Engine Testing

Engine Testing
Theory and Practice

Third edition

A.J. Martyr
M.A. Plint

Published on behalf of:
SAE International
400 Commonwealth Drive,
Warrendale, PA 15096-0001

Butterworth-Heinemann is an imprint of Elsevier
Linacre House, Jordan Hill, Oxford OX2 8DP
30 Corporate Drive, Suite 400, Burlington, MA 01803

First edition 1995
Reprinted 1996 (twice), 1997 (twice)
Second edition 1999
Reprinted 2001, 2002
Third edition 2007

Library of Congress Cataloging in Publication Data
A catalog record for this book is available from the Library of Congress

SAE ISBN 978-0-7680-1850-9

SAE Edition
R-382
SAE International
400 Commonwealth Drive
Warrendale, PA 15096-0001 U.S.A.
Phone: (724) 776-4841
Fax: (724) 776-0790
E-mail: publications@sae.org
http://www.sae.org

Printed and bound in UK

Contents

Preface

The preface of this book is probably the least read section of all; however, it is the only part in which I can pay tribute to my friend and co-author of the first two editions, Dr Michael Plint, who died suddenly in November 1998, only four days after the publication of the second edition.

All the work done by Michael in the previous editions has stood up to the scrutiny of our readers and my own subsequent experience. In this edition, I have attempted to bring our work up to date by revising the content to cover the changing legislation, techniques and some of the new tools of our industry. In a new Chapter 1, I have also sought to suggest some good practices, based on my own 40 years of experience, aimed at minimizing the problems of project organization that are faced by all parties involved in the specification, modification, building and commissioning of engine test laboratories.

The product of an engine test facility is data and byproduct is the experience gained by the staff and hopefully retained by the company. These data have to be relevant to the experiments being run, and every component of the test facility has to play its part, within an integrated whole, in ensuring that the test data are as valid and uncorrupted as possible, within the sensible limits of the facility's role. It was our intention when producing the first edition to create an eclectic source of information that would assist any engineer faced with the many design and operational problems of both engine testing and engine test facilities. In the intervening years, the problems have become more difficult as the nature of the engine control has changed significantly, while the time and legislative pressures have increased. However, it is the laws of physics that rule supreme in our world and they can continue to cause problems in areas outside the specialization of many individual readers. I hope that this third edition helps the readers involved in some aspect of engine testing to gain a holistic view of the whole interactive package that makes up a test facility and to avoid, or solve, some of the problems that they may meet in our industry.

Having spoken to a number of readers of the two proceeding editions of this book I have reorganized the contents of most of the chapters in order to reflect the way in which the book is used.

Writing this edition has, at times, been a lonely and wearisome task that would not have been completed without the support of my wife Diana and my friends. Many people have assisted me with their expert advice in the task of writing this third edition. I have to thank all my present AVL colleagues in the UK and Austria, particularly Stuart Brown, David Moore and Colin Freeman who have shared many of my experiences in the test industry over the last 20 years, also Dave Rogers, Craig Andrews, Hans Erlach and finally Gerhard Müller for his invaluable help with the complexities of electrical distribution circuits.

A.J. Martyr

Inkberrow

22 September 2006

Acknowledgements

Figures 3.6, 3.13, 3.15 and *3.16* Reprinted from *Industrial Noise Control*, Fader, by permission of John Wiley and Sons Ltd

Figure 3.9 Reprinted from technical literature of Type TSC by courtesy of Christie and Grey Ltd, UK

Figure 3.14 Reprinted from *Encyclopaedia of Science and Technology*, Vol. 12, 1987, by kind permission of McGraw-Hill Inc., New York

Figure 5.3 Reprinted from CIBSE Guide C, section C4, by permission of the Chartered Institution of Building Services Engineers

Figure 5.10 Reprinted from I.H.V.E. Psychometric Chart, by permission of the Chartered Institution of Building Services Engineers

Figure 7.1 Reprinted from BS799. Extracts from British Standards are reprinted with the permission of BSI. Complete copies can be obtained by post from BSI Sales, Linford Wood, Milton Keynes, MK14 6LE, UK

Figure 7.2 Reprinted from *The Storage and Handling of Petroleum Liquids*, Hughes and Swindells, with the kind permission of Edward Arnold Publishers

Figure 7.3 Reprinted from 'Recommendations for pre-treatment and cleaning of heavy fuel oil' with the kind permission of Alfa Laval Ltd

Figure 8.3 Reprinted from Drawing no. GP10409 (Carl Shenck AG, Germany)

Figures 8.4 and *8.5* Reprinted from Technical Documentation T 32 FN, with the kind permission of Hottinger Baldwin Messtechnik GmbH, Germany

Figures 8.6 and *8.7* Illustration courtesy of Ricardo Test Automation Ltd

Figures 8.9, 8.10 and *8.11* Reprinted with permission of Froude Consine, UK

Figure 8.12 Reprinted from technical literature, Wichita Ltd

Figure 9.4 Reprinted from *Practical Solution of Torsional Vibration Problems*, 3rd edition, W. Ker-Wilson, 1956

Figure 9.7 Reprinted from literature, with the kind permission of British Autoguard Ltd

Figures 9.8 and *9.11* Reprinted from sales and technical literature, with the kind permission of Twilfex Ltd

Figure 12.2 Reprinted from Paper ISATA, 1982, R.A. Haslett, with the kind permission of Cussons Ltd

Figure 12.4 Reprinted from Technology News with the kind permission of Petroleum Review

Figure 14.7 Reprinted from SAE 920 462 (SAE International Ltd)

Figure 17.1 From *Schmiertechnik und Tribologie* 29, H. 3, 1982, p. 91, Vincent Verlag Hannover (now: *Tribologie und Schmierungstechnik*)

Figures 17.2 and *17.3* Reprinted by permission of the Council of the Institution of Mechanical Engineers from 'The effect of viscosity grade on piston ring wear', S.L. Moore, *Proc. I. Mech.* E C184/87

Figure 18.2 Illustration courtesy of Ricardo Test Automation Ltd

Figures 2.3, 5.4, 5.5, 5.6, 5.7, 6.8, 7.5, 7.6, 10.5 10.8, 14.9, 14.10, 14.11, 16.1, 16.2, 16.5, 16.7, 16.8 and *18.4* Reprinted by kind permission of AVL List GmbH

Introduction

Over the working lifetime of the authors the subject of internal engine development and testing has changed, from being predominantly within the remit of mechanical engineers, into a task that is well beyond the remit of any one discipline that requires a team of specialists covering, in addition to mechanical engineering, electronics, power electrics, acoustics, software, computer sciences and chemical analysis, all supported by expertise in building services and diverse legislation.

It follows that the engineer concerned with any aspect of engine testing, be it fundamental research, development, performance monitoring or routine production testing, must have at his fingertips a wide and ever-broadening range of knowledge and skills.

A particular problem he must face is that, while he is required to master ever more advanced experimental techniques – such areas as emissions analysis and engine calibration come to mind – he cannot afford to neglect any of the more traditional aspects of the subject. Such basic matters as the mounting of the engine, coupling it to the dynamometer and leading away the exhaust gases can give rise to intractable problems, misleading results and even on occasion to disastrous accidents. More than one engineer has been killed as a result of faulty installation of engines on test beds.

The sheer range of machines covered by the general term *internal combustion engine* broadens the range of necessary skills. At one extreme we may be concerned with an engine for a chain saw, a single cylinder of perhaps 50 c.c. capacity running at 15 000 rev/min on gasoline, with a running life of a few hours. Then we have the vast number of passenger vehicle engines, four, six or eight cylinder, capacities ranging from one litre to six, expected to develop full torque over speeds ranging from perhaps 1500 rev/min up to 7000 rev/min (the upper limit rising continually), and with an expected life of perhaps 6000 hours. The motor-sport industry continues to push the limits of both engine and test plant design with engines revving at speeds approaching 20 000 r.p.m. and, in rally cars, engine control systems having to cope with cars leaving the ground, then requiring full power when they land. At the other extreme is the cathedral type marine engine, a machine perhaps 10 m tall and weighing 1000 tonnes, running on the worst type of residual fuel, and expected to go on turning at 70 rev/min for more than 50 000 hours.

The purpose of this book is to bring together the information on both the theory and practice of engine testing that any engineer responsible for work of this kind must have available. It is naturally not possible, in a volume of manageable size, to give all the information that may be required in the pursuit of specialized lines of development, but it is the intention of the authors to make readers aware of the many tasks they may face and to give advice based on experience; a range of references for more advanced study has been included.

Throughout the book accuracy will be a recurring theme. The purpose of engine testing is to produce information, and inaccurate information can be useless or worse. A feeling for accuracy is the most difficult and subtle of all the skills required of the test engineer. Chapter 19, dealing with this subject, is perhaps the most important in the book and the first that should be read.

Experience in the collaboration with architects and structural engineers is particularly necessary for engineers involved in test facility design. These professions follow design conventions and even draughting practices that differ from those of the mechanical engineer. To give an example, the test cell designer may specify a strong floor on which to bolt down engines and dynamometers that has an accuracy approaching that of a surface plate. To the structural engineer this will be a startling concept, not easily achieved.

The internal combustion engine is perhaps the best mechanical device available for introducing the engineering student to the practical aspects of engineering. An engine is a comparatively complicated machine, sometimes noisy and alarming in its behaviour and capable of presenting many puzzling problems and mystifying faults. A few hours spent in the engine testing laboratory are perhaps the best possible introduction to the real world of engineering, which is remote from the world of the lecture theatre and the computer simulation in which, inevitably, the student spends much of his time.

While it contains some material only of interest to the practising test engineer, much of this book is equally suitable as a student text, and this purpose has been kept very much in mind by the authors. In response to the author's recent experience, the third edition has a new Chapter 1 dedicated to the problems involved in specifying and managing a test facility build project.

A note of warning: the general management of engine tests

What may be regarded as traditional internal combustion engines had in general very simple control systems. The spark ignition engine was fitted with a carburettor controlled by a single lever, the position of which, together with the resisting torque applied to the crankshaft, set all the parameters of engine operation. Similarly, the performance of a diesel engine was dictated by the position of the fuel pump rack, either controlled directly or by a relatively simple speed governor.

The advent of engine control units (ECUs) containing ever more complex maps and taking signals from multiple vehicle transducers has entirely changed the situation. The ECU monitors many aspects of powertrain performance and makes continuous adjustments. The effect of this is effectively to take the control of the test conditions out of the hands of the engineer conducting the test. Factors entirely extraneous to the investigation in hand may thus come into play.

The introduction of exhaust gas recirculation (EGR) under the control of the ECU is a typical example. The only way open to the test engineer to regain control of his test is to devise means of bypassing the ECU, either mechanically or by intervention in the programming of the control unit.

A note on references and further information

It would clearly not be possible to give all the information necessary for the practice of engine testing and the design of test facilities in a book of this length. References suitable for further study are given at the end of most chapters. These are of two different kinds:

*   a selection of fundamental texts or key papers
*   relevant British Standards and other reference standard specifications.

The default source of many students is now the world wide web which contains vast quantities of information related to engines and engine testing, much of which is written by and for the automotive after-market where a rigorous approach to experimental accuracy is not always evident; for this reason and due to the transient nature of many websites, there are very few web-based references.

Units and conversion factors

Throughout this book use is made of the metric system of units, variously described as:

The MKS (metre-kilogram-second) System

SI (Système International) Units

These units have the great advantage of logical consistency but the disadvantages of still a certain degree of unfamiliarity and in some cases of inconvenient numerical values.

Fundamental Units

Mass	kilogram (kg)	$1\,kg = 2.205\,lb$
Length	metre (m)	$1\,m = 39.37\,in$
Force	newton (N)	$1\,N = 0.2248\,lbf$

Derived Units

Area	square metre (m²)	$1\,m^2 = 10.764\,ft^2$
Volume	cubic metre (m³), litre (l)	$1\,m^3 = 10001 = 35.3\,ft^3$
Velocity	metre per second (m/s)	$1\,m/s = 3.281\,ft/s$
Work, Energy	joule (J)	$1\,J = 1\,Nm = 0.7376\,ft\text{-}lbf$
Power	watt (W)	$1\,W = 1\,J/s$
		$1\,horsepower\,(hp) = 745.7\,W$
Torque	newton metre	$1\,Nm = 0.7376\,lbf\text{-}ft$

The old metric unit of energy was the calorie (cal), the heat to raise the temperature of 1 gram of water by 1°C.

$1\,cal = 4.1868\,J$. 1 kilocalorie (kcal) $= 4.1868\,kJ$.

Temperature degree Celsius °C $\qquad\qquad\qquad\qquad\qquad\qquad\qquad\qquad \theta$

Absolute temperature kelvin K $\qquad\qquad\qquad\qquad\qquad\qquad\qquad\qquad T$

$$T = \theta + 273.15$$

Pressure pascal (Pa)
$$1\,Pa = 1\,N/m^2 = 1.450 \times 10^{-4}\,lbf/in^2$$
$$1\,MPa = 10^6\,Pa = 145\,lbf/in^2$$

This unit is commonly used to denominate stress.

Throughout this book the bar is used to denominate pressures:

1 bar (bar) = 10^5 Pa = 14.5 lbf/in^3

Standard test conditions for i.c. engines as defined in BS 5514/ISO 3046[1] specify:

Standard atmospheric pressure = 1 bar = 14.5 lbf/in^2

Note: 'Standard atmosphere' as defined by the physicist[2] is specified as a barometric pressure of 760 millimetres of mercury (mmHg) at 0°C.

1 standard atmosphere = 1.01325 bar = 14.69 lbf/in^2

The difference between these two standard pressures is a little over 1 per cent. This can cause confusion. Throughout this book 1 bar is regarded as standard atmospheric pressure.

The torr is occasionally encountered in vacuum engineering.

1 torr = 1 mmHg = 133.32 Pa

In measurements of air flow use is often made of water manometers.

1 mm of water (mmH$_2$O) = 9.81 Pa

References

1.   BS 5514 *Reciprocating Internal Combustion Engines: Performance.*
2.   Kaye, G.W.C. and Laby, T.H. (1973) *Tables of Physical and Chemical Constants,* Longmans, London.

Further reading

BS 350 Pt 1 *Conversion factors and tables*
BS 5555 *Specification for SI units and recommendations for the use of their multiples and of certain other units*

1 Test facility specification, system integration and project organization

Introduction

An engine test facility is a complex of machinery, instrumentation and support services, housed in a building adapted or built for its purpose. For such a facility to function correctly and cost-effectively, its many parts must be matched to each other while meeting the operational requirements of the user and being compliant with various regulations.

Engine and vehicle developers now need to measure improvements in engine performance that are frequently so small as to require the best available instrumentation in order for fine comparative changes in performance to be observed. This level of measurement requires that instrumentation is integrated within the total facility such that their performance and data are not compromised by the environment in which they operate and services to which they are connected.

Engine test facilities vary considerably in power rating and performance; in addition there are many cells designed for specialist interests, such as production test or study of engine noise, lubrication oils or exhaust emissions.

The common product of all these cells is data that will be used to identify, modify, homologate or develop performance criteria of all or part of the tested engine. All post-test work will rely on the relevance and veracity of the test data, which in turn will rely on the instrumentation chosen to produce it and the system within which the instruments work.

To build or substantially modify a modern engine test facility requires co-ordination of a wide range of specialized engineering skills; many technical managers have found it to be an unexpectedly complex task.

The skills required for the task of putting together test cell systems from their many component parts have given rise, particularly in the USA, to a specialized industrial role known as system integration. In this industrial model, a company or more rarely a consultant, having one of the core skills required, takes contractual responsibility for the integration of all of the test facility components from various sources. Commonly, the integrator role has been carried out by the supplier of test cell control systems and the role has been restricted to the integration of the dynamometer and control room instrumentation.

In Europe, the model is somewhat different because of the long-term development of a dynamometry industry that has given rise to a very few large test plant contracting companies.

However, the concept of *systems integrator* is useful to define that role, within a project, that takes the responsibility for the final functionality of a test facility; so the term will be used, where appropriate, in the following text.

This chapter covers the vital importance of good user specification and the various organizational structures required to complete a successful test facility project.

Test facility specification

Without a clear and unambiguous specification no complex project should be allowed to proceed.

This book suggests the use of three levels of specification:

1. *Operational specification*: describing 'what it is for', created by the user prior to any contract to design or build a test facility.
2. *Functional specification*: describing 'what it consists of and where it goes', created either by the user group having the necessary skills, as part of a feasibility study by a third party, or by the main contractor as part of the first phase of a contract.
3. *Detailed functional specification*: describing 'how it all works' created by the project design authority within the supply contract.

Creation of an operational specification

This chapter will tend to concentrate on the operational specification which is a user-generated document, leaving some aspects of the more detailed levels of functional specification to subsequent chapters covering the design process. The operational specification should contain a clear description of the task for which the facility is being created. It need not specify in detail the instruments required, nor does it have to be based on a particular site. The operational specification is produced by the end user; its first role will normally be to support the application for budgetary support and outline planning; subsequently, it remains the core document on which all other detailed specifications are based. It is sensible to include a brief description of envisaged facility acceptance tests within the document since there is no better means of developing and communicating the user's requirement than to describe the results to be expected from described work tasks.

- It is always sound policy to find out what is available on the market at an early stage, and to reconsider carefully any part of the specification that makes demands that exceed what is commonly offered.
- A general cost consciousness at this stage can have a permanent effect on capital and subsequent running costs.

Because of the range of skills required in the design and commissioning of a 'green field' test laboratory it is remarkably difficult to produce a succinct specification that is entirely satisfactory, or even mutually comprehensible, to all specialist participants.

The difficulty is compounded by the need for some of the building design details that determine the final shape, such as roof penetrations or floor loadings, to be determined before the detailed design of internal plant has been finalized. It is appropriate that the operational specification document contains statements concerning the general 'look and feel' and any such pre-existing conditions or imposed restrictions that may impact on the facility layout. It should list any prescribed or existing equipment that has to be integrated, the level of staffing and any special industrial standards the facility is required to meet. In summary, it should at least address the following questions:

- What are the primary and secondary purposes for which the facility is intended and can these functions be condensed into a sensible set of acceptance procedures to prove the purposes that may be achieved?
- What is the realistic range of units under test (UUT)?
- How are test data (the product of the facility) to be displayed, distributed, stored and post-processed?
- What possible extension of specification or further purposes should be provided for in the initial design and to what extent would such 'future proofing' distort the project phase costs?
- May there be a future requirement to install additional equipment and how will this affect space requirement?
- Where will the UUT be prepared for test?
- How often will the UUT be changed and what arrangements will be made for transport into and from the cells?
- How many different fuels are required and must arrangements be made for quantities of special or reference fuels?
- What up-rating, if any, will be required of the site electrical supply and distribution system?
- To what degree must engine vibration and exhaust noise be attenuated within the building and at the property border?
- Have all local regulations (fire, safety, environment, working practices, etc.) been studied and considered within the specification?

Feasibility studies and outline planning permission

The work required to produce a site-specific operational specification, or statement of intent, may produce a number of alternative layouts each with possible first-cost or operational problems. In all cases an environmental impact report should be produced covering both the facility's impact of its surroundings and, in the case of low emission measuring laboratories, the locality's impact on the facility.

Complex technocommercial investigatory work may be needed so a feasibility study might be considered, covering the total planned facility or that part that gives rise to doubt or the subject of radically differing strategies. In the USA, this type of work is often referred to as a proof design contract.

The secret of success of such studies is the correct definition of the required 'deliverable' which must answer the technical and budgetary dilemmas, give clear and costed recommendations and, so far as is possible, be supplier neutral. The final text should be capable of easy incorporation into the Operation and Functional Specification documents.

A feasibility study will invariably be concerned with a specific site and, providing appropriate expertise is used, should prove supportive to gaining budgetary and outline planning permission; to that end, it should include within its content a site layout drawing and graphical representation of the final building works.

Benchmarking

Cross-referencing with other test facilities or test procedures is always useful when specifying your own. Benchmarking is merely a modern term for an activity that has been practised by makers of products intended for sale, probably ever since the first maker of flint axes went into business: it is the act of comparing your product with competing products and your production and testing methods with those of your competitors. The difference today is that it is now highly formalized and practised without compunction. Once it is on the market any vehicle or component thereof can be bought and tested by the manufacturer's competitors, with a view to taking over and copying any features that are clearly in advance of the competitor's own products. There are test facilities built and run specifically for benchmarking.

This evidently increases the importance of patent cover, of preventing the transfer of confidential information by disaffected employees and of maintaining confidentiality during the development process; such concerns need to have preventative measures built into the specification of the facility rather than added as an afterthought.

Safety regulations and planning permits covering test cells

Feasibility not only concerns the technical and commercial viability, but also whether one will be allowed to create the new or altered test laboratory; therefore, the responsible person should consider discussion at an early stage with the following agencies:

- local planning authority;
- local petroleum officer and fire department;
- local environmental officer;
- building insurers;

- local electrical supply authority;
- site utility providers.

Note the use of the word 'local'. There are very few regulations specifically mentioning engine test cells, much of the European legislation is generic and frequently has unintended consequences for the automotive test industry. Most legislation is interpreted locally and the nature of that interpretation will depend on the industrial experience of the officials concerned, which can be highly variable. There is always a danger that inexperienced officials will over-react to applications for engine test facilities and impose unrealistic restraints on the design. It may be found useful to keep in mind one basic rule that has had to be restated over many years:

> An engine test cell, using liquid fuels, is a 'zone 2' hazard containment box. It is not possible to make its interior inherently safe since the test engine worked to the extremes of its performance is not inherently safe; therefore the cell's function is to contain and minimise the hazards and to inhibit human access when they are present. (See Chapter 4, Test cell and control room design: an overall view)

Most of the operational processes carried out within a typical automotive test cell are generally no more hazardous than those hazards experienced by garage mechanics, motorists or racing pit staff in real life. The major difference is that in the cell the running engine is stationary in a space that is different from that for which it was designed and therefore humans may be able to gain close and potentially dangerous access to it.

It is more sensible to interlock the cell doors to prevent access to an engine running above 'idle' state, than to attempt to make the rotating elements 'safe' by the use of close fitting guarding that will inhibit operations and fall into operational disuse.

The authors of the high level operational specification need not concern themselves with some of such details, but simply state that industrial best practice and compliance with current legislation is required. The arbitrary imposition of existing operational practices on a new test facility should be avoided at the operational specification stage until confirmed as appropriate, since they may restrict the inherent benefits of the technological developments available.

One of the restraints commonly imposed on the facility buildings concerns the number and nature of chimney stacks or ventilation ducts. This is often a cause of tension between the architect, planning authority and facility designers. With some ingenuity these essential items can be disguised, but the resulting designs will inevitably require more space than the basic vertical inlet and outlet ducts. Similarly, noise break-out via such ducting may be reduced to the background at the facility border but the space required for attenuation will complicate the plant room layout (see Chapter 3, Vibration and noise).

Note that the use of gaseous fuels will impose special restrictions on the design of test facilities and, if included in the operational specification, the relevant authorities and specialist contractors must be involved from the planning stage. Modifications may include blast pressure relief panels in the cell structure and exhaust ducting, which need to be included from design inception.

Specification for a control and data acquisition system

The choice of test automation supplier need not be part of the operation specification but, since it will form part of the functional specification, and since the choice of test cell software may be the singularly most important technocommercial decision in placing a contract for a modern test facility, it would seem sensible to consider the factors that should be addressed in making that choice. The test cell automation software lies at the core of the facility operation therefore its supplier will take an important role within the final system integration. The choice therefore is not simply one of a software suite but of a key support role in the design and ongoing development of the new facility.

Laboratories where the systems are to be fully computerized should consider the

- local capability of each software/hardware supplier;
- installed base of each possible supplier, relevant to the industrial sector;
- level of operator training and support required for each of the short-listed systems;
- compatibility of the control system with any intended, third party hardware;
- modularity or upgradeability of both hardware and software;
- requirements to use pre-existing data or to export data from the new facility to existing databases;
- ease of creating test sequences;
- ease of channel calibration and configuration;
- flexibility of data display and post-processing options.

A methodical approach allows for a 'scoring matrix' to be drawn up whereby competing systems may be objectively judged.

Anyone charged with producing specifications is well advised to carefully consider the role of the test cell operators. Significant upgrades in test control and data handling will totally change the working environment of the cell operator. There are many cases of systems being imposed on users which never reach their full potential because of inadequacy of training or inappropriate specification of the system.

Use of supplier's specifications

It is all too easy for us to be influenced by headline speed and accuracy numbers in the specification sheets for computerized systems. The effective time constants of many engine test processes are not limited by the data handling rates of the computer system, but rather of the physical process being measured and controlled. Thus the speed at which a dynamometer can make a change in torque absorption is governed more by the rate of magnetic flux generation in its coils, or the rate at which it can change the mass of water in its internals, rather than the speed at which its control algorithm is being recalculated. The skill in using such information is to identify the numbers that are relevant to task for which the item is required.

Faster is not necessarily better and it is often more expensive.

Functional specifications: some common difficulties

Building on the operational specification, which describes what the facility has to do, the *functional specification* describes how the facility is to perform its defined tasks and what it will contain. If the functional specification is to be used as the basis for competitive tendering then it should avoid being unnecessarily prescriptive. Overprescriptive specifications, or those including sections that are technically incompetent, are problems to specialist contractors. The first type may prevent better or more cost-effective solutions being quoted, while the later mean that a company who, through lack of experience, claims compliance wins the contract, then inevitably fails to meet the customer's expectations.

Overprescription may range from ill matching of instrumentation to unrealistically wide range of operation of subsystems.

A classic problem in facility specification concerns the range of engines that can be tested in one test cell using common equipment and a single shaft system. Clearly there is a great cost advantage for the whole production range of a manufacturer's engines to be tested in one cell. However, the detailed design problems and subsequent maintenance implications that such a specification may impose can be far greater than the cost of creating two or more cells that are optimized for narrower ranges of engines. Not only is this a problem inherent in the 'turn-down' ratio of fluid services and instruments having to measure the performance of a range of engines from say 500 to 60 kW, but the range of vibratory models produced may defy the capability of any one shaft system to handle.

This issue of dealing with a range of vibratory models may require that cells be dedicated to particular types or that alternative shaft systems are provided for particular engine types. Errors in this part of the specification and the subsequent design strategy are often expensive to resolve after commissioning. Not even the most demanding customer can break the laws of physics with impunity.

Before and during the specification and planning stage of any test facility, all participating parties should keep in mind the vital question: By what cost- and time-effective means do we prove that this complex facility meets the requirement and specification of the user? It is never too early to consider the form and content of acceptance tests, since from them the designer can infer much of the detailed functional specification. Failure to incorporate these into contract specifications from the start leads to delays and disputes at the end.

Interpretation of specifications

Employment of contractors with the relevant industrial experience is the best safe-guard against overblown contingencies or significant omissions in quotations arising from user-generated specifications.

Provided with a well written operational and functional specification any competent subcontractor experienced in the engine or vehicle test industry should be able to provide a detailed specification and quote for their module or service within the total project. Subcontractors who do not have experience in the industry will not be able to appreciate the special, sometimes subtle, requirements imposed upon their designs by the transient conditions, operational practices and possible system interactions inherent in the industry.

In the absence of a full appreciation of the project based on previous experience they will search the specification for 'hooks' on which to hang their standard products or designs, and quote accordingly. This is particularly true of air or fluid conditioning plant where the bare parameters of temperature range and heat load can lead the inexperienced to equate test cell conditioning with that of a chilled warehouse. An escorted visit to an existing test facility should be the absolute minimum experience for subcontractors quoting for systems such as chilled water, electrical installation and HVAC.

General project organization

In all but the smallest test facility projects, there will be three generic types of contractor with whom the customer's project manager has to deal. They are

- civil contractor;
- building services contractors;
- test instrumentation contractor.

How the customer decides to deal with these three industrial groups and integrate their work will depend on the availability of in-house skills and the skills and experience of any preferred contractors.

The normal variations in project organization, in ascending order of customer involvement in the process, are

- a consortium working within a design and build or 'turnkey'* contract based on the customer's operational specification and working to the detailed functional specification and fixed price produced by the consortium;
- guaranteed maximum price (GMP) contracts where a complex project management system, having a 'open' cost accounting system, is set up with the mutual intent to keep the project within a mutually agreed maximum value. This requires joint project team cohesion of a high order;

---

*The term *turnkey* is now widely misused. The original turnkey contract was one carried out to an agreed specification by a contractor taking total responsibility for the site and all associated works with virtually no involvement by the end user until the keys were handed over so that acceptance tests can be performed.

- a customer-appointed main contractor employing supplier chain and working to customer's functional specification;
- a customer appointed civil contractor followed by services and system integrator contractor each appointing specialist subcontractors, working with customer's functional specification and under the customer's project management and budgetary control;
- a customer controlled series of subcontract chains working to the customer's detailed functional specification, project engineering, site and project management.

Whichever model is chosen the two vital roles of project manager and design authority (systems integrator) have to be clear to all and provided with the financial and contractual authority to carry out their allotted roles. It should be noted that in the UK, all but the smallest contracts involving construction or modification of test facilities will fall under the control of a specific section of health and safety legislation known as Construction Design and Management Regulations 1994 (CDM Regulations) which require nomination of these and other project roles.

Project roles and management

The key role of the client, or user, is to invest great care and effort into the creation of a good operational and functional specification. Once permission to proceed has been given, based on this specification, and the main contractor has been appointed, the day to day role of the client user group should, ideally, reduce to that of attendance at review meetings and being 'on-call'.

Nothing is more guaranteed to cause project delays and cost escalation than ill-considered or informal changes of detail by the client's representatives. Whatever the project model, the project management system should have a formal system of notification of change and an empowered group within both the customer's and contractor's organization to deal with such changes quickly. The type of form shown in Fig. 1.1 allows individual requests for project change to be recorded and the implications of the change to be discussed and quantified. Change can have negative or positive effect on project costs and can be requested by both the client and the contractor; with the right working relationship in a joint project team, change notes can be the mechanism by which cost- or time-saving alterations can be raised during the course of work.

All projects have to operate within the three restraints of time, cost and quality (content) (Fig. 1.2). The relative importance of these three criteria has to be understood by the client and project manager. The model is different for each client and each project and however much a client may protest that all three criteria have equal weighting and are fixed; if change is introduced, one has to be a variable.

Customer:	Variation No:
Project name:	Project No:

Details of proposed change:
(include any reference to supporting documentation)

Requirement (tick and initial as required)
Design authority required design change
Customer instruction:
Customer request:
Contractor request:
Urgent quotation required:
Customer agreed to proceed at risk:
Work to cease until variation agreed:
Requested to review scope of supply:
Other (specify):

	Name	Email/Phone No.
Contractor representative		
Authorized		
Customer representative	Name	Email/Phone No.
Authorized		

Actions:
Sales quote submitted: (date and initial)
Authorized for action: (date and initial)
Implemented: (date and initial)

Figure 1.1 *A sample contract variation record sheet*

If one of the points of the triangle is moved there is a consequential change in one or both of the others and that the later in the program the change is required, the greater the consequential effect.

One oft-repeated error, which is forced by time pressure on the overall programme, is to deliver instrumentation and other equipment into a facility building before internal environmental conditions are suitable. It is always better to deliver such plant late and into a suitable environment, then make up time by increasing installation man hours, than it is to have incompatible trades working in the same building space and suffering the almost inevitable damage to the test equipment.

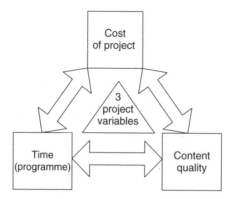

Figure 1.2   *Project constraints: in real life, if two are fixed the third will be variable*

Key project management tools

The techniques of project management are outside the remit of this book but some important tools and skills suitable for any project manager in charge of an automotive test facility are suggested.

Communications and responsibility matrix

Any multicontractor and multidisciplinary project creates a complex network of communications. These networks between suppliers, contractors and personnel within the customer's organization may pre-exist or be created during the course of the project; the danger is that informal communications may give rise to unauthorized variations in project content or timing. Good project management is only possible with a disciplined communication system and this should be designed into and maintained during the project. The arrival of email as the standard communication method has increased the need for communication discipline and introduced the need, within project teams, of creating standardized, computer-based filing systems.

If competent staff are available to create and maintain a project specific intranet website then the project manager has a good means of maintaining control over formal communications. Such a network can give access permission, such as 'read only', 'submit' and 'modify', as appropriate to individuals' roles to all of the groups and nominated staff having any commercial or technical interest in the project.

The creation of a responsibility matrix is most useful when it covers the important minutiae of project work, that is, not only who supplies a given module but who insures, delivers, off-loads, connects and commissions the module.

Use of 'master drawing' in project control

The use of a facility layout or schematic drawing graphically showing facility modules that can be used by all tendering contractors and continually updated by the main contractor or design authority can be a vital tool in any multidisciplinary project where there may be little detailed appreciation between specialized contractors for the spatial requirements of each other.

Constant, vigilant site management is required during the final building 'fit out' phase of an automotive test facility if clashes over space allocation are to be avoided, but good preparation and contractor briefing can reduce the inherent problem. If the systems integrator or main contractor takes ownership of project floor layout plans and these plans are used at every subcontractor meeting, kept up to date to record the layout of all services and major modules, then most of the space utilization, service route and building penetration problems will be resolved before work commences. Where possible and appropriate, contractors method statements should use the common 'table top' project plans to show the area of their own installation in relation to the building and installations of others.

Project timing chart

Most staff involved with a project will recognize a classic Gantt chart; not all will understand their role or the interactions of their tasks within that plan. It is the task of the project manager to ensure that each contractor and all key personnel work within the project plan structure. This is not served by sending an electronic version of a large and complex Gantt chart, but by early contract briefing and preinstallation progress meetings.

There are some key events in every project that are absolutely time critical and these have to be given special attention by both client and project manager. Consider, for example, the arrival of a chassis dynamometer and the site implications:

- One or more large trucks will have to arrive on the client's site, in the correct order, and require suitable site access for manoeuvring.
- The chassis dynamometer will require a large crane to off-load. The crane's arrival and site positioning will have to be coordinated within an hour of the trucks' arrival.
- Access into the chassis dynamometer pit area will have to be kept clear for special heavy handling equipment until the unit is installed, the access thereafter will be closed up by deliberately delayed building work.
- Other contractors will have to be kept out of the effected work and access areas, as will client's and contractor's vehicles and equipment.

Preparation for such an event takes detailed planning, good communications and authoritative management. The non- or late arrival of one of the key players because 'they did not understand the importance' clearly causes acute problems in the example above, but the same ignorance of programmed roles causes delays and overspends that are less obvious throughout any project where detailed planning and communications are left to take care of themselves.

A note on documentation

Test cells and control room electrical systems are, in the nature of things, subject to detailed modification during the build and commissioning process. The documentation, representing the 'as-commissioned' state of the facility, must be of a high standard and easily accessible to maintenance staff and contractors. The form and due delivery of documentation should be specified within the functional specification and form part of the acceptance criteria. Subsequent responsibility for keeping records and schematics up to date within the operator's organization must be clearly defined and controlled.

Summary

Like all complex industrial projects, the creation, or significant modification, of an engine test laboratory should start with the creation of an operational specification involving all those departments and individuals having a legitimate interest. The specification of the engine testing tasks and definition of suitable acceptance tests are essential prerequisites of such a project.

The roles of system integrator, design authority and project manager must be well defined and empowered. Specifically, ensure that the design authority for systems integration is explicitly given to a party technically competent and contractually empowered to carry it out.

There are at least three levels of specification that should be considered and it is essential to invest time in their preparation if a successful project is to be achieved. Paper is cheaper to change than concrete.

The project management techniques required are those of any multidisciplinary laboratory construction but require knowledge of the core testing process so that the many subtasks are integrated appropriately.

The statement made early in this chapter, 'Without a clear and unambiguous specification no complex project should be allowed to proceed', seems self-evident; yet many companies, within and outside our industry, continue either to allocate the task inappropriately or underestimate its importance and subject it to post-order change. The consequence is that project times are extended by an iterative quotation period and from the point at which the users realize that their (unstated or misunderstood) expectations are not being met, usually during commissioning.

2 The test cell as a thermodynamic system

The energy of the world is constant; the entropy strives towards a maximum.[1]

Rudolph Clausius (1822–1888)

Introduction

The closed engine test cell system makes a suitable case for students to study an example of the flow of heat and change in entropy. In almost all engine test cells the vast majority of the energy comes into the system as highly concentrated 'chemical energy' entering the cell via the smallest penetration in the cell wall, the fuel line. It leaves the cell as lower grade heat energy via the largest penetrations: the ventilation duct, engine exhaust pipe and the cooling water pipes. In the case of cells fitted with electrically regenerative dynamometers, almost one-third of the energy supplied by fuel will leave the cell as electrical energy able to slow down the electrical energy supply meter. Many problems are experienced in test cells worldwide when the thermodynamics of the cell have not been correctly catered for in the design of cooling systems. The most common problem is high air temperature within the test cell, either generally or in critical areas. The practical effects of such problems will be covered in detail in Chapter 5, Ventilation and air conditioning, but it is vital for the cell designer to have a general appreciation of the contribution of the various heat sources and the strategies for their control.

In the development of the theory of thermodynamics much use is made of the concept of the open system. This is a powerful tool and can be very helpful in considering the total behaviour of a test cell. It is linked to the idea of the control volume, a space enclosing the system and surrounded by an imaginary surface, the control surface (Fig. 2.1).

The great advantage of this concept is that once one has identified all the mass and energy flows into and out of the system it is not necessary to know exactly what is going on inside the system in order to draw up a 'balance sheet' of inflows and outflows.

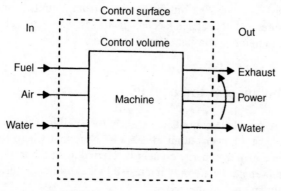

Figure 2.1 *An open thermodynamic system*

The various inflows and outflows to and from a test cell are as follows:

*In**	*Out*
Fuel	
Ventilation air (some may be used by the engine as combustion air)	Ventilation air
Combustion air (treated)	Exhaust (includes air used by engine)
	Engine cooling water
Charge air (when separately supplied)	Dynamometer cooling water
Cooling water	Electricity from dynamometer
Electricity for services	Losses through walls and ceiling

Balance sheets may be drawn up for fuel, air, water and electricity, but by far the most important is the *energy balance*, since every one of these quantities has associated with it a certain quantity of energy. The same concept may be applied to the engine within the cell. This may be pictured as surrounded by its own control surface, through which the following flows take place:

In	*Out*
Fuel	Power
Air used by the engine	Exhaust
Cooling water	Cooling water
Cooling air	Cooling air
	Convection and radiation

---

* Compressed air may be a further energy input; however, usage is generally intermittent and unlikely to make a significant contribution.

Measurement of thermal losses from the engine is dealt with in Chapter 13, Thermal efficiency, measurement of heat and mechanical losses, where the value of the method in the analysis of engine performance is made clear.

The energy balance of the engine

Table 2.1 shows a possible energy balance sheet for a cell in which a gasoline engine is developing a steady power output of 100 kW. Note that where fluids (air, water, exhaust) are concerned, the energy content is referred to an arbitrary zero, the choice of which is unimportant: we are only interested in the difference between the various energy flows into and out of the cell.

Given sufficient detailed information on a fixed engine/cell system it is possible to carry out a very detailed energy balance calculation (see Chapter 5, Ventilation and air conditioning, for a more detailed treatment). Alternatively there are some commonly used 'rule of thumb' calculations available to the cell designer; the most common of these relates to the energy balance of the engine which is known as the '30–30–30–10 rule'. This refers to the energy balance shown in Table 2.2.

The key lesson to be learnt by the non-specialist reader is that any engine test cell has to be designed to deal with energy flows that are at least three times greater than the 'headline' engine rating. To many this will sound obvious but a common fixation on engine power and a casual familiarity with, but lack of appreciation of, the energy density of petroleum fuels has misled many people in the past to significantly

Table 2.1 *Simplified energy flows for a test cell fitted with a hydraulic dynamometer and 100 kW gasoline engine*

Energy balance			
In		*Out*	
Fuel	300 kW	Exhaust gas	60 kW
Ventilating fan power	5 kW	Engine cooling water	90 kW
		Dynamometer cooling water	95 kW
		Ventilation air	70 kW
Electricity for cell services	25 kW	Heat loss, walls and ceiling	15 kW
	330 kW		330 kW

The energy balance for the engine, see Chapter 11, is as follows:

In		*Out*	
Fuel	300 kW	Power	100 kW
		Exhaust gas	90 kW
		Engine cooling water	90 kW
		Convection and radiation	20 kW
	300 kW		300 kW

Table 2.2 *Example of the 30–30–30–10 rule*

In via	Out via
Fuel 300 kW	Dynamometer 30% (90 + kW)
	Exhaust system 30% (90 kW)
	Engine fluids 30% (90 kW)
	Convection and radiation 10% (30 kW)

under-rate cell cooling systems. Like any rule of thumb this is crude but does provide a starting point for the calculation of a full energy balance and a datum from which we can evaluate significant differences in balance caused by the engine itself and its mounting within the cell.

Firstly, there are differences inherent in the engine design. Diesels will tend to transfer less energy into the cell than petrol engines of equal physical size. For example, testers of rebuilt bus engines often notice that different models of diesels with the same nominal power output will show quite different distribution of heat into the test cell air and cooling water.

Secondly, there are differences in engine rigging in the cell which will vary the temperature and surface area of engine ancillaries, such as exhaust pipes. Finally, there is the amount and type of equipment within the test cell, all of which makes a contribution to the convection and radiation heat load to be handled by the ventilation system.

Specialist designers have developed their own versions of a software model, based both on empirical data and theoretical calculation, all of which is used within this book, and which produces the type of energy balance shown in Fig. 2.2. Such tools are useful but cannot be used uncritically as the final basis of design, particularly when a range of engines are to be tested or the design has to cover two or more cells, then the energy diversity factor has to be considered.

Diversity factor and the final specification of a facility energy balance

To design a multicell test laboratory able to control and dissipate the maximum theoretical power of all its prime movers on the hottest day of the year will lead to an oversized system and possibly poor temperature control at low heat outputs. The amount by which the thermal rating of a facility is reduced from that theoretical maximum is the diversity factor. In Germany it is called the *Gleichzeitigkeits Faktor* and is calculated from zero heat output upwards, rather than 100 per cent heat output downwards, but the results should be the same, providing the same assumptions are made.

The diversity factor often lies between 60 and 85 per cent of maximum rating, but individual systems will vary from endurance beds with high rating down to anechoic beds with very low rating.

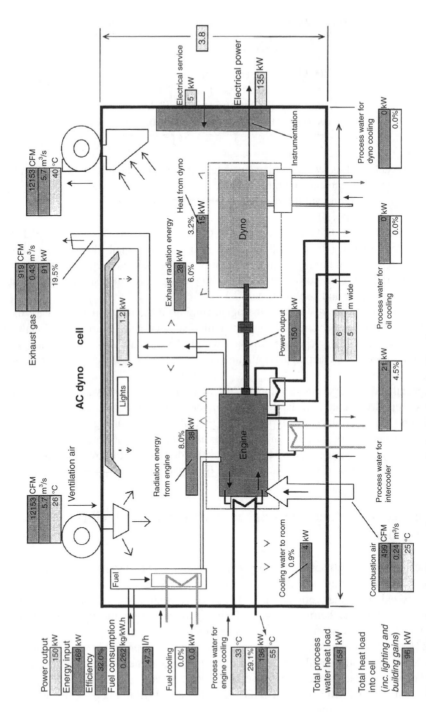

Figure 2.2 Output diagram from test cell thermal analysis software

Power output   150 kW
Energy input   469 kW
Efficiency   32.0%
Fuel consumption   0.262 kg/kW.h
  47.3 l/h
Fuel cooling   0.0%   0.0 kW
Process water for engine cooling   33 °C   29.1%   136 kW   55 °C
Total process water heat load   158 kW
Total heat load into cell (inc. lighting and building gains)   96 kW

Electrical service   5 kW
Electrical power   135 kW
Instrumentation
Heat from dyno   3.2%   15 kW
Exhaust radiation energy   28 kW   6.0%
Dyno
Power output   150 kW
Process water for dyno cooling   0 kW   0.0%
Process water for oil cooling   0 kW   0.0%
6 m   5 m wide
Process water for intercooler   21 kW   4.5%
Combustion air   499 CFM   0.24 m³/s   25 °C
Cooling water to room   0.9%   4 kW
Engine

Exhaust gas   919 CFM   0.43 m³/s   91 kW   19.5%
Lights   1.2 kW
Radiation energy from engine   8.0%   38 kW

Ventilation air   12153 CFM   5.7 m³/s   26 °C
12153 CFM   5.7 m³/s   40 °C

AC dyno   cell

3.8

Fuel

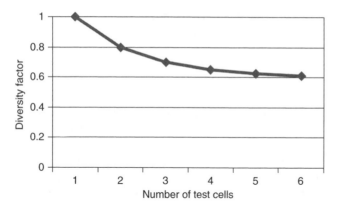

Figure 2.3 *Diversity factor of thermal rating of facility services, plotted against number of test cells, based on empirical data from typical automotive test facilities*

In calculating, or more correctly estimating, the diversity factor it is essential that the creators of the operational and functional specifications use realistic values of actual engine powers, rather than extrapolations based on possible future developments. A key consideration is the number of cells included within the system. It is clear that one cell may at some time run at its maximum rating, but it may be considered less likely that four cells will all run at maximum rating at the same time: the possible effect of this is shown in Fig. 2.3. There is a degree of bravery and confidence, based on relevant experience, required to significantly reduce the theoretical maximum to a contractual specification, but very significant savings in first and running costs may be possible if it is done correctly.

Once a realistic maximum power rating for the facility has been calculated, the facility design team can use information concerning the operating regime, planned test sequences, past records, engine type variation, etc., to draw up diversity factors for heat energy balance and electrical power requirements. Future proofing may be better designed into the facility by incremental addition of plant rather than oversizing at the beginning.

Common or individual services in multicell laboratories?

When considering the thermal loads and the diversity factor of a facility containing several test cells, it is sensible to consider the strategy to be adopted in the design of the various services. The choice has to be based on the operation requirements rather than just the economies of purchasing and running the service modules. Services, such as cooling water (raw water), are always common and treated via a central cooling tower system. Services, such as cell ventilation and engine exhaust gas extraction, may either serve individual cells or are shared. In these cases sharing may show cost savings and simplify the building design by reducing penetrations;

however, it is prudent to build in some standby or redundancy to prevent total facility shut-down in the event of, for example, a fan failure.

A problem that must be avoided in the design of common services is 'cross talk' between cells where the action in one cell, or other industrial plant, disturbs the control achieved in another. This is a particular danger when a service, for example chilled water, has to serve a wide range of thermal loads. In this case a central plant may be designed to circulate glycol/water mix at 6°C through two or more cells wherein the coolant is used by devices ranging from large intercoolers to small fuel conditioners; any sudden increase in demand may significantly increase the system return temperature and cause an unacceptable disturbance in the control temperatures. In such systems there needs to be individual control loops per instrument or a very high thermal inertia gained through the installation of a sufficiently large cold buffer tank.

Summary

The energy balance approach outlined in this chapter will be found helpful in analysing the performance of an engine and in the design of test cell services (Chapters 5 and 6). It is recommended that at an early stage in the design of a new test cell, diagrams such as Fig. 2.2 should be drawn up and labelled with flow and energy quantities appropriate to the capacity of the engines to be tested.

The large quantities of ventilation air, cooling water, electricity and heat that are involved will often come as a surprise. Early recognition can help to avoid expensive wasted design work by ensuring that

- the general proportions of cell and services do not depart too far from accepted practice (any large departure is a warning sign);
- the cell is made large enough to cope with the energy flows involved;
- sufficient space is allowed for such features as water supply pipes and drains, air inlet grilles, collecting hoods and exhaust systems. Note that space is not only required within the test cell but also in any service spaces above or below the cell and the penetrations within the building envelope.

Reference

1. Eastop, T.D. and McConkey, A. (1993) *Applied Thermodynamics for Engineering Technologists*, Longmans, London.

Further reading

Heywood, J.B. (1988) *Internal Combustion Engine Fundamentals*, McGraw-Hill, Maidenhead.

3   Vibration and noise

Introduction

Vibration is considered in this chapter with particular reference to the design and operation of engine test facilities, engine mountings and the isolation of engine-induced disturbances. Torsional vibration is covered as a separate subject in Chapter 9, Coupling the engine to the dynamometer.

The theory of noise generation and control is briefly considered and a brief account given of the particular problems involved in the design of anechoic cells.

Vibration and noise

Almost always the engine itself is the only significant source of vibration and noise in the engine test cell.[1-5] Secondary sources such as the ventilation system, pumps and circulation systems or the dynamometer are usually swamped by the effects of the engine.

There are several aspects to this problem:

- The engine must be mounted in such a way that neither it nor connections to it can be damaged by excessive movement or excessive constraint.
- Transmission of engine-induced vibration to the cell structure or to other buildings must be controlled.
- Excessive noise levels in the cell should be avoided or contained as far as possible and the design of alarm signals should take in-cell noise levels into account.

Fundamentals: sources of vibration

Since the vast majority of engines likely to be encountered are single- or multi-cylinder in-line vertical engines, we shall concentrate on this configuration.

An engine may be regarded as having six degrees of freedom of vibration about orthogonal axes through its centre of gravity: linear vibrations along each axis and rotations about each axis (see Fig. 3.1).

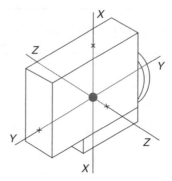

Figure 3.1   *Internal combustion engine: principle axes and degrees of freedom*

In practice, only three of these modes are usually of importance:

- vertical oscillations on the X axis due to unbalanced vertical forces;
- rotation about the Y axis due to cyclic variations in torque;
- rotation about the Z axis due to unbalanced vertical forces in different transverse planes.

Torque variations will be considered later. In general, the rotating masses are carefully balanced but periodic forces due to the reciprocating masses cannot be avoided. The crank, connecting rod and piston assembly shown in Fig. 3.2 is subject to a periodic force in the line of action of the piston given approximately by:

$$f = m_p \omega_c^2 r \cos\theta + \frac{m_p \omega_c^2 r \cos 2\theta}{n} \qquad \text{where } n = l/r \qquad (1)$$

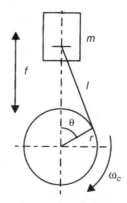

Figure 3.2   *Connecting rod crank mechanism: unbalanced forces*

Here m_p represents the sum of the mass of the piston plus, by convention, one-third of the mass of the connecting rod (the remaining two-thirds is usually regarded as being concentrated at the crankpin centre).

The first term of eq. (1) represents the first-order inertia force. It is equivalent to the component of centrifugal force on the line of action generated by a mass m_p concentrated at the crankpin and rotating at engine speed. The second term arises from the obliquity of the connecting rod and is equivalent to the component of force in the line of action generated by a mass $m/4n$ at the crankpin radius, but rotating at twice engine speed.

Inertia forces of higher order ($3\times$, $4\times$, etc., crankshaft speed) are also generated but may usually be ignored.

It is possible to balance any desired proportion of the first-order inertia force by balance weights on the crankshaft, but these then give rise to an equivalent reciprocating force on the Z axis, which may be even more objectionable.

Inertia forces may be represented by vectors rotating at crankshaft speed and twice crankshaft speed. Table 3.1 shows the first- and second-order vectors for engines having from one to six cylinders.

Table 3.1 *First- and second-order forces, multicylinder engines*

Note the following features:

- In a single cylinder engine, both first- and second-order forces are unbalanced.
- For larger numbers of cylinders, first-order forces are balanced.
- For two and four cylinder engines, the second-order forces are unbalanced and additive.

This last feature is an undesirable characteristic of a four cylinder engine and in some cases has been eliminated by counter-rotating weights driven at twice crankshaft speed.

The other consequence of reciprocating unbalance is the generation of rocking couples about the transverse or Z axis and these are also shown in Fig. 3.1.

- There are no couples in a single cylinder engine.
- In a two cylinder engine, there is a first-order couple.
- In a three cylinder engine, there are first- and second-order couples.
- Four and six cylinder engines are fully balanced.
- In a five cylinder engine, there is a small first-order and a larger second-order couple.

Six cylinder engines, which are well known for smooth running, are balanced in all modes.

Variations in engine turning moment are discussed in Chapter 9, coupling the engine to the dynamometer. These variations give rise to equal and opposite reactions on the engine, which tend to cause rotation of the whole engine about the crankshaft axis. The order of these disturbances, i.e. the ratio of the frequency of the disturbance to the engine speed, is a function of the engine cycle and the number of cylinders. For a four-stroke engine, the lowest order is equal to half the number of cylinders: in a single cylinder there is a disturbing couple at half engine speed while in a six cylinder engine the lowest disturbing frequency is at three times engine speed. In a two-stroke engine, the lowest order is equal to the number of cylinders.

The design of engine mountings and test bed foundations

The main problem in engine mounting design is that of ensuring that the motions of the engine and the forces transmitted to the surroundings as a result of the unavoidable forces and couples briefly described above are kept to manageable levels. In the case of vehicle engines it is sometimes the practice to make use of the same flexible mounts and the same location points as in the vehicle; this does not, however, guarantee a satisfactory solution. In the vehicle, the mountings are carried on a comparatively light structure, while in the test cell they may be attached to a massive pallet or even to a seismic block. Also in the test cell the engine may be fitted with additional equipment and various service connections. All of these factors alter the dynamics of the system when compared with the situation of the engine in service and can give rise to fatigue failures of both the engine support brackets and those of auxiliary devices, such as the alternator.

Truck diesel engines usually present less of a problem than small automotive engines, as they generally have fairly massive and well-spaced supports at the fly-wheel end. Stationary engines will in most cases be carried on four or more flexible mountings in a single plane below the engine and the design of a suitable system is a comparatively simple matter.

We shall consider the simplest case, an engine of mass m kg carried on undamped mountings of combined stiffness k N/m (Fig. 3.3). The differential equation defining the motion of the mass equates the force exerted by the mounting springs with the acceleration of the mass:

$$\frac{m d^2 x}{dt^2} + kx = 0 \tag{2}$$

a solution is

$$x = cons \tan t \times \sin \sqrt{\frac{k}{m}} \cdot t$$

$$\frac{k}{m} = \omega_0^2 \qquad \text{natural frequency} = \eta_0 = \frac{\omega_0}{2\pi} = \frac{1}{2\pi} \sqrt{\frac{k}{m}} \tag{3}$$

the static deflection under the force of gravity $= mg/k$ which leads to a very convenient expression for the natural frequency of vibration:

$$\eta_0 = \frac{1}{2\pi} \sqrt{\frac{g}{\text{static deflection}}} \tag{4a}$$

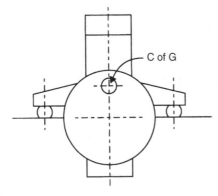

Figure 3.3 *Engine carried on four flexible mountings*

or, if static deflection is in millimetres:

$$\eta_0 = \frac{15.76}{\sqrt{\text{static deflection}}}$$ (4b)

This relationship is plotted in Fig. 3.4

Next, consider the case where the mass m is subjected to an exciting force of amplitude f and frequency $\omega/2\pi$. The equation of motion now reads:

$$m\frac{d^2x}{dt^2} + kx = f \sin \omega t$$

the solution includes a transient element; for the steady state condition amplitude of oscillation is given by:

$$x = \frac{f/k}{(1 - \omega^2/\omega_0^2)}$$ (5)

here f/k is the static deflection of the mountings under an applied load f. This expression is plotted in Fig. 3.5 in terms of the amplitude ratio x divided by static deflection. It has the well-known feature that the amplitude becomes theoretically infinite at resonance, $\omega = \omega_0$.

If the mountings combine springs with an element of viscous damping, the equation of motion becomes:

$$m\frac{d^2x}{dt^2} + c\frac{dx}{dt} + kx = f \sin \omega t$$

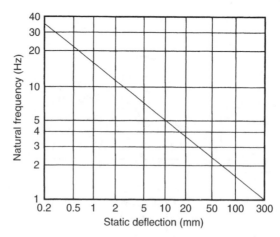

Figure 3.4   *Relationship between static deflection and natural frequency*

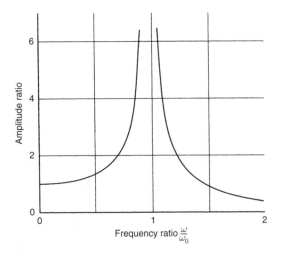

Figure 3.5   *Relationship between frequency and amplitude ratio (transmissibility) undamped vibration*

where c is a damping coefficient. The steady state solution is:

$$x = \frac{f/k}{\sqrt{\left(1 - \frac{\omega^2}{\omega_0^2}\right)^2 + \frac{\omega^2 c^2}{mk\omega_0^2}}} \sin(\omega t - A) \tag{6a}$$

If we define a dimensionless damping ratio:

$$C^2 = \frac{c^2}{4mk}$$

this equation may be written:

$$x = \frac{f/k}{\sqrt{\left(1 - \frac{\omega^2}{\omega_0^2}\right)^2 + 4C^2 \frac{\omega^2}{\omega_0^2}}} \sin(\omega t - A) \tag{6b}$$

(if $C = 1$ we have the condition of critical damping when, if the mass is displaced and released, it will return eventually to its original position without overshoot).

The amplitude of the oscillation is given by the first part of this expression:

$$\text{amplitude} = \frac{f/k}{\sqrt{\left(1 - \frac{\omega^2}{\omega_0^2}\right)^2 + 4C^2 \frac{\omega^2}{\omega_0^2}}}$$

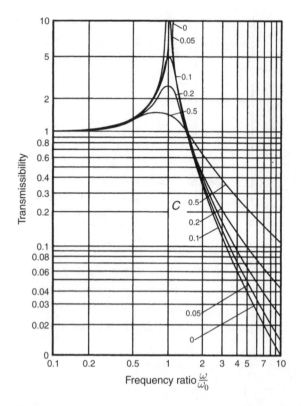

Figure 3.6 *Relationship between transmissibility (amplitude ratio) and frequency, damped oscillations for different values of damping ratio C (logarithmic plot)*

This is plotted in Fig. 3.6, together with the curve for the undamped condition, Fig. 3.5, and various values of *C* are shown. The phase angle *A* is a measure of the angle by which the motion of the mass lags or leads the exciting force. It is given by the expression:

$$A = \tan^{-1} \frac{2C}{\dfrac{\omega_0{}^0}{\omega} - \dfrac{\omega}{\omega_0}} \tag{7}$$

At very low frequencies, *A* is zero and the mass moves in phase with the exciting force. With increasing frequency the motion of the mass lags by an increasing angle, reaching 90° at resonance. At high frequencies the sign of *A* changes and the mass leads the exciting force by an increasing angle, approaching 180° at high ratios of ω to ω_0.

Natural rubber flexible mountings have an element of internal (hysteresis) damping which corresponds approximately to a degree of viscous damping corresponding to $C = 0.05$.

The essential role of damping will be clear from Fig. 3.6: it limits the potentially damaging amplitude of vibration at resonance. The ordinate in Fig. 3.6 is often described as the transmissibility of the mounting system: it is a measure of the extent to which the disturbing force f is reduced by the action of the flexible mounts. It is considered good practice to design the system so that the minimum speed at which the machine is to run is not less than three times the natural frequency, corresponding to a transmissibility of about 0.15. It should be noticed that once the frequency ratio exceeds about 2 the presence of damping actually has an adverse effect on the isolation of disturbing forces.

Practical considerations in the design of engine and test bed mountings

In the above simple treatment we have only considered oscillations in the vertical direction. In practice, as has already been pointed out, an engine carried on flexible mountings has six degrees of freedom (Fig. 3.1). While in many cases a simple analysis of vibrations in the vertical direction will give a satisfactory result, under test cell conditions a more complete computer analysis of the various modes of vibration and the coupling between them may be advisable. This is particularly the case with tall engines with mounting points at a low level, when cyclic variations in torque may induce transverse rolling of the engine.

Reference 6 lists the design factors to be considered in planning a system for the isolation and control of vibration and transmitted noise:

- specification of force isolation
    - as attenuation, dB
    - as transmissibility
    - as efficiency
    - as noise level in adjacent rooms
- natural frequency range to achieve the level of isolation required
- load distribution of the machine
    - is it equal on each mounting?
    - is the centre of gravity low enough for stability?
    - exposure to forces arising from connecting services, exhaust system, etc.
- vibration amplitudes – low frequency
    - normal operation
    - fault conditions
    - starting and stopping
    - is a seismic block or sub-base needed?

- higher-frequency structure-borne noise (100 Hz+)
  - is there a specification?
  - details of building structure
  - sufficient data on engine and associated plant
- transient forces
  - shocks, earthquakes, machine failures
- environment
  - temperature
  - humidity
  - fuel and oil spills.

Detailed design of engine mountings for test bed installation is a highly specialized matter, see Ker-Wilson[1] for guidance on standard practice. In general, the aim is to avoid 'coupled' vibrations, e.g. the generation of pitching forces due to unbalanced forces in the vertical direction, or the generation of rolling moments due to the torque reaction forces exerted by the engine. These can give rise to resonances at much higher frequencies than the simple frequency of vertical oscillation calculated in the following section and to consequent trouble, particularly with the engine-to-brake connecting shaft.

Massive foundations and supported bedplates

The analysis and prediction of the extent of transmitted vibration to the surroundings is a highly specialized field. The theory is dealt with in Ref. 1, the starting point being the observation that a heavy block embedded in the earth has a natural frequency of vibration that generally lies within the range 1000 to 2000 c.p.m. There is thus a possibility of vibration being transmitted to the surroundings if exciting forces, generally associated with the reciprocating masses in the engine, lie within this frequency range. An example would be a four cylinder four-stroke engine running at 750 rev/min. We see from Table 3.1 that such an engine generates substantial second-order forces at twice engine speed or 1500 c.p.m. Figure 3.7, redrawn from Ref. 1, gives an indication of acceptable levels of transmitted vibration from the point of view of physical comfort.

Figure 3.8 is a sketch of a typical seismic block. Reinforced concrete weighs roughly 2500 kg/m³ and this block would weigh about 4500 kg. Note that the surrounding tread plates must be isolated from the block, also that it is essential to electrically earth (ground) the mounting rails. The block is shown carried on four combined steel spring and rubber isolators, each having a stiffness of 100 kg/mm (Fig. 3.9). From eq. 2a, the natural frequency of vertical oscillation of the bare block would be 4.70 Hz or 282 c.p.m., so the block would be a suitable base for an engine running at about 900 rev/min or faster. If the engine weight were, say, 500 kg

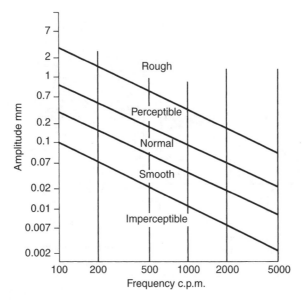

Figure 3.7 *Perception of vibration*

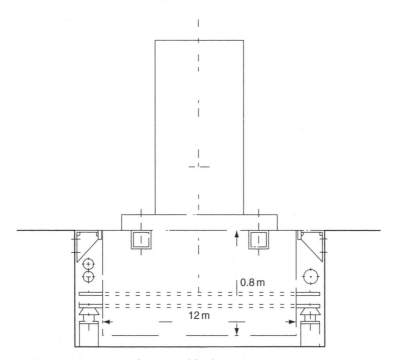

Figure 3.8 *Spring-mounted seismic block*

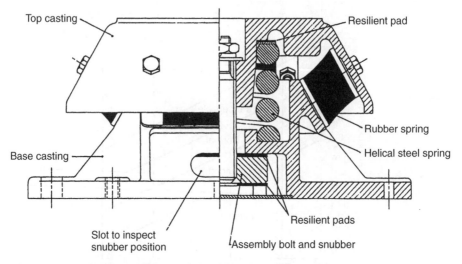

Figure 3.9   *Combined spring and rubber flexible mount*

the natural frequency of block and engine would be reduced to 4.46 Hz, a negligible change. An ideal design target for the natural frequency is considered to be 3 Hz.

Heavy concrete foundations (seismic blocks) carried on a flexible membrane are expensive to construct, calling for deep excavations, complex shuttering and elaborate arrangements, such as tee-slotted bases, for bolting down the engines. With the wide range of different types of flexible mounting now available, it is questionable whether, except in special circumstances, such as a requirement to install test facilities in close proximity to offices, their use is economically justified. The trough surrounding the block may be of incidental use for installing services, if the gap is small then there should be means of draining out contaminated fluid spills.

It is now common practice for automotive engines to be rigged on vehicle type engine mounts, then on trolley systems, which are themselves mounted on isolation feet therefore less engine vibration is transmitted to the building floor. In these cases a more modern alternative to the deep seismic block is shown in Fig. 3.10a and is sometimes used where the soil conditions are suitable. Here the test bed sits on a thickened and isolated section of the floor cast in situ on the compacted native ground. The gap between the floor and block is almost filled with expanded polystyrene boards and sealed at floor level with a flexible, fuel resistant, sealant. A damp-proof membrane should be inserted under both floor and pit or block.

Where the subsoil is not suitable for the arrangement shown in Fig. 3.10a then a pit is required, cast to support a concrete block that sits on a mat or pads of a material such as cork/nitrile rubber composite which is resistant to fluid contamination (Fig. 3.10b), alternatively a cast iron bedplate supported by air springs, as shown in Figure 3.11, may be installed.

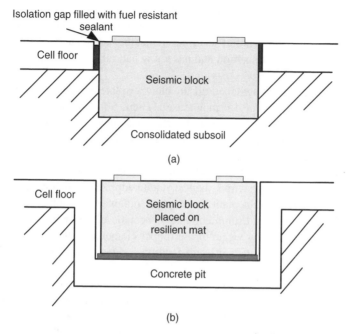

Figure 3.10 *(a) Isolated foundation block for test stand set on to firm subsoil; (b) seismic block on to resilient matting in a shallow pit*

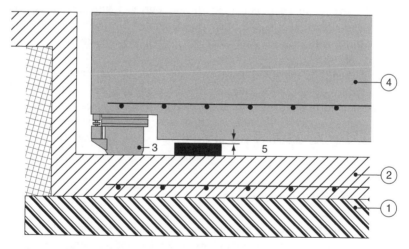

Figure 3.11 *Seismic block or cast bedplate (not shown) (4) mounted on air springs (3) within a shallow concrete pit (2) on consolidated subsoil (1). When the air supply is off the self-levelling air springs allow the block to settle on support blocks (5) the rise and fall typically being 4 to 6 mm. Maintenance access to the springs is not shown*

The use of a cast iron or steel fabricated bedplate mounted on springs is widespread and air-spring support system of four or more units is connected so that, after commissioning, the system is self-levelling so that the base plate remains in a constant load-independent vertical position and has a low natural frequency of between 3 and 2.5 Hz.

When the air supply is switched off the block or plate will settle down and rest on packers, this allows removal for maintenance (item 5 in Fig. 3.11). The air springs require a reliable, low flow, condensate free air supply that is not always easy to provide to the lowest point in the cell system. Sufficient room must be left around the bed plate to allow maintenance access to the air-spring units.

In cases where there is no advantage in having the bedplate face at ground level (as is required for pallet systems), the test stand bedplate can be mounted on the flat cell floor. It is possible that plant and engines mounted on rubber viscous mounts or air-spring systems could, unintentionally, become electrically isolated from the remainder to the facility by virtue of the rubber elements; it is vital that a common grounding scheme is included in such facility designs (see Chapter 10, Electrical design considerations).

A special application concerns the use of seismic blocks for supporting engines in anechoic cells. It is, in principle, good practice to mount engines undergoing noise testing as rigidly as possible, since this reduces noise radiated as the result of movement of the whole engine on its mountings. Lowering of the centre of gravity is similarly helpful since the engines have to be mounted with a shaft centre height of at least 1 metre to allow for microphone placement.

While the dynamometer is not a significant source of vibration, it is common practice to mount both engine and brake on a common block; if they are separated the relative movement between the two must be within the capacity of the connecting shaft and its guard.

Finally it should be remarked that there is available on the market a bewildering array of different designs of isolator or flexible mounting, based on steel springs, air springs, natural or synthetic rubber of widely differing properties used in compression or shear, and combinations of these materials. For the non-specialist the manufacturer's advice should be sought and design sheets obtained (Fig. 3.12).

Summary of vibration section

This section should be read in conjunction with Chapter 9, which deals with the associated problem of torsional vibrations of engine and dynamometer. The two aspects – torsional vibration and other vibrations of the engine on its mountings – cannot be considered in complete isolation.

The exciting forces arising from (inevitable) unbalance in a reciprocating engine are considered and a description of the essential features of mounting design is given, together with a check list of points to be considered.

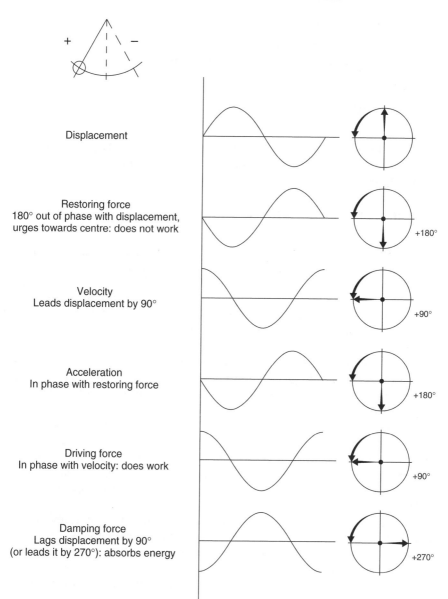

Vibrations of a pendulum or spring and mass

Displacement

Restoring force
180° out of phase with displacement,
urges towards centre: does not work

+180°

Velocity
Leads displacement by 90°

+90°

Acceleration
In phase with restoring force

+180°

Driving force
In phase with velocity: does work

+90°

Damping force
Lags displacement by 90°
(or leads it by 270°): absorbs energy

+270°

Figure 3.12 *Diagram of phase relationship of vibrations*

Noise: fundamentals

Sound intensity

The starting point in the definition of the various quantitative aspects of noise measurement is the concept of sound intensity, defined as:

$$I = \frac{p^2}{\rho c} \text{W/m}^2$$

where p^2 is the mean square value of the acoustic pressure, i.e. the pressure variation due to the sound wave, ρ the density of air and c the velocity of sound in air.

Intensity is measured in a scale of decibels (dB):

$$dB = 10 \ \log_{10} \left(\frac{I}{I_0}\right) = 20 \ \log_{10} \left(\frac{\overline{p}}{\overline{p_0}}\right)$$

where I_0 corresponds to the average lower threshold of audibility, taken by convention as $I_0 = 10^{-12}\text{W/m}^2$, an extremely low rate of energy propagation.

From these definitions it is easily shown that a doubling of the sound intensity corresponds to an increase of about 3 dB ($\log_{10} 2 = 0.301$). A tenfold increase gives an increase of 10 dB, while an increase of 30 dB corresponds to a factor of 1000 in sound intensity. It will be apparent that intensity varies through an enormous range.

The value on the decibel scale is often referred to as the *sound pressure level* (SPL). In general, sound is propagated spherically from its source and the inverse square law applies. Doubling the distance results in a reduction in SPL of about 6 dB ($\log_{10} 4 = 0.602$).

The human ear is sensitive to frequencies in the range from roughly 16 Hz to 20 kHz, but the perceived level of a sound depends heavily on its frequency structure. The well-known Fletcher–Munson curves, Fig. 3.13, were obtained by averaging the performance of a large number of subjects who were asked to decide when the apparent loudness of a pure tone was the same as that of a reference tone of frequency 1 kHz.

Loudness is measured in a scale of *phons*, which is only identical with the decibel scale at the reference frequency. The decline in the sensitivity of the ear is greatest at low frequencies. Thus at 50 Hz an SPL of nearly 60 dB is needed to create a sensation of loudness of 30 phons.

Acoustic data are usually specified in frequency bands one octave wide. The standard mid-band frequencies are:

31.5   62.5   125   250   500   1000   2000   4000   8000   16 000 Hz

e.g. the second octave spans 44–88 Hz. The two outer octaves are rarely used in noise analysis.

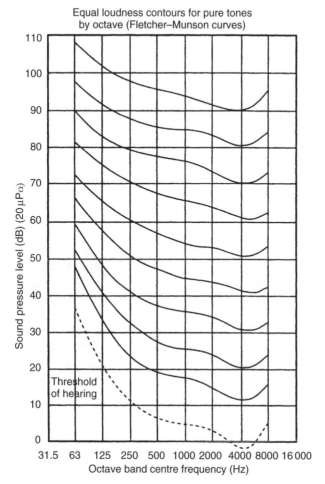

Figure 3.13   *Fletcher–Munson curves of equal loudness*

Noise measurements

Most instruments for measuring sound contain weighted networks which give a response to frequency which approximates to the Fletcher–Munson curves. In other words, their response to frequency is a reciprocal of the Fletcher–Munson relationship (Fig. 3.13). For most applications the A-weighting curve gives satisfactory results and the corresponding SPL readings are given in *dBA*. *B*- and *C*-weightings are sometimes used for high sound levels, while a special *D*-weighting is used primarily for aircraft noise measurements, Fig. 3.14.

The dBA value gives a general 'feel' for the intensity and discomfort level of a noise, but for analytical work the unweighted results should be used. The simplest type of sound level meter for diagnostic work is the octave band analyser. This instrument

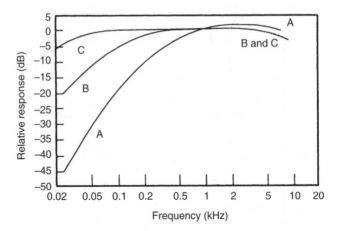

Figure 3.14 *Noise weighting curves*

can provide flat or *A*-weighted indications of SPL for each octave in the standard range. For more detailed study of noise emissions an instrument capable of analysis in one-third octave bands is more effective. With such an instrument it may, for example, be possible to pinpoint a particular pair of gears as a noise source.

For serious development work on engines, transmissions or vehicle bodies much more detailed analysis of noise emissions is provided by the discrete Fourier transform (DFT) or fast Fourier transform (FFT) digital spectrum analyser. The mathematics on which the operation of these instruments is based is somewhat complex, but for-tunately they may be used effectively without a detailed understanding of the theory involved. It is well known that any periodic function, such as the cyclic variation of torque in an internal combustion engine, may be resolved into a fundamental frequency and a series of harmonics (see Chapter 8, Dynamometers and the mea-surement of torque). General noise from an engine or transmission does not repeat in this way and it is accordingly necessary to record a sample of the noise over a finite interval of time and to process these data to give a spectrum of SPL against frequency. The Fourier transform algorithm allows this to be done.

Permitted levels of noise

Noise levels actually within an engine test cell nearly always exceed the levels per-mitted by statutes, while the control room noise level must be kept under observation and appropriate measures taken.

Member European states were required to bring into force the laws, regulations and administrative provisions necessary to comply with the Physical Agents Directive before 15 February 2006. This directive lays down the minimum requirements for the protection of workers from risks including to their health arising or likely to arise from exposure to noise and in a particular risk of permanent damage.

The Directive places the first action daily personal exposure level at 80 dBA (Noise at Work Act 85 dBA in 2005) and the second action level at 85 dBA (Noise of Work Act 90 dBA in 2005). The employer's obligations under prevailing regulations include the undertaking of an assessment when any employee is likely to be exposed to the first action level. This means that hearing protection equipment must be available to all staff having to enter a running cell, where sound levels will often exceed 85 dBA, and that control room noise levels should not exceed 80 dBA.

Noise cancelling earphones, in place of the noise attenuating type, should be considered for use in high noise areas where vocal communication between staff is required.

Noise external to test facility and planning regulations

Almost all planning authorities will impose some form of restriction on the building of a new engine test laboratory that will seek to restrict the noise pollution caused. In the case of an existing industrial site, these restrictions will often take the form of banning any increase in sound levels at the nominated boundaries of the site. It is vital therefore that in the early planning and design stages a set of boundary noise readings are taken by both the system integrator and the user. The survey should be based on readings taken at important points, such as the boundary position nearest to residential buildings and located on an authorized map or with GPS readings. The readings should be taken in as normal working situation as possible without exceptional occurrences and at several times throughout the period the facility is to be used. Such a datum set of readings should be agreed as relevant with any interested party taken before work starts and will provide a key reference for any dispute post-commissioning.

Noise in the test cell environment

The measured value of SPL in an environment such as an engine test cell gives no information as to the power of the source: a noisy machine in a cell having good sound-absorbent surfaces may generate the same SPL as a much quieter machine surrounded by sound-reflective walls. The absorption coefficient is a measure of the sound power absorbed when a sound impinges once upon a surface. It is quite strongly dependent on frequency and tends to fall as frequency falls below about 500 Hz.

Information on absorption coefficients for a wide range of structural materials and sound insulators is given in IHVE Guide B 12. A few approximate values are given in Table 3.2 and indicate the highly reverberatory properties of untreated brick and concrete.

Table 3.2 *Absorption coefficients at an octave band centre frequency of 1 kHz*

Concrete, brickwork	0.03
Glass	0.1
Breeze blocks	0.5
Acoustic tiles	0.9
Open window	1.0

It should be remembered that the degree to which sound is absorbed by its surroundings makes no difference to the intensity of the sound received directly from the engine.

'Cross talk' between test cell and control room and other adjacent rooms can occur through any openings in the partition walls and also through air conditioning ducts and other service pipes, when there is a common system.

Vehicle noise measurements

A number of regulations lay down permitted vehicle noise levels and the methods by which they are to be measured. In the EC it is specified that A-weighted SPL should be measured during vehicle passage at a distance of 7.5 m from the centre line of the vehicle path. The EEC rules on the permissible sound level and the exhaust system of motor vehicles (70/157 adapted by 1999/101) covers all motor vehicles with a maximum speed of more than 25 km/h. The limits range from 74 dB(A) for motor cars to 80 dB(A) for high-powered goods vehicles.

Many motor sport venues worldwide have introduced their own noise limits for vehicles using their tracks, often in the range of 95–98 dBA measured at a fixed percentage of engine speed and a set distance from the exhaust relying on the use of hand-held meters operated by scrutineers. Results of such tests can be highly misleading if care is not taken to avoid sound reflection from surroundings, nor do they presently measure noise levels coming from induction systems. Some tracks are now fitted with complex 'drive-by' monitoring systems which more closely represent the 'noise pollution' created.

Anechoic test cells

The testing of engines and vehicles for noise, vibration and harshness (NVH) forms a substantial element of vehicle development programmes, and calls for specialized (and very expensive) test facilities. Essentially, such a cell should provide an environment approximating as closely as possible to that of the vehicle on the road.

Such cells are of two kinds: *semi-anechoic* (in the USA, commonly and correctly 'hemi-anechoic') in which walls and ceiling are lined with sound-absorbent materials while the floor is reflective, generally of concrete, and *full anechoic cells,* in which all surfaces including the floor are sound absorbent.

A full anechoic cell is adapted to measuring the noise radiated in every direction from a source, while the semi-anechoic cell simulates the situation where the source is located in the open, but resting on a reflective horizontal plane; clearly, the latter is appropriate for land vehicle testing.

The design of anechoic cells is a matter for the specialist, but BS 4196, while specifically concerned with methods of determining sound power levels of noise sources, gives some useful guidance on certain aspects of anechoic cell design, including

- desirable shape and volume of the cell;
- desirable absorption coefficient of surfaces;
- specification of absorptive treatment;
- guidance regarding avoidance of unwanted sound reflections.

A key part of the technical specification of an anechoic cell is known as the 'cut-off' frequency which is the frequency below which the rate of sound level decay in the chamber no longer replicates the outdoors or, more precisely, a 'free-field' environment. The cut-off frequency normally required for automotive anechoic cells (both engine and vehicle) is around 120 Hz.

The acceptance test of the anechoic characteristic of the cell is carried out using a broadband generator placed at the geometric centre of the cell. The level of sound decay is measured by a microphone physically moved along chords tensioned between the noise source and the high level corners of the cell. The decay measured is compared with the 6 dB distance doubling characteristic of a free-field. The task of certification of an anechoic cell and equipment required is a specialist field.

There are certain other points that should be borne in mind by the non-specialist if he is required to take responsibility for setting up a test facility of this kind:

- The structure of an anechoic cell should be isolated from any environment in which noise is generated, such as production plant. Common ventilation systems should be avoided as should any other paths to facility spaces that act for noise or vibration transmission.
- The internal volume of the building shell will be roughly twice that of the usable space because of the space required for the acoustic lining. This lining often takes the form of foam shapes which may protrude some 700 mm from the supporting structure, which may itself be some 300 mm from the cell wall.
- Access doors and windows affect the acoustic performance adversely and must be kept to a minimum commensurate with safety. The acoustic lining covering exits should be coloured to make their location obvious. Windows should preferably be avoided since glass has poor sound-absorbent properties.

- Acoustic research makes use of highly sensitive instruments and correspondingly sensitive measurement channels. The design of the signal cabling must be carefully considered at the design stage to avoid interference and a multitude of trailing leads in the cell.
- The ventilation of acoustic cells presents particular problems. In most ordinary test work, the noise contributed by the ventilation system may be ignored but this is not the case with acoustic cells, where the best solution is usually to dispense with a forced draught fan and to draw in the air by way of suitable filters and silencers, using only an induced draught fan. However, full power running in anechoic cells is normally of short duration and the precise control of inlet air temperature is not so critical as in cells concerned with power measurement.
- Acoustic cell linings usually absorb liquids and may be inflammable, increasing fire hazards. Particular attention should be paid to smoke detection and to the fire quenching system.
- The dynamometer must clearly be isolated acoustically from the cell proper. For work on engines and transmissions, the dynamometer is located outside the cell, calling for an overlong coupling shaft, the design of which will present problems. In many designs the shaft system is split into two sections with an intermediate bearing mounted on a support block just inside the anechoic chamber.
- For work on complete vehicles a rolling road dynamometer, Chapter 18, will be the usual solution: this will transmit a certain amount of noise to the cell which may be measured by motoring the rolls in the absence of a vehicle.
- The engine in an anechoic cell must be raised above floor level, typically by 1.0–1.2 m to shaft centre line, to permit microphones to be located below it. This calls for non-standard engine mounting arrangements.

Exhaust noise

The noise from test cell exhaust systems can travel considerable distances and be the subject of complaints from neighbouring premises, particularly if running takes place at night or during weekends. The design of test cell exhaust systems is largely dictated by the requirement that the performance of engines under test should not be adversely affected and by restraints on noise emission, both internally and external to the building. Essentially, there are two types of device for reducing the noise level in ducts: resonators and absorption mufflers. A resonator, sometimes known as a reactive muffler, is shown in Fig. 3.15. It consists of a cylindrical vessel divided by partitions into two or more compartments. The exhaust gas travels through the resonator by way of perforated pipes which themselves help to dissipate noise. The device is designed to give a degree of attenuation, which may reach 50 dB, over a range of frequencies, see Fig. 3.16 as a typical example.

Absorption mufflers, which are the type most commonly used in test facilities, consist essentially of a chamber lined with sound-absorbent material through which the exhaust gases are passed in a perforated pipe. Absorption mufflers give broadband

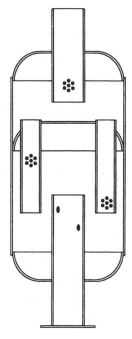

Figure 3.15 *Exhaust resonator or reactive muffler*

damping but are less effective than resonators in the low-frequency range. However, they offer less resistance to flow.

Selection of the most suitable designs for a given situation is a matter for the specialist. Both types of silencer are subject to corrosion if not run at a temperature above the dew point of the exhaust gas and condensation in an absorption muffler is particularly to be avoided.

As with a number of engine attached devices, the exhaust silencer may be considered as part of the engine rigging when it is engine model specific, or it may be considered as part of the cell, or both may be used. Modern practice tends to use the vehicle exhaust system complete with silencers within the cell to extract the gas mixed with a proportion of cell air into a duct fitted, before exiting the building, via absorption attenuators. Fans used in the extraction are rated to work up to gas/air temperatures of 200°C and the effect of the temperature on the density of the mixture and therefore the fan power must be remembered in the design process.

Tail pipes

The noise from the final pipe section is directional and therefore often points skywards, although the ideal would be to terminate in a wide radius 90° bend away

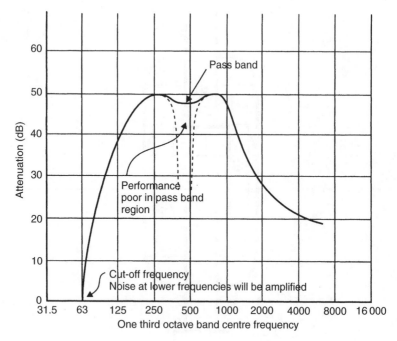

Figure 3.16   *Sample performance of a reactive muffler*

from the prevailing wind. The pipe end can, if incorrectly positioned and with a plain 90° end, suffer from wind-induced pressure effects; this has led to difficulties in getting correlation of results between cells. It has to be remembered that the condensate of exhaust gases is very corrosive, also that rain, snow, etc., should not be allowed to run into undrainable catch points or the engine. Where there is a need to minimize any smoke plume, a tail pipe of the design shown in Fig. 6.7a may assist in mixing and dispersal.

Notation

Vibration

Mass of piston +1/3 connecting rod	m_p kg	
Angular velocity of crankshaft	$\omega_c S^{-1}$	
Crank radius	r m	
Crank angle from t.d.c.	θ	
Connecting rod length/r	n	

Unbalanced exciting force	f N
Mass of engine	m kg
Combined stiffness of mountings	k N/m
Amplitude of vibration	$\times$ m
Angular velocity of vibration	ω rad/s
Angular velocity at resonance	ω_0 rad/s
Natural frequency	n_0 Hz
Phase angle	A rad
Damping coefficient	cN s/m
Damping ratio	C
Acceleration due to gravity	g m/s^2

Notation

r.m.s. value of acoustic pressure	$\bar{p}$ N/m^2
Density of air	ρ kg/m^3
Velocity of sound in air	c m/s
Sound intensity	I W/m^2
Threshold sound intensity	I_o W/m^2

References

1. Ker-Wilson, W. (1959) *Vibration Engineering*, Griffin, London.
2. Thompson, W.T. (1988) *Theory of Vibration*, 3rd edn, Prentice-Hall, London.
3. Fader, B. (1981) *Industrial Noise Control*, Wiley, Chichester.
4. Turner, J.D. and Pretlove, A.J. (1991) *Acoustics for Engineers*, Macmillan Education, London.
5. Warring, R.H. (1983) *Handbook of Noise and Vibration Control*, 5th edn, Trade and Technical Press, Morden, UK.
6. Maw, A.N. (1992) The design of resilient mounting systems to control machinery noise in buildings. *Plastics, Rubber and Composites Processing and Applications* 18, 9–16.

Further reading

BS 3425 *Method for the Measurement of Noise emitted by Motor Vehicles.*
BS 3539 *Specification for Sound Level Meters for the Measurement of Noise Emitted by Motor Vehicles.*
BS 4196 Parts 0 to 8 *Sound Power Levels of Noise Sources.*
BS 2475 *Specification for Octave and One-third Octave Band-pass Filters.*
BS 3045 *Method of Expression of Physical and Subjective Magnitudes of Sound or Noise in Air.*
BS 4198 *Method for Calculating Loudness.*

BS 4675 Parts 1 and 2 *Mechanical Vibration in Rotating Machinery.*

BS 5330 *Method of Testing for Estimating the Risk of Hearing Handicap due to Noise Exposure.*

I.H.V.E. Guide B12: *Sound Control,* Chartered Institution of Building Services, London.

Noise at Work Regulations (1990) Health and Safety Executive, London.

4 Test cell and control room design: an overall view

Introduction

This chapter is intended for readers who may lack experience of engine testing or test engineers having to consider the specification of test cells and deals in broad outline with salient features of the main types of engine test cell and the associated operators control space, ranging from the simplest possible area for the user with an occasional requirement to run a test, to complex multiple cell installations of vehicle manufacturer and research organizations. Questions involved in the sizing of test cells, control rooms, the provision of services, engine handling and safety and fire suppression are discussed. Following a description of typical cell types, the subject of optimum space required for the cell, control room and services is covered.

It is assumed that those involved in the design process have an operational specification (see Chapter 1) on which to base their discussions.

Overall size of individual test cells

One of the early considerations in planning a new test facility will be the space required. The areas to be separately considered are

- the engine or powertrain test cell;
- the control room;
- the space required for services and support equipment;
- the support workshop or engine rig and derig area;
- the storage area required for engine rig items and consumables.

A cramped cell, in which there is not room to move around in comfort, is a permanent source of danger and inconvenience. The smaller the volume of the cell the more difficult it is to control the ventilation system under conditions of varying load (see Chapter 5). As a rule of thumb, there should be an unobstructed walkway 1 metre wide all round the test engine. Cell height is determined by a number of factors including the provision or not of a crane beam in the structure. In practice, most modern automotive cells are between 4 and 4.5 metres internal height.

Table 4.1   *Some actual cell dimensions found in UK industry*

6.5 m long × 4 m wide × 4 m high	QA test cell for small automotive diesels fitted with eddy-current dynamometer
7.8 m long × 6 m wide × 4.5 m high	ECU development cell rated for 250 kW engines, containing work bench and some emission equipment
6.7 m long × 6.4 m wide × 4.7 m high	Gasoline engine development cell with a.c. dynamometer, special coolant and intercooling conditioning
9.0 m wide × 6 m × 4.2 m high (to suspended ceiling)	Engine and gearbox development bed with two dynamometers in 'Tee' configuration. Control room runs along 9.0 metre wall

It is often necessary, when testing vehicle engines, to accommodate the exhaust system as used on the vehicle, and this may call for space and extra length in the cell. It must be remembered that much of the plant in the cell requires calibration from time to time and there must be adequate access for the calibration engineer, his instruments and in some cases ladder. The major layout problem may be caused by the calibration of the dynamometer, involving accommodation of a torque arm and dead weights.

Control room space depends significantly upon the individual operations management; some are spaces where the engine operator stands and others are work space and conference area for a development team (Table 4.1).

Some typical test cell designs

The basic minimum

There are many situations in which there is a requirement to run engines under load, but so infrequently that there is no economic justification for building a permanent test cell. Examples are organizations concerned with engine overhaul and rebuilding, and specialist engine tuners.

To meet needs of this sort all that is required is a suitable area provided with

- water supplies and drains;
- portable fuel supply system;
- adequate ventilation;
- arrangements to take engine exhaust to exterior;
- minimum necessary sound insulation;
- adequate fire and safety precautions.

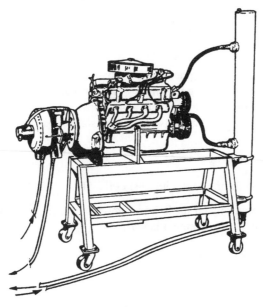

Figure 4.1 *Portable engine test stand*

Figure 4.1 shows a typical installation, consisting of the following elements:

- portable test stand for engine and dynamometer;
- engine cooling system;
- engine exhaust system;
- control console.

A 'bolt-on' dynamometer (see Chapter 8) requires no independent foundation, and is particularly suitable for a simple mobile installation. This type of dynamometer is bolted to the engine bell housing, using an adaptor plate, with a splined shaft connection engaging the clutch plate. Occasionally, the dynamometer may be installed without removal of the engine from a truck chassis by dropping the propeller shaft and mounting the dynamometer on a hanger frame bolted to the chassis.

The dynamometer cooling water can be simply run from mains supply and run to waste or it can be circulated by a pump within a sump into which the water drains. The engine cooling system includes a cooling column that maintains the top-hose temperature to a set figure and minimizes the quantity of cooling water run to waste or back to the sump. In some cases the radiator associated with the engine under test and placed in front of a constant speed fan may be used.

The control console requires, as a minimum, indicators for dynamometer torque and speed. The adjustment of the dynamometer flow control valve and engine throttle may be by cable linkage or simple electrical actuators. The console should also house the engine stop control and an oil pressure gauge. A simple manually operated fuel consumption gauge of the 'burette' type is adequate for this type of installation.

Typical applications are

- proving of truck or bus engines after overhaul or rebuild;
- tuning of engines for racing in same workshops, etc.;
- military vehicle overhaul in portable workshops;
- checking, post-rebuild, emissions against legislative requirements.

General purpose automotive engine test cells, 50 to 450 kW

This category represents by far the largest number of test cells in service (see Figs 4.2 and 4.3 for variants of this common layout).

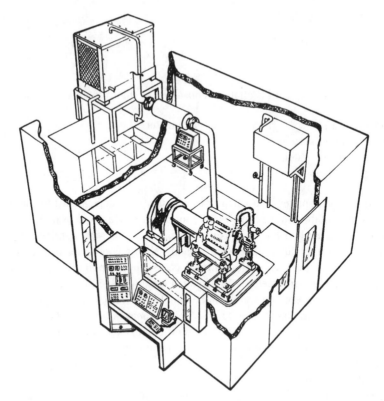

Figure 4.2 *General purpose test cell arranged against an outside wall with control desk 'side on' to engine*

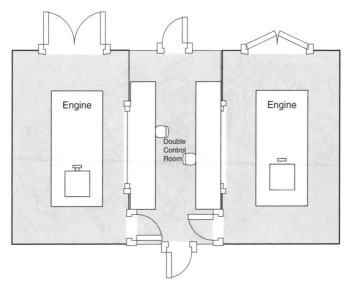

Figure 4.3 *Engine test cells with shared control rooms running down the length of the cell*

Such cells are often built in multiples, side-by-side and in a line with a common control room that is shared by two cells on either side. Engines enter the cell by way of a large door in the rear wall while the operator may enter by way of a door in the front wall to one side of the control desk. Typically there is a double-skinned toughened glass window in front of the control desk. Most wall-mounted instrumentation, smoke meters, fuel consumption meters, etc., is carried on the side wall remote from the cell access door.

In cells rated at above about 150 kW where engine changes and rigging were carried out in the cell it was usual to provide a crane rail located above the test bed axis with a hoist of sufficient capacity to handle engines and dynamometer. The penalty of such a lifting beam is that the structure has to be built to take the full rated crane load plus its plant support load. In modern cells where the engine is rigged outside the cell and trolley or pallet mounted, in-cell cranes are not usually included as the cost benefit of a crane structure may be judged as marginal.

Figure 4.4 shows a dynamometer and engine support mounted on a common cast iron tee-slotted bedplate; an assembly following this basic format forms the 'island' at the centre of most engine test cells.

Signal conditioning units for the various engine transducers (pressure, temperature, etc.) are usually housed in a box carried by an adjustable boom cantilevered and hinged from the cell wall. The boom provides a route for segregated cable routes from engine instrumentation to the control room.

Boom boxes can be of considerable size, with contents that increase in the process of cell development. The ability of the wall to take the stresses imposed by the

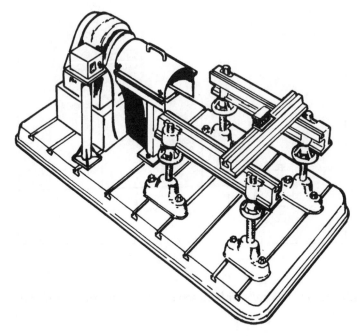

Figure 4.4 *Dynamometer and engine mounting stand on a common tee-slot bedplate. This type of stand, requiring crane loading of the engine, is being replaced by pallet or trolley mounting systems*

cantilevered load of the boom and its attachments should be carefully checked. Where headroom above the boom is restricted and it is not possible to fit a sufficiently angled tie-wire then the boom may have to be attached directly to a steel pillar fixed at floor and ceiling.

In the cell illustrated (Fig. 4.2), exhaust pipes lead from the engine manifold to the in-cell exhaust and silencing system via a flexible connection section. Each exhaust outlet can include an exhaust back pressure control valve.

Such a general purpose cell should be adaptable to take a wide range of engines within its thermal rating, but the number of changes is typically not more than one or two per week. An adjustable engine mounting as shown in Fig. 4.4 is common; such a universal stand can give maximum flexibility but set-up times may be rather long.

The layout shown in Figs 4.2 and 4.3, with the test bed axis coinciding with the control desk, gives good control window visibility, but is less easy to stack in a multiple cell layout than that of Fig. 4.5 which has the engine access and control area at the engine end of the cell. This layout works best when the control desk is within a wide corridor, but compromises the layout of the control space and makes confidentiality of individual control areas difficult to arrange.

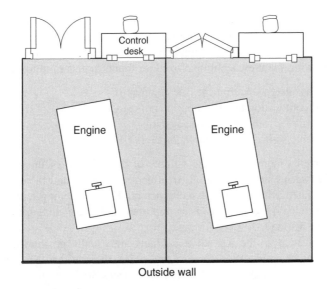

Figure 4.5   *Engine test cells having engine entry from the control corridor allowing side-by-side 'stacking' of cells*

Typical applications are

- development departments of engine and component manufacturers;
- independent testing and development laboratories;
- fuel and lubricant development and standard test procedures;
- training and educational facilities;
- military workshops;
- specialized test cells for research and development.

Research and development engine and powertrain test cells

In this book, the term 'powertrain' testing is used to cover engine plus direct mounted transmission, but does not cover vehicle transmissions complete with vehicle shafting; these are considered as 'transmission test rigs'.

Powertrain rigs are often designed to be able to take up different configurations, on a large bedplate, as required by the unit under test (UUT). The typical configurations catered for are

- transverse engine plus gearbox driving two dynamometers;
- in-line engine and gearbox driving one dynamometer;
- in-line engine plus gearbox driving two dynamometers ('Tee' set-up).

To allow fast transition times, the various UUT have to be pallet mounted in a system that presents a common height and alignment to the cell interface points.

Automotive engine test and development facilities built since the mid-1990s will have the following features, additional to those found in general purpose cells:

1.   Exhaust gas analysis equipment (see Chapter 16)
2.   Dedicated combustion air treatment plant (see Chapter 6)
3.   Ability to run 'in vehicle' exhaust systems
4.   High dynamic four-quadrant dynamometers (see Chapter 8).

Such requirements increase both the volume of the cell and of the space required to house plant such as combustion air treatment equipment and the electrical drive cabinets associated with the a.c. dynamometers. The positioning and condition in which such 'service' plant operates is as important and as demanding as those of the contents of the test cell.

While plant such as the a.c. drive cabinets are usually positioned outside the cell, for reasons of ambient temperature and noise, they need to be as close to the dynamometer as practical to avoid higher than necessary costs for the connecting power cable.

Modern dynamometer drive cabinets tend to be large, heavy and difficult to manoeuvre within restricted building spaces. Therefore, expert planning is required, as is anti-condensation heating if there is a long period between installation and commissioning. Combustion air units also need to be close to the engine to prevent heat loss or gain through the delivery trunking. They are often positioned in the service room immediately above the engine, with the delivery duct running via a fire damper through the cell ceiling to a flexible duct attached to the engine as part of the rigging process. If the humidity of the combustion air is being controlled, the unit will require condensation and steam drains out of the building or into foul water drains.

Typical applications are

• development departments of vehicle and engine manufacturers and major oil companies;
• motor-sport developers;
• specialist consultancy companies;
• government testing and monitoring laboratories.

Inclined, static or dynamic automotive engine test beds

With the increased use of 'off-road' vehicles there is likely to be an occasional requirement for test beds capable of handling engines running with the crankshaft centre line inclined to the horizontal. This will present problems with hydraulic dynamometers with open water outlet connections. Closed circuit cooling systems, as are usual with eddy current machines, or electrical motor-based machines are easily adapted to inclined running.

Some rigs have been built where the whole engine and dynamometer bedplate is mounted upon a system of hydraulic cylinders allowing for dynamic movement in three planes while the engine is running. The cost and operational complexity of such installations will be obvious to most readers.

For the very occasional case of the engine with vertical crankshaft, e.g. outboard boat engines tested without dummy transmissions, the electrical dynamometer is the obvious choice and may generally be used without modification, although dry gap eddy current machines have also been used. Special arrangements need to be made for torque calibration.

Automotive engine production test cells (hot test)

These cells are highly specialized installations forming part of an automation system lying outside the scope of this book. The objective is to check, in the minimum possible process time, that the engine is complete and runs. Typical 'floor-to-floor' times for small automotive engines range between 5 and 8 minutes.

The whole procedure – engine handling, rigging, clamping, filling, starting, draining and the actual test sequence – is highly automated, with interventions, if any, by the operator limited to dealing with fault identification. Leak detection may be difficult in the confines of a hot test stand therefore it is often carried out at a special (black-light) station following test.

The test cell is designed to read from identity codes on the engine and recognize variants to adjust the pass or fail criteria accordingly.

Typical measurements made during a production test include

- time taken for engine to start;
- cranking torque;
- time taken for oil pressure to reach normal level;
- exhaust gas composition.

Most gasoline engines are no longer loaded by any form of dynamometer during a hot test, but in the case with diesel engines load is applied with power output measured and recorded.

Cold testing in production

In addition to rotational testing 'in process' of subassemblies, cold testing is sometimes applied to (near) completely built engines. This is a highly automated process (see Chapter 19, section Data processing in cold testing).

Cold test rigs are invariably situated within the assembly process line fed by a 'power and free' or similar conveyor system. The mechanical layout, based on docking on to a drive motor and transducer pick-up system, is physically simpler than a hot test cell since it does not require fuel supplies or hot gas and fluid evacuation.

Cold test areas also have the cost advantage of being without any significant enclosure other than safety guarding.

Production hot test cell layout

In the design phase of a production hot test facility a number of fundamental decisions have to be made, including

- layout, e.g. conveyor loop with workstations, carousel;
- what remedial work, if any, to be carried out on test stand;
- processing of engines requiring minor/major rectification;
- engine handling system, e.g. bench height, conveyor and pallets, 'J' hook conveyor, automated guided vehicle;
- engines rigged and derigged at test stand or remotely;
- storage and recycling of rigging items;
- maintenance facilities and system fault detection;
- measurements to be made, handling and storage of data.

Production testing imposes heavy wear and tear, particularly on engine rigging components which need constant monitoring and spares available.

Automatic shaft docking systems may represent a particularly difficult design problem where multiple engine types are tested, and where faulty engines are cranked or run for periods giving rise to unusual torsional vibration and torque reversals.

Shaft docking splines need adequate lubrication and, like any automatic docking item, can become a maintenance liability if one damaged component is allowed to travel round the system, causing consequential damage to mating parts.

Modular construction of key subassemblies will allow repairs to be carried out quickly by replacement of complete units thus minimizing production downtime.

Large- and medium-speed diesel engine test areas

As the physical size of an engine increases the logistics of handling them becomes more significant, therefore the test area is often located within the production plant close to the final build area.

Above a certain size engines are tested within an open shop in which they have been finally assembled. The dynamometers designed to test engines in the ranges of about $\geq 20\,MW$ are small in comparison to the prime mover, therefore the test equipment is brought to the engine rather than the more usual arrangement where the engine is taken to a cell.

Cells for testing medium-speed diesels require access platforms along the sides of the engine to enable rigging of the engine and inspection of the top mounted equipment, including turbochargers, during test. There is a design temptation to install services under these platforms but these spaces can be difficult and unpleasant to access, therefore the maintenance items such as control valves should, where possible, be wall or boom mounted.

Rig items can be heavy and unwieldy. Rigging of engines of this size is a design exercise in itself but the common technique is to prerig engines of differing configurations in such a way that they present a common interface when put in the cell. This allows the cell to be designed with permanently installed semi-automated or power-assisted devices to connect exhausts, intercooler and engine coolant piping. The storage of rig adaptors will need careful layout in the rig/derig area.

Shaft connection is usually manual with some form of assisted shaft lift and location system.

Special consideration should be given in these types of test areas to the draining, retention and disposal of liquid spills or wash-down fluids.

Large engine testing often exceeds the normal working day, therefore running at night or weekends is not uncommon and may lead to complaints of exhaust noise or smoke from residential areas nearby. Each cell or test area will have an exhaust system dedicated to a single engine; traditionally and successfully the silencers have been of massive construction built from the ground at the rear of the cells. Modern versions may be fitted with smoke dilution cowls. These exhaust systems require condensate and rain drains to prevent accelerated corrosion.

Seeing and hearing the engine

Except in the case of production test beds, large diesel engines and portable dynamometer stands, it is almost universal practice to separate the control space from the cell proper. This space may be in a corridor joining several similar workstations. It can be a separate, cell specific, control room or shared and sandwiched between the lengths of two cells.

Thanks to modern instrumentation and closed circuit television, a window between the control room and test cell is rarely absolutely necessary, although many users will specify one. The cell window causes a number of design problems; it compromises the fire rating and sound attenuation of the cell and uses valuable space in the control room. The choice to have a window or not is often made on quite subjective grounds. Cells in the motor sport industry all tend to have large windows because they are visited by sponsors and press, whereas OEM multicell research facilities increasingly rely on remote monitoring and closed-circuit television. With the advent of the thin, flat computer screen many cell windows have become compromised by banked arrays of displays.

The importance or otherwise of engine visibility from a window is linked with a fundamental question: which way round is the cell to be arranged?

It is often more convenient from the point of view of engine handling and installation to have the engine at the rear, adjacent to a rear access door, but this gives the worst visibility if the design relies only on a window.

The experienced operator will be concentrating attention on the indications of instruments and display screen, and will only catch changes happening in the cell out of the corner of an eye. It is important to avoid possible distractions, such as dangling labels or identity tags that can flutter in the ventilation wind.

Hearing has always been important to the experienced test engineer, who can often detect an incipient failure by ear well before it manifests itself in any other way.

Unfortunately, modern test cells, with their generally excellent sound insulation, cut off this source of information and consideration should be given to the provision of in-cell microphones with external loudspeakers or earphones.

Flooring and subfloor construction

The floor, or seismic block when fitted (see Chapter 3, Vibration and noise), must be provided with arrangements for bolting down the engine and dynamometer. A good solution is precisely level and cast in two or more cast-iron T-slotted rails. The machined surfaces of these rails form the datum for all subsequent alignments and they must be set and levelled with great care. Any twist could lead to serious distortion. The use of fabricated box beams is a false economy.

Sometimes complete cast-iron floor slabs with multiple T-slots are used as the cell floor rather than as a spring supported floor plate within a pit; these tend to trap liquids and are highly sound-reflective. They can also be very slippery.

Floors should have a surface finish that does not become unduly slippery when fluids have been spilt; special marine deck paints are recommended over the more cosmetically pleasing smooth finishes; however, any surface finish used must be able to resist the fluids, including fuels, used in the cell.

It is good practice to provide floor channels on each side of the bed, as they are particularly useful for running 'cold' services and drains in an uncluttered manner; however, regulations in most countries call for spaces below floor level to be scavenged by the ventilating system to avoid any possibility of the build-up of explosive vapours. Fuel services should preferably not be run in floor trenches.

Floor channels should be covered with well-fitting chequer plates, not weighing more than about 20 kg each, and provided with lifting holes. The plates can be cut as necessary to accommodate service connections.

Doors

Doors that meet the requirements of noise attenuation and fire containment are inevitably heavy and require more than normal effort to move them; this is a safety consideration to be kept in mind when designing the cell. Forced or induced ventilation fans can give rise to pressure differences across doors, possibly making it dangerous or impossible to open a large door. The recommended cell depression for ventilation control is 50 Pa.

All test cell doors must be either on slides or be outward opening. There are designs of both sliding and hinged doors that are suspended and drop to seal in the closed position.

Sliding doors have the disadvantage of creating 'dead' wall space when open. Doors should be provided with small observation windows and may be subject to regulations regarding the provision of exit signs.

The cell operator should give consideration to interlocking the cell doors with the control system to prevent human access during chosen operating conditions. A common strategy is to force the engine into a 'no load/idle' state as soon as a door is opened.

Walls and roof

Test cell walls are required to meet certain special demands in addition to those normally associated with an industrial building. They, or the frame within which they are built, must support the load imposed by any crane installed in the cell, plus the weight of any equipment mounted on or suspended below the roof. They must be of sufficient strength and suitable construction to support wall-mounted instrumentation cabinets, fuel systems and any equipment carried on booms cantilevered out from the walls. They should provide the necessary degree of sound attenuation and must comply with requirements regarding fire retention (usually a minimum of one hour containment).

High density building blocks provide good sound insulation which may be enhanced by filling the voids with dried casting sand, but this may cause difficulties in creating wall penetrations after the original construction. Walls of whatever construction usually require some form of internal acoustic treatment, such as 50-mm thick sound absorbent panels, to reduce the level of reverberation in the cell. Such panels can be effective on walls and ceilings, even if some areas are left uncovered for the mounting of equipment. The alternative, easier to clean, option of having ceramic tiled walls is not uncommon, where test operators are not allowed in the cell, to be exposed to high noise levels, during engine running.

The choice of cell wall material has been complicated in recent years by the availability of special construction panels made of sound absorbent material sandwiched between metal sheets, of which the inner (cell) side is usually perforated. These 100-mm thick panels may be used with standardized structural steel frames for the construction of test cells and, while not offering the same level of sound attenuation as dense block walls, give a quick and clean method of construction with a pleasing finished appearance; this is particularly useful when cells are to be built in an existing building. However, proper planning is necessary if heavy equipment is to be fixed as internal 'hard-mounting' points must be built into the steel structure.

The roof of a test cell often has to support the services housed above which may include large and heavy electrical cabinets. Modern construction techniques such as the use of 'rib-decking' (Fig. 4.6), which consists of a corrugated metal ceiling that

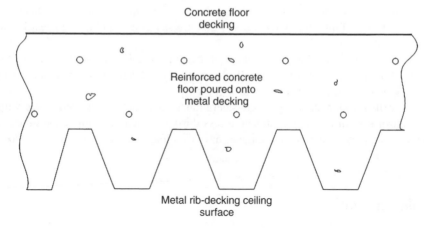

Figure 4.6 *Section through metal and concrete 'rib-decking' construction of a cell roof*

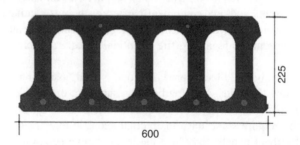

Figure 4.7 *Section through a concrete plank which gives a quick and high strength cell roof construction, but can be difficult to modify post-installation*

provides the base of a reinforced concrete roof pouring in situ, are commonly used. An alternative is hollow core concrete planking (Fig. 4.7), but the internal voids in this material mean that a substantial topping of concrete screed is required to obtain good sound insulation.

If penetrations have to be made in the structure after installation, these systems are not particularly easy to modify, therefore good preplanning and well-integrated design of services is important.

False, fire retardant, ceilings suspended from hangers fixed into the roof can be fitted if the 'industrial' look of concrete or corrugated metal is unacceptable, but this has to be shaped around all of the roof penetrating service ducts and pipes and can not always be financially justified.

Lighting

The typical test cell ceiling may be cluttered with fire sprinkler systems, exhaust outlets, ventilation ducting and a lifting beam. The lights are often the last consideration, but are of vital importance. They must be securely mounted so as not to move in the ventilation 'wind' and give a high and even level of lighting without causing glare from the control room window. Unless special and unusual conditions or regulations apply, cell lighting does not need to be explosion proof; however, lights may be working in an atmosphere of soot-laden fumes and so must be sufficiently robust, be easily cleaned and operate at moderate exposed surface temperatures.

The detailed design of a lighting system is a matter for the specialist. The 'lumen' method of lighting design gives the average level of illumination on a working plane for a particular number of 'luminaries' (light sources) of specified power arranged in a symmetrical pattern. Factors such as the proportions of the room and the reflectivity of walls and ceiling are taken into account.

The unit of illumination in the International System of Units is known as the lux, in turn defined as a radiant power of one lumen per square metre. The unit of luminous intensity of a (point) source is the candela, defined as a source which emits one lumen per unit solid angle or 'steradian'. The efficiency of light sources in terms of candela per watt varies widely, depending on the type of source and the spectrum of light that it emits.

The IES Code lays down recommended levels of illumination in lux for different visual tasks. A level of 500 lux in a horizontal plane 500 mm above the cell floor should be satisfactory for most cell work, but areas of deep shadow must be avoided. In special cells where in situ inspections take place, the lighting levels may be variable between 500 and 1000 lux.

Emergency lighting with a battery life of at least one hour and an illumination level in the range of 30 to 80 lux should be provided in both test cell and control room.

It is sometimes very useful to be able to turn off the cell lights from the control desk, whether to watch for sparks or red-hot surfaces or simply to hide the interior of the cell from unauthorized eyes. An outlet for a 'wandering' 110 V inspection lamp should be included in the layout.

Engine handling systems

A considerable number of connections must be made to any engine before testing can proceed. These include the coupling shaft, fuel, cooling water, exhaust and a wide range of transducers and instrumentation. It is obviously cost-effective when dealing with high throughput of small automotive units to carry out such work in a properly equipped workshop rather than use expensive test cell space. Once rigged the engines have to be transported to the cell.

The degree of complexity of the system adopted for transporting, installing and removing the engine within the cell naturally depends on the frequency with which the engine is changed. In some research or lubricant test cells the engine is more or less a permanent fixture, but at the other extreme the test duration for each engine in a production cell will be measured in minutes, and the time taken to change engines must be cut to an absolute minimum.

There is a corresponding variation in the handling systems:

- Simple arrangements when engine changes are comparatively infrequent. The engine is either craned into the cell and rigged in situ or mounted on a suitable stand such as that shown in Fig. 4.4 which is then lifted, by crane or forklift, into the cell. All connections are made subsequently by skilled staff with workshop back-up.
- The prerigged engine is mounted on a wheeled trolley or truck manoeuvred pallet carrying various transducers and service connections which are coupled to the engine before it is moved into the cell. An engine rigging workshop needs to be fitted with a dummy test cell station that presents datum connection points identical in position and detail to those in the cells, particularly the dynamometer shaft in order for the engine to be pre-aligned. Clearly to gain maximum benefit, each cell in the facility needs to be built with critical fixed interface items in identical positions from a common datum. These positions should be repeated by dummy interface points in any prerigging stand, in the workshop.
- For production test beds it is usual to make all engine to pallet connections prior to the combined assembly entering the cell. An automatic docking system permits all connections (including in some cases the driving shaft) to be made in seconds and the engine to be filled with liquids.

Workshop support in the provision of suitable fittings and adaptors should always be made available. All pipe connections should be flexible and exhaust connections can be particularly troublesome and short-lived at high temperatures; they should be treated as consumable items.

Rigging of exhaust systems often requires a welding bay; this must obey regulations concerning shielding. Electrical arc welding must not be allowed within the test cell because of the danger of damage to instrumentation by stray currents.

The workshop area used for derigging should be designed to deal with the inevitable spilled fluids and engine wash activities. Floor drains should run into an oil intercept unit.

Cell support services space

The design criteria of the individual services are covered in other chapters but at the planning stage suitable spaces have to be reserved for the following systems listed in order of space required:

- ventilation plant including fans, inlet and outlet louvers and ducting including sound attenuation section (see Chapter 5);
- engine exhaust system (see Chapter 6);

- electrical power distribution cabinets, including large drive cabinets in the case of a.c. dynamometers (see Chapter 10);
- fluid services including cooling water, chilled water, fuel and compressed air (see Chapters 6 and 7).

In addition to these standard services, space in some facilities needs to include

- combustion air treatment unit (see Chapter 5);
- exhaust gas emissions equipment (see Chapter 16).

It is very common to mount these services above the cell on the roof slab and it is often desirable, although rarely completely possible, for the services of individual cells to be contained within the footprint of the cell and control room below.

Control room design

As the operator will spend a great deal of time in the control room its layout should be as convenient and comfortable as possible. There should be a working surface away from the control station, where discussions may be held. A keyboard will be needed, but not all the time and in production cells it should be kept in a drawer when not in use.

Multiple computers can lead to several identical cordless keyboards and 'mice' mixed together on a dangerously confusing desk space.

Some equipment often installed in control rooms, such as emissions instrumentation, can have a quite large power consumption. This must be taken into account when specifying the ventilation system, which should conform to office standards.

The control console itself calls for careful design on sound ergonomic principles, and should, initially, include space for later additions. Figure 4.2 shows a typical general purpose control desk consisting of a work surface and an angled 19-in rack instrumentation tower. By definition, the primary purpose of the control desk is either to permit the operator to control the engine directly, or to allow him to manage and monitor a control system that may be automated to a greater or less degree. Pride of place will thus be given to the controls that govern engine torque and speed, the prime variables, and to the corresponding indicators of these variables, which may take various forms.

Of almost equal importance are secondary indications and warning signals covering such features as coolant temperature, lubricating oil pressure, exhaust temperature, etc.

Many modern systems are now integrated in a pedestal desk layout with the desk space supporting multiple screen displays that display all the test and facility condition data required.

The size and position of the observation window greatly affects the control room layout. In special climatic or anechoic cells windows are not fitted because they compromise the integrity of the structure and function.

If a window is fitted, it will normally be supplied by a specialist contractor and consist of two or three sheets of toughened glass, typically about 10 cm apart, in a configuration designed to reduce internal reflection.

Displays and controls for secondary equipment such as cell services can be relegated to lower levels in the cabinets, while instruments used only intermittently, such as fuel or smoke meters, should be installed towards the top and operated if necessary from a standing position. The positioning of graphic displays or computer monitors can be a problem, as they occupy considerable space and the solution may be to use articulated support booms.

The choice of indicator depends on the type of control and the kind of work to be done. If, as is sometimes the case, the operator needs to manage the engine manually by a T-handle or rotary dial type throttle control he should have an analogue speed indicator prominently in view.

In general the analogue indicator, in which a pointer moves over a graduated dial, is the appropriate choice when the quantity indicated is under manual control. If the indicator is required to show the stage reached in a test, current levels of such quantities as power and speed and any departures from desired values, a digital readout, perhaps accompanied by warning signals, is appropriate.

Presentation of readings in the form of a row of coloured bars on a screen can be useful if the balance of a number of inter-related quantities is to be kept under review, as is sometimes the case in gas turbine testing.

The operator's desk should be designed so that liquids spilled on the working surface cannot run down into electronic equipment mounted beneath. Control rooms should have two emergency escape routes. The floors should have non-slip and anti-static surfaces.

Instruments and controls: planning the layout

Even if the above principles are kept firmly in mind, it requires clear design management to prevent the test cell and control room from degenerating over time into a jumble of different devices positioned in available spaces. This is not only inefficient but unsafe.

Computerization has greatly complicated the problem. Before total computer dependency, it was unusual to see in test cells a direct display of more than perhaps 10 or 12 quantities with possibly 10 more on a multi-pen chart recorder. In computer-controlled cells, particularly those associated with engine mapping, it is not unusual to have six display screens mounted on a frame, many of them giving rapidly changing indications.

This can present the control system designer with a major problem in seeing 'the wood for the trees' and organizing a coherent display. In most cases the suppliers have passed the problem to the user by allowing them freedom to produce their own customized display screens. The problem, while transferred, has not been solved; some form of optimization and discipline of data display are required within the user organization if serious misunderstandings or operational errors are to be avoided.

It may be found helpful, in planning the layout of instruments and controls, to classify them under the following headings:

- operational and safety instrumentation;
- primary instrumentation;
- secondary instrumentation.

If the status or output of most or all of the instrumentation is shown on one or more display screens, these displays must be arranged so that those of high priority are visually prominent.

Primary instrumentation

This covers the essential data relating to the experimental purpose of the cell. The display of this information will be arranged in a prominent position where it is comfortably seen from the operator's desk.

The definition of primary as opposed to secondary instrumentation depends entirely on the purpose of the cell. In a cell intended for noise measurement studies, for example, sound level indicators and recorders have pride of place while indications of engine torque, exhaust temperature or fuel consumption are of secondary importance.

A test cell devoted to diesel engine exhaust emissions has as primary instrumentation the required gas analysis system, with all the associated controls of the gas dilution and sampling system. If, as is often the case, the engine is being put through a computer-controlled transient drive cycle, information regarding engine state is of secondary importance and displayed accordingly.

Secondary instrumentation

Information not of direct relevance to the test should be displayed away from the operator's normal field of attention, or it may be recorded on selectable pages of the main VDU display. Typical secondary information concerns slowly changing quantities such as barometric pressure, relative humidity and cooling water temperature.

In-cell controls

There are a number of tasks that may require an operator to be in the cell while the engine is running. Examples are the tuning of racing car engines, the setting of idling adjustments and checking for leaks. It is very desirable, for safety reasons, that the engine is brought via door interlock to idle when the cell door is opened and that there should always be a second operator at the control desk while a colleague is within the running cell.

A simple control station within the cell, enabling the operator there to control the engine and observe torque and speed, is often useful but must be interlocked to prevent control actions being taken outside the cell at the same time. Duplication of all control desk displays within the cell is seldom justified.

Emergency stop and engine shutdown

The engine test cell control system must include a robust method of shutting down the engine and dynamometer (see Chapter 11, under sections Safety systems and Emergency stop function). All test facilities must be provided with a number of emergency stop buttons, of large size and conspicuously marked, both at the control desk and at strategic positions in the cell. An emergency stop button is intended for use in the case of a local emergency; it operates only in the cell affected.

Since spark ignition engines may be stopped by interrupting the supply to the ignition circuit this is usually linked to the emergency stop system. A diesel engine calls for a more robust shutdown system: while many automotive diesels are fitted with a shutdown solenoid, which should be wired into the emergency stop system, it should be backed up by some means of independently moving the fuel pump rack to the stop position. Some test plant suppliers produce standard devices or modified throttle controllers which perform this function, using stored pneumatic power as a precaution against electrical failure.

There is always the possibility of a diesel engine running away, usually as a result of being fuelled by lubricating oil: this puts it out of the control of the fuel pump. Such an occurrence is rare but not unknown, even with modern engines. Horizontal bus engines, if inadvertently overfilled with lubricating oil, are prone to this trouble. If the emergency stop system is activated when an engine is running away all load is removed from electrical or eddy current dynamometers and the situation is made worse.

The two ways of stopping the engine in this situation are either by closing off the combustion air supply or applying sufficient load to stall the engine. Unless the engine has been specially rigged with a closure valve or a means of injecting CO_2, the former method is often not practicable. Stalling the engine in such an emergency calls for a cool head and is best arranged by providing a special 'fast stop' button which ramps the dynamometer to full torque and de-energizes the engine systems. Such rapid stalling of the engine can be damaging: a fast stop button should only be located at the control desk under a protective flap and operators should be carefully briefed on its use.

Fire control

This subject has two separate system design subjects:

- the detection of flammable, explosive or dangerous gases in the cell and the associated actions and alarms;
- the suppression of fire and the associated actions and alarms.

The gas detection system, usually supplied by a specialist subcontractor, will be linked to the cell control system and building management system as required by the operator and in conformity to local and national regulation.

The gas hazard alarm and fire extinguishing systems are always entirely independent of and hierarchical to the emergency stop arrangements. They are linked to the ventilation control system since in the event of a fire, the cell needs to be isolated from sources of air and the fire needs to be contained.

The operation of both systems should be specified within the control and safety interlock matrix mentioned in Chapter 10, Electrical design considerations.

There is much legislation relevant to industrial fire precautions and also a number of British Standards. The Health and Safety Executive, acting through the Factory Inspectorate, is responsible for regulating such matters as fuel storage arrangements, and should be consulted, as should the local fire authority.

Safety regulations and European ATEX codes applied to engine test cells

Atmospheric Explosion (ATEX) regulations were introduced as part of harmonization of European regulations for such industries as mines and paper mills where explosive atmospheres occur; the engine and vehicle test industry was not explicitly identified within the wording, therefore the European industry has had to negotiate the conditions under which test cells may be excluded. As with all matters relating to regulation, it is important for the local and industrial sector specific interpretation to be checked.

The classification of zones is as shown in Table 4.2.

As has been determined over many years, secondary explosion protection measures such as using EX-rated equipment (even if it were available) in the engine test cell makes little sense since the ignition source is invariably the engine itself. Therefore it is necessary to use primary explosion prevention methods that prevent the space from ever containing an explosive atmosphere covered by ATEX.

Table 4.2 *ATEX and US zone classification for occurrence of explosive atmospheres*

USA divisions	ATEX zone designation	Explosive gas atmosphere exists	Remark
	Zone 0	Continuously, or for long periods	>1000 h/year
Division 1 = zone 0 and 1	Zone 1	Occasionally	10–1000 h/year
Division 2 = zone 2	Zone 2	For a short period only	<10 h/year

These primary precautions, which cover both gasoline- and diesel-fuelled beds without distinction, are certified in Europe by the relevant body TUF and are as follows:

- The cell space must be sufficiently ventilated both by strategy and volume flow to avoid an explosive atmosphere.
- There has to be continuous monitoring and alarming of hydrocarbon concentration (normally 'warning' at 20 per cent of lower explosive mixture and 'shutdown' at 40 per cent).
- Leak-proof fuel piping using fittings approved for use with the liquids contained.
- The maximum volume of fuel 'available' in the cell in the case of an emergency or alarm condition is 10 litres.

With these conditions fulfilled, the only EX-rated electrical devices that need to be included in the cell design are the gas detection devices and any purge extraction fan.

The requirement for the purge fan, which may have a dual purge/ventilation extract role can be achieved by use of a copper shroud ring for the impeller which prevents sparks in the case of a rotor/stator contact.

Fire extinguishing systems

In the period between the second and third edition of this book, some gas-based fire suppressant systems have been banned for use in new building systems having fallen foul of environmental protection legislation.

Table 4.3 lists the common fire suppression technologies and summarizes their characteristics which are then covered in more detail in the following paragraphs.

Microfog water systems

Microfog, or high pressure mist systems, unlike other water-based fire extinguishing systems, have the great advantage that they remove heat from the fire source and its surroundings and thus reduce the risk of reignition when it is switched off.

Microfog systems use very small quantities of water and discharge it as a very fine spray. They are particularly efficient in large cells, such as vehicle anechoic chambers, where the fire source is likely to be of small dimensions relative to the size of the cell.

Other advantages of these high pressure mist systems are that they tend to entrain the black smoke particles that are a feature of engine cell fires and prevent or considerably reduce the major clean-up of ceiling and walls. Such systems are used in facilities containing computers and high-powered electrical drive cabinets with minimum damage after activation and fast restart.

Table 4.3 *Characteristics of major fire suppression systems*

	Water sprinkler	Inert gas (CO_2)	Chemical gases	High pressure water mist
Cooling effect on fire source	Some	None	None	Considerable
Effect on personnel in cell	Wetting	Hazardous/fatal	Minor health hazards	None
Effect on environment	High volume of polluted water	Greenhouse and ozone layer depleting	Greenhouse and ozone layer depleting	None
Damage by extinguishing agent	Water damage	None	Possible corrosive/hazardous byproducts	Negligible
Warning alarm time before activation	None required	Essential	Essential	None required
Effect on electrical equipment	Extensive	Small	Possible corrosive by products	Small
Oxygen displacement	None	In entire cell space	In entire cell space	At fire source

Carbon dioxide

Carbon dioxide (CO_2) was used extensively in the industry until the environmental impact was widely understood. While they can be used against flammable liquid fires, they are hazardous to life, therefore a warning alarm period must be given before activation to ensure staff have evacuated the cell. Breathing difficulties become apparent above a concentration of 4 per cent and a concentration of 10 per cent can lead to unconsciousness or, after prolonged exposure, to death.

CO_2 is about 1.5 times heavier than air and it will tend to settle at ground level in enclosed spaces. The discharge of a CO_2 flood system is likely to be very noisy and produce a sudden drop of cell temperature causing dense misting of the atmosphere to take place that obscures any remaining vision possible through smoke.

Dry powder

Powders are designed for high-speed extinguishment of highly flammable liquids such as petrol, oils, paints and alcohol; they can also be used on electrical or engine fires. It must be remembered that dry chemical powder does not cool nor does it have a lasting smothering effect and therefore care must be taken against reignition.

Halons

Following the Montreal Protocol, some halons (halogenated hydrocarbons) are being phased out. These contain chlorine or bromine, thought to be damaging to the ozone layer, and their production has been banned. Unfortunately, this ban embraces Halon 1211 and Halon 1301, both hitherto used for total flooding and in portable extinguishers for dealing with flammable liquid fires. It is still feasible to use existing stocks, but clearly there are strong arguments for seeking replacements. Some are available, mostly fluorinated hydrocarbons, but care must be exercised as they may have toxic effects at their design concentrations.

Inergen

Inergen is the trade name for the extinguishing gas mixture of composition 52 per cent nitrogen, 40 per cent argon, 8 per cent carbon dioxide. There are other alternative fire suppressants, including pure argon, many of which may be used in automatic mode even when the compartment is occupied, provided the oxygen concentration does not fall below 10 per cent and the space can be quickly evacuated.

With all gaseous systems, precautions should be taken to ensure that accidental or malicious activation is not possible. In carbon dioxide systems, automatic mode should only be used when the space is unoccupied.

Total flood systems, of whatever kind, are usable only after the area has been evacuated and sealed. They must be interlocked with the doors and special warning signs must be provided.

Foam

Foam extinguishers could be used on engine fires but they are more suited to flammable liquid spill fires or fires in containers of flammable liquid. If foam is applied to the surface or subsurface of a flammable liquid it will form a protective layer. Some powders can also be used to provide rapid knock-down of the flame. Care must be taken, however, since some powders and foams are incompatible.

Warning: Wall penetrations between control space and cell should be sealed by the use of Hawke Gland™ boxes or similar devices. Poor maintenance or incomplete closure of these fire-breaks, commonly caused by frequent installation of temporary looms, can allow fire or extinguishant such as CO_2 to escape explosively.

Vapour detection

Explosion proof devices are available for detection of preset levels of oil mist, hydrocarbon vapour and the gases used in engine test cells. It is common to set the alarm system handling these devices with a two-stage setting since their use presents a practical problem in setting the sensitivity to a level that warns of real danger without continuously tripping at a level which experience shows to be safe. Once again, the strategy controlling the alarms and interlocks should form part of the alarm and interlock matrix within the functional specification (see Chapter 10, Electrical design considerations).

Summary

The potentially hazardous nature of the test cell environment is emphasized and emergency stop and fire precautions are described. There are a number of different basic types of test cell, depending on the use for which they are intended. Over- and underprovision of facilities should both be avoided.

The special features of test cell construction, which differs from that of any other kind of industrial building, are described and references given to chapters of this book covering the design of specific services.

References

1. http://www.planningportal.gov.uk/england/professionals/

5   Ventilation and air conditioning

Introduction

In this chapter, the strategies of test cell ventilation are reviewed and the concept of the test cell as an open system is applied to the analysis of thermal loadings and ventilation requirements.[1,2] The design process for ventilation systems is described with a worked example.

It must be remembered that an internal combustion (IC) engine is essentially an air engine and that the air used by the engine may come from the cell ventilation air or from a treatment unit outside the cell. Whatever its source, the performance and power output of the engine is affected by the condition, temperature, pressure and humidity of the ingested air.

The purpose of air conditioning and ventilation is the maintenance of an acceptable environment in an enclosed space. This is a comparatively simple matter where only human activity is taking place, but becomes progressively more difficult as the energy flows into and out of the space increase. An engine test cell represents perhaps the most demanding environment encountered in industry. Large amounts of power will be generated in a comparatively small space, surfaces at high temperature are unavoidable and large flows of cooling water, air and electrical power have to be accommodated, together with rapid variations in thermal load.

Test cell ventilation strategies

While much of this chapter covers the most commonly used method of removing engine-generated heat from the test cell, by forced ventilation using ambient (outside) air, it is possible to recirculate cell air through a conditioning system.

The final choice will be influenced by the range of ambient conditions, building space restraints and the type of testing being carried out.

Some authorities still use 'room volume changes per hour' as a measure of ventilation requirements; in this book the measure is only used to deal with legislative or safety guidelines such as in the following table.

Ventilation systems for test cells not only remove heat but also prevent the build-up of dangerous levels of gases and vapours. Such requirements are dealt with by specifically designed vapour purge systems (see below) and by ensuring sufficient air flow through the cell. The flow required to remove the heat generated by an

Table 5.1 *Suggested air changes per hour, 'rule of thumb'*

Banks	2–4	Laboratories	5–10
Cafés	10–12	Offices	5–7
Shower rooms	15–20	Toilets	6–10
Garages	6–8	Factories	8–10

engine or vehicle usually provides sufficient air flow to clear gasoline vapour and exhaust gases under most normal circumstances and the correct control strategy will ensure that the cell space runs slightly below ambient pressure thus preventing such gases being pushed into other work areas.

Purge fans: safety requirements to reduce explosion risk

In cells using volatile fuels the ventilation system will have to incorporate a purge fan, the purpose of which is to remove heavier than air and potentially explosive fumes from the lowest points of the cell complex. The purge fan plays an essential role in reducing the fire and explosion risk in a test cell (for relevance to ATEX regulations, see Chapter 4). The purge system should be integrated into the test cell control system in such a way as to ensure, on cell start-up, that it has run for a minimum of 10 min, with no hydrocarbon sensors showing an alarm state, before engine ignition and therefore fuel inlet valves are allowed to be energized.

The rating of the purge fan may be covered by local legislation, but as a minimum should provide for 30 air changes of the enclosed cell space per hour.

The purge suction duct should extend to the lowest point of the cell within any services trench system; in some designs the purge fan is also used for providing exhaust dilution. Purge fan flow is additive to the outgoing flow of the main extract fan which has to be balanced by the replenishment volume in balanced systems. The source of the replenishment may be the combined flow of the combustion air system and the main inlet air fan or the main inlet fan alone. Whatever strategy is used, the cell should run, when the doors are closed, at a negative pressure of around 30–50 Pa compared with ambient. Any greater than this will make it difficult to initiate door opening.

If doors of the cell open for more than a few seconds, then the cell can be considered as operating in 'workshop mode' with fuel system turned off, then it is possible to balance the treated combustion air flow with that of the purge fan so as to provide a space heating system. In cells with no combustion air system a separate 'comfort' mode air heating and cooling system may have to be provided for the cell.

Air handling units

The use of an air handling unit (AHU) capable of recirculation of the majority, or variable proportion, of test cell air through a temperature conditioning system can have operational advantages over a forced ventilation system. They may occupy less building space and may greatly reduce the problem of engine noise break-out. Designs are usually based on packaged units, supplied with building services such as chilled and medium pressure hot water (MPHW) supply. They are normally mounted above the cell in the services room and aligned on the long axis of the cell to minimize ducting. A more rare alternative arrangement, in high roofed cells, uses ceiling-mounted units that are based on a direct evaporative cooling and electrical heating units.

Air handling units, particularly the direct evaporative type, can be energy efficient but a detailed and site-specific calculation of initial and running costs is needed to make a valid comparison. Since closed systems will recirculate entrained pollutants, a separate combustion air system or alternative make-up source will always be required to prevent build-up and to balance the loss through the low level purge.

The heat capacity of cooling air

By definition, the test cell environment is mainly controlled by regulating the quantity, temperature and in some cases the humidity of the air passing through it.

Air is not the ideal heat transfer medium: it has low density and specific heat and is transparent to radiant heat, while its ability to cool hot surfaces is much inferior to that of liquids.

The main properties of air of significance in air conditioning may be summarized as follows:

the gas equation:

$$p_a \times 10^5 = \rho R \, (t_a + 273) \tag{1}$$

where:

p_a = atmospheric pressure, bar
ρ = air density, kg/m^3
R = gas constant for air = 287 J/kg K
t_a = air temperature, °C

Under conditions typical of test cell operation, with $t_a = 25°C$ (77°F), and standard atmospheric pressure (see Units and conversion factors), the density of air, from eq. (1), is:

$$\rho = \frac{1.01325 \times 10^5}{287 \times 298} = 1.185 \, kg/m^3$$

or about 1/850 that of water.

The specific heat at constant pressure of air at normal atmospheric conditions is approximately:

$$C_p = 1.01 \, \text{kJ/kg K}$$

or less than one quarter that of water.

The air flow necessary to carry away 1 kW of power with a temperature rise of 10°C is:

$$m = \frac{1}{1.01 \times 10} = 0.099 \, \text{kg/s} = 0.084 \, \text{m}^3/\text{s} = (2.9 \, \text{ft}^3/\text{s})$$

This is a better basis for design than any rule of thumb regarding number of cell air changes per hour.

Heat transfer from the engine

It is useful to gain a feel for the relative significance of the elements that make up the total of heat transferred from a running engine to its surroundings by considering rates of heat transfer from bodies of simplified form under test cell conditions.[3]

Consider a body of the shape sketched in Fig. 5.1. This might be regarded as roughly equivalent, in terms of projected surface areas in horizontal and vertical directions, to a gasoline engine of perhaps 100 kW maximum power output, although the total surface area of the engine could be much greater. Let us assume the surface temperature of the body to be 80°C and the temperature of the cell air and cell walls 30°C.

Heat loss occurs as a result of two mechanisms: natural convection and radiation. The rate of heat loss by natural convection from a vertical surface in *still air* is given approximately by:

$$Q_v = 1.9 \, (t_s - t_a)^{1.25} \, \text{W/m}^2 \qquad (2)$$

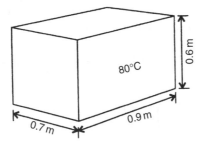

Figure 5.1 *Simplified model of exampled 100 kW engine, for analysis of heat transfer to surroundings*

The total area of the vertical surfaces in Fig. 5.1 = 1.9 m². The corresponding convective loss is therefore:

$$1.9 \times 1.9 \times (80 - 30)^{1.25} = 480\,\text{W}$$

The rate of heat loss from an upward facing horizontal surface is approximately:

$$Q_h = 2.5\,(t_s - t_a)^{1.25}\,\text{W/m}^2 \tag{3}$$

giving in the present case a convective loss of:

$$0.63 \times 2.5\,(80 - 30)^{1.25} = 210\,\text{W}$$

The heat loss from a downward facing surface is about half that for the upward facing case, giving a loss = 110 W.

We thus arrive at a rough estimate of convective loss of 800 W. For a surface temperature of 100°C, this would increase to about 1200 W.

However, this is the heat loss in still air: the air in an engine test cell is anything but still and very much greater rates of heat loss can and probably will occur. This effect must never be forgotten in considering cooling problems in a test cell: an increase in air velocity greatly increases the rate of heat transfer to the air and may thus aggravate the problem.

As a rough guide, doubling the velocity of air flow past a hot surface increases the heat loss by about 50 per cent. The air velocity due to natural convection in our example is about 0.3 m/s.

An air velocity of 3 m/s would be moderate for a test cell with ventilating fans producing a vigorous circulation, and such a velocity past the body of Fig. 5.1 would increase convective heat loss four-fold, to about 3.2 kW at 80°C and 4.8 kW at 100°C.

The rate of heat loss by radiation from a surface depends on the emissivity of the surface (the ratio of the energy emitted to that emitted by a so-called black body of the same dimensions and temperature) and on the temperature difference between the body and its surroundings. Air is essentially transparent to radiation, which thus serves mainly to heat up the surfaces of the surrounding cell; this heat must subsequently be transferred to the cooling air by convection, or conducted to the surroundings of the cell.

Heat transfer by radiation is described by the Stefan–Boltzmann equation, a form of which is:

$$Q_r = 5.77\varepsilon \left[\left(\frac{t_s + 273}{100} \right)^4 - \left(\frac{t_w + 273}{100} \right)^4 \right] \tag{4}$$

A typical value of emissivity (ε) for machinery surfaces would be = 0.9, t_s = temperature of hot body (°C) and t_w = temperature of enclosing surface (°C).

In the present case:

$$Q_r = 5.77 \times 0.9 \left[\left(\frac{353}{100} \right)^4 - \left(\frac{303}{100} \right)^4 \right] = 370 \, \text{W/m}^2$$

In our present example, total surface area $= 3.16 \, \text{m}^2$, giving a radiation heat loss of 1170 W.

For a surface temperature of 100°C, this would increase to 1800 W.

Heat transfer from the exhaust system

The other main source of heat loss associated with the engine is the exhaust system. In the case of turbocharged engines this can be particularly significant. Assume in the present example that exhaust manifold and exposed exhaust pipe are equivalent to a cylinder 80 mm diameter × 1.2 m long at a temperature of 600°C, surface area 0.3 m². Heat loss at this high temperature will be predominantly by radiation and equal to:

$$0.3 \times 5.77 \times 0.9 \left[\left(\frac{873}{100} \right)^4 - \left(\frac{303}{100} \right)^4 \right] = 8900 \, \text{W}$$

from eq. (4).

Convective loss is:

$$0.3 \times 1.9 \times (600 - 30)^{1.25} = 1600 \, \text{W}$$

from eq. (2).

It is clear that this can heavily outweigh the losses from the engine and points to the importance of reducing the run of unlagged exhaust pipe as much as possible.

Heat losses from the bodies sketched in Fig. 5.1 and described above representing an engine and exhaust system, surroundings at 30°C, may be summarized as follows:

	Convection	*Radiation*
engine, jacket and crankcase at 80°C	3.2 kW	1.2 kW
engine, jacket and crankcase at 100°C	4.8 kW	1.8 kW
exhaust manifold and tailpipe at 600°C	1.6 kW	8.9 kW

Heat transfer from walls

Most of the heat radiated from engine and exhaust system will be absorbed by the cell walls and ceiling, also by instrument cabinets and control boxes, and subsequently transferred to the ventilation air by convection.

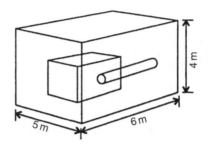

Figure 5.2 *Simplified test cell for heat transfer calculation*

Imagine the 'engine' and 'manifold' considered above to be installed in a test cell of the dimensions shown in Fig. 5.2. The total wall area is $88\,m^2$. Assuming a wall temperature 10°C higher than the mean air temperature in the cell and an air velocity of 3 m/s the rate of heat transfer from wall to air is in the region of $100\,W/m^2$, or 8.8 kW for the whole wall surface, roughly 90 per cent of the heat radiated from engine and exhaust system; the equilibrium wall temperature is perhaps 15°C higher than that of the air.

While an attempt to make a detailed analysis on these lines, using exact values of surface areas and temperatures, would not be worthwhile, this simplified treatment may clarify the principles involved.

Sources of heat in the test cell

The engine

As introduced in Chapter 2, various estimates of the total heat release to the surroundings from a water-cooled engine and its exhaust system have been published. One authority[4] quotes a maximum of 15 per cent of the heat energy in the fuel, divided equally between convection and radiation. This would correspond to about 30 per cent of the power output of a diesel engine and 40 per cent in the case of a gasoline engine.

In the experience of the authors, a figure of 40 per cent (0.4 kW/kW engine output) represents a safe upper limit to be used as a basis for design for water-cooled engines. This is divided roughly in the proportion 0.1 kW/kW engine to 0.3 kW/kW exhaust system. It is thus quite sensitive to exhaust layout and insulation.

In the case of an air-cooled engine, the heat release from the engine will increase to about 0.7 kW/kW output in the case of a diesel engine and to about 0.9 kW/kW output for a gasoline engine. The proportion of the heat of combustion that passes to the cooling water in a water-cooled engine in an air-cooled unit must of course pass directly to the surroundings.

The dynamometer

A water-cooled dynamometer, whether hydraulic or eddy current, runs at a moderate temperature and heat losses to the cell are unlikely to exceed 5 per cent of power input to the brake. Usually a.c. and d.c. machines are air cooled and heat loss to surroundings is in the region of 6–10 per cent of power input.

Other sources of heat

Effectively, all the electrical power to lights, fans and instrumentation in the test cell will eventually appear as heat transmitted to the ventilation air. The same applies to the power taken to drive the forced-draught ventilating fans: this is dissipated as heat in the air handled by the fans.

The secondary heat exchangers used in coolant and oil temperature control will similarly make a contribution to the total heat load. Here, long lengths of unlagged pipe work holding the primary fluid will not only add heat to the ventilation load but will adversely effect control of temperature.

Heat losses from the cell

The temperature in an engine test cell is generally higher than usual for an industrial environment. There is thus in some cases appreciable transfer of heat through cell walls and ceiling, depending on the configuration of the site but, except in the case of a test cell forming an isolated unit, these losses may probably be neglected.

Recommended values as a basis for the design of the ventilation system are given in Table 5.2. In all cases they refer to the maximum rated power output of the engines to be installed.

Calculation of ventilation load

The first step is to estimate the various contributions to the heat load from engine, exhaust system, dynamometer, lights and services. This information should be summarized in a single flow diagram. Table 5.2 shows typical values.

Table 5.2 *Heat transfer to ventilation air*

	kW/kW power output
Engine, water cooled	0.1
Engine, air cooled	0.7–0.9
Exhaust system (manifold and silencer)	0.3
Hydraulic dynamometer	0.05
Eddy current dynamometer	0.05
a.c./d.c. dynamometer	0.15

Heat transfer to the ventilation air is to a degree self-regulating: the cell temperature will rise to a level at which there is an equilibrium between heat released and heat carried away. The amount of heat carried away by a given air flow is clearly a function of the temperature rise ΔT from inlet to outlet.

If the total heat load is H_L kW then the required air flow rate is:*

$$Q_A = \frac{H_L}{1.01 \times 1.185\,\Delta T} = 0.84\frac{H_L}{\Delta T}\ \text{m}^3/\text{s} \tag{5}$$

A temperature rise $\Delta T = 10°C$ is a reasonable basis for design. Clearly the higher the value of ΔT, the smaller the corresponding air flow. However, a reduction in air flow has two influences on general cell temperature: the higher the outlet temperature the higher the mean level in the cell, while a smaller air flow implies lower air velocities in the cell, calling for a greater temperature difference between cell surfaces and air for a given rate of heat transfer.

Design of ventilation ducts and distribution systems

Pressure losses: fundamentals

The velocity head or pressure associated with air flowing at velocity V is given by:

$$p_v = \frac{\rho V^2}{2}\ \text{Pa}$$

This represents the pressure necessary to generate the velocity. A typical value for ρ, density of air, is $1.2\,\text{kg/m}^3$, giving:

$$p_v = 0.6\,V^2\,\text{Pa}$$

The pressure loss per metre length of a straight duct is a fairly complex function of air velocity, duct cross-section and surface roughness. Methods of derivation with charts are given in Ref. 5. In general, the loss lies within the range 1–10 Pa/m, the larger values corresponding to smaller duct sizes. For test cells with individual ventilating systems duct lengths are usually short and these losses are small compared with those due to bends and fittings such as fire dampers.

The choice of duct velocity is a compromise depending on considerations of size of ducting, power loss and noise. If design air velocity is doubled the size of the ducting is clearly reduced, but the pressure losses are increased roughly fourfold, while the noise level is greatly increased (by about 18 dB for a doubling of velocity).

Maximum duct velocities recommended in Ref. 5 are given in Table 5.3.

---

* For air at standard atmospheric conditions, $\rho = 1.185\,\text{kg/m}^3$. A correction may be applied when density departs from this value but is probably not worthwhile.

Table 5.3 *Maximum recommended duct velocities*

Volume rate of flow (m³/s)	Maximum velocity (m/s)	Velocity pressure (Pa)
<0.1	8–9	38–55
0.1–0.5	9–11	55–73
0.5–1.5	11–15	73–135
>1.5	15–20	135–240

It is general practice to aim for a flow rate of 12 m/s through noise attenuators built within ducting.

The total pressure of an air flow p_t is the sum of the velocity pressure and the static pressure p_s (relative to atmosphere):

$$p_t = p_s + p_v = p_s + \frac{\rho V^2}{2}$$

The design process for a ventilating system includes the summation of the various pressure losses associated with the different components and the choice of a suitable fan to develop the total pressure required to drive the air through the system.

Ducting and fittings

Various codes of practice have been produced covering the design of ventilation systems.[6,7] The following brief notes are based on the *Design Notes* published by the Chartered Institution of Building Services.

Galvanized sheet steel is the most commonly used material, and ducting is readily available in a range of standard sizes in rectangular or circular sections. Rectangular section ducting has certain advantages: it can be fitted against flat surfaces and expensive round-to-square transition lengths for connection to components of rectangular section such as centrifugal fan discharge flanges, filters and coolers are avoided.

Table 5.4 *Maximum recommended duct velocities*

Volume of flow (m³/s)	Maximum velocity (m/s)	Velocity pressure (Pa)
<0.1	8–9	38–55
0.1–0.5	9–11	55–73
0.5–1.5	11–15	73–135
>1.5	15–20	135–240

Pressures in ventilation systems are often measured in mmH_2O:
$1\,mmH_2O = 9.81\,Pa$.

Once the required air flow rate has been settled and the general run of the ducting decided, the next step is to calculate the pressure losses in the various elements in order to specify the pressure to be developed by the fan. In most cases a cell will require both a forced draught fan for air supply and an induced draught fan to extract the air. The two fans must be matched to maintain the cell pressure as near as possible to atmospheric. For control purposes cell pressure is usually set at 50 Pa below atmospheric which gives the safest set of conditions concerning door pressurization and fume leakage.

Figure 5.3 shows various components in diagrammatic form and indicates the loss in total pressure associated with each. This loss is given by:

$$\Delta p_t = K_e \frac{\rho V^2}{2}$$

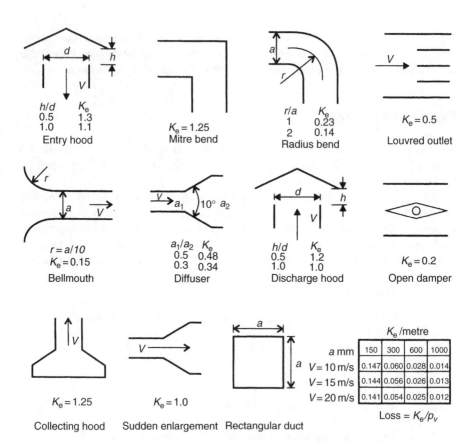

Figure 5.3 *Pressure losses in components of ventilation systems*

Information on pressure losses in plant items such as filters, heaters and coolers is generally provided by the manufacturer.

The various losses are added together to give the cumulative loss in total pressure (static pressure + velocity pressure) which determines the required fan performance.

Inlet and outlet ducting

The ducts also have to be designed so that noise created in the test cell does not break out into the surrounding environment. The problem has to be solved by the use of special straight attenuating sections of duct which are usually of a larger section and which should be designed in such a way as to give an air velocity around an optimum value of 12 m/s. Such silencer sections may tend to artificially increase the length of ducting within the service space. Vertical intake and discharge ducting is usually more effective than horizontal wall mounted louvres in reducing noise 'break-out' at the site border. The arrangement of air inlets and outlets calls for careful consideration if short-circuiting and local areas of stagnant air within the cell are to be avoided.

There are many possible layouts based on combinations of high level, low level and above engine direction ducts; four commonly used layouts are shown in Figs 5.4–5.7.

Figure 5.4 shows a commonly used design in cells without large subfloor voids; designers have to ensure that the air flow is directed over the engine rather than

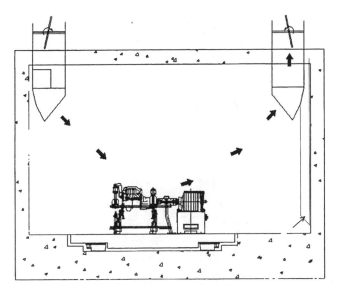

Figure 5.4 *Variable speed fans in a pressure balanced system using high level inlet and outlet ducts*

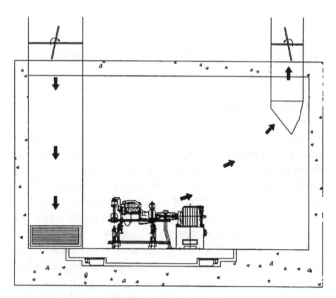

Figure 5.5   *Low level inlet and high level outlet ducted system; the low level duct can either be part of a balanced fan system as in Fig. 5.4 or drawing outside air through an inlet silencer*

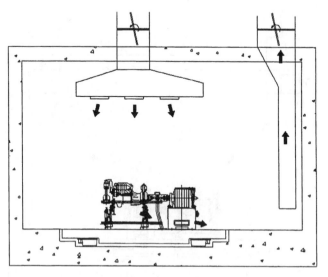

Figure 5.6   *A balanced fan system as in Fig. 5.4, but inlet ducted through nozzles from an over engine plenum and extracted at low level*

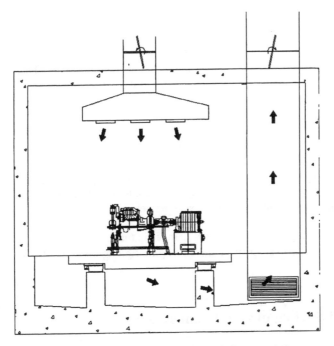

Figure 5.7 *A balanced fan system in a cell with large subfloor service space where ventilation air is drawn over the engine and extracted from low level*

taking a higher level short-cut between inlet and outlet. In this and the next design the purge of floor level gases would be carried out via the exhaust gas dilution duct (see comments concerning exhaust systems in Chapter 6).

Figure 5.5 shows an alternative layout where the intake air comes into the cell at low level, either by forced ventilation, as shown, or through an attenuated duct drawing ambient air from outside. In both cases the air is drawn over the engine and exits at high level.

An extraction inlet hood above the engine, as shown in Fig. 5.6, may inhibit the use of service booms and other engine access.

Where a substantial subfloor space is present the ventilation air can be drawn out below floor level and over the engine from an overhead inlet ventilation plenum.

It will be clear by now that the choice of system layout has a major influence on the layout of equipment both in the cell and the service space above.

Where an AHU is not fitted but heat energy needs to be conserved in the ventilation air then a recirculation duct, complete with flow control dampers, can be fitted between the outlet and inlet ducts above fans. This can be set to recirculate between 0 to over 80 per cent of the total air flow.

Fire dampers

Designs of inlet and outlet ducting should obey the rule of hazard containment; therefore fire dampers must be fitted in ducts at the cell boundary. These devices act to close off the duct and act as a fire barrier at the cell boundary.

The two types commonly used are:

- Normally closed, motorized open. These are formed by stainless steel louvres within a special framework having their own motorized control gearing and closure mechanism. They are fitted with switches allowing the BMS or cell control system to check their positional status.
- Normally open, thermal link retained steel curtain. These only operate when the gas temperature reaches their release temperature and have to be reset manually.

A major advantage of the motorized damper is that when the cell is not running or being used as a workshop space during engine rigging the flow of outside (cold) air can be cut off.

External ducting of ventilation systems

The external termination of ducts can take various forms, often determined by architectural considerations, but always having to perform the function of protecting the ducts from the ingress of rain, snow and external debris such as leaves (see diagrams of exhaust cowls in Chapter 6). To this end, the inlet often takes the form of a motorized set of louvres, interlocked with the fan start system, immediately in front of coarse then finer filter screens.

In some cases, in temperate climates, the whole service space above the cell may be used as the inlet plenum for an air-handling unit; this allows more freedom in position of the inlet louvres.

In exceptionally dusty conditions such as exist in arid desert areas the external louvres should be motorized to close and, with vertical louvre sections, to prevent dust build-up on horizontal surfaces.

The use of 'spot fans' for supplementary cooling

A common error made by operators facing high cell temperatures is to bring in auxiliary, high speed fans which can make the original problem worse and introduce new air flow-related problems. Spot fans can greatly increase the total amount of heat transferred to the ventilation air, with a consequent increase in overall cell temperature. A jet or eddy of hot air can have undesirable effects: for example, it can raise the temperature of the engine inlet air and upset the calibration of force transducers that are sensitive to a Δt across their surface.

Control of ventilation systems

The simplest practical ventilation control system is that based on a two speed extraction fan in the roof of the cell drawing air though a low level duct at one end of the cell (a variant of Fig. 5.5). The fan runs at low speed during times when the cell is used as a workshop and at high speed during engine running. In this case, the cell pressure is ambient unless the inlet filter is blocked.

Closed cells require both inlet and extract fans to be fitted with variable speed drives and a control system to balance them. There are several alternatives, but under a commonly used control strategy operating with a non-recirculation system, one fan operates under temperature control and the other under pressure. Thus, with both fans running and with a cell pressure of about 50 Pa below ambient, when the cell air temperature rises the extract fan increases in speed to increase air flow; this speed increase creates a drop in cell pressure which is detected by the control system, so the inlet fan speed is increased until equilibrium is restored. The roles of the fans are somewhat interchangeable, the difference being that in the case above the control transients tend to give a negative cell pressure and if the control roles are reversed it gives positive cell pressure.

The control fans have to be able to deal with the additional flows in and out of the cell via purge and combustion air flows where these systems are fitted. The control system also has to be sufficiently damped to prevent surge and major disturbance in the case of a cell door being opened.

Fans

Methods of testing fans and definitions of fan performance are given in BS 848. This is not an entirely straightforward matter; a brief summary follows.

Once again the control volume technique will be found useful. Figure 5.8 shows a fan surrounded by a control surface.

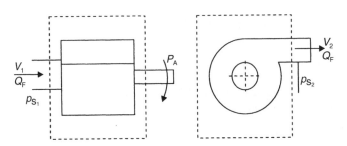

Figure 5.8 *Centrifugal fans as an open system*

The various flows into and out of the control volume are as follows.

In   air flow Q_F at velocity V_1 and pressure p_{s1} power input P_A
Out   air flow Q_F at velocity V_2 and pressure p_{s2}, where p_{s1} and p_{s2} are
          the static pressures at inlet and outlet, respectively.

Total pressure at inlet

$$P_{t1} = p_{s1} + \frac{\rho V_1^2}{2}$$

Total pressure at outlet

$$P_{t2} = p_{s2} + \frac{\rho V_2^2}{2}$$

Fan total pressure

$$P_{tF} = P_{t2} - P_{t1}$$

Air power (total)

$$P_{tF} = Q_F P_{tF}$$

Most fan manufacturers quote fan static pressure which is defined as:

$$P_{sF} = P_{tF} - \frac{\rho V_2^2}{2}$$

This ignores the velocity pressure of the air leaving the fan and does not equal the pressure difference $p_{s2} - p_{s1}$ between the inlet and outlet static pressures. (It is worth noting that, in the case of an axial flow fan with free inlet and outlet, fan static pressure as defined above will be zero.)

The total air power P_{tF} is a measure of the power required to drive the fan in the absence of losses.

The air power (static) is given by:

$$P_{SF} = Q_f p_{SF}$$

Manufacturers quote either fan static efficiency η_{SA} or fan total η_{tA}.

The shaft power is given by:

$$P_A = \frac{P_{SF}}{\eta_{SA}} = \frac{P_{tF}}{\eta_{tA}}$$

Fan noise

In a ventilation system, the fan is usually the main source of system noise. The noise generated varies as the square of the fan pressure head so that doubling the system resistance for a given flow rate will increase the fan sound power fourfold, or by about 6 dB.

As a general rule, to minimize noise, ventilation fans should operate as close to the design point as possible.

Ref. 6 gives examples of good and bad designs of fan inlet and discharge; a poor design gives rise to increased noise generation. Ref. 8, Part 2, gives guidance on methods of noise testing fans.

Classification of fans[8]

1. *Axial flow fans.* For a given flow rate, an axial flow fan is considerably more compact than the corresponding centrifugal fan and fits very conveniently into a duct of circular cross-section. The fan static pressure per stage is limited, typically to a maximum of about 600 Pa at the design point, while the fan dynamic pressure is about 70 per cent of the total pressure. Fan total efficiencies are in the range 65–75 per cent. Axial flow fans mounted within a bifurcated duct, so that the motor is external to the gas flow, are a type commonly used in individual exhaust dilution systems. An axial flow fan is a good choice as a spot fan, or for mounting in a cell wall without ducting. Multistage units are available, but tend to be fairly expensive.

2. *Centrifugal fans, flat blades, backward inclined.* This is probably the first choice in most cases where a reasonably high pressure is required, as the construction is cheap and efficiencies of up to 80 per cent (static) and 83 per cent (total) are attainable. A particular advantage is the immunity of the flat blade to dust collection. Maximum pressures are in the range 1–2 kPa.

3. *Centrifugal fans, backward curved.* These fans are more expensive to build than the flat bladed type. Maximum attainable efficiencies are 2–3 per cent higher, but the fan must run faster for a given pressure and dust tends to accumulate on the concave faces of the blades.

4. *Centrifugal fans, aerofoil blades.* These fans are expensive and sensitive to dust, but are capable of total efficiencies exceeding 90 per cent. There is a possibility of discontinuities in the pressure curve due to stall at reduced flow. They should be considered in the larger sizes where the savings in power cost are significant.

5. *Centrifugal fans, forward curved blades.* These fans are capable of a delivery rate up to 2.5 times as great as that from a backward inclined fan of the same size, but at the cost of lower efficiency, unlikely to exceed 70 per cent total. The power curve rises steeply if flow exceeds the design value.

The various advantages and disadvantages are summarized in Table 5.5.

Table 5.5 *Fans: advantages and disadvantages*

Fan type	Advantages	Disadvantages
Axial flow	Compact Convenient installation Useful as free-standing units	Moderate efficiency Limited pressure Fairly expensive
Centrifugal, flat blades, backward inclined	Cheap Capable of high pressure Good efficiency Insensitive to dust	
Centrifugal, backward curved	Higher efficiency	More expensive Higher speed for given pressure Sensitive to dust
Centrifugal aerofoil	Very high efficiency	Expensive Sensitive to dust May stall
Centrifugal forward curved	Small size for given duty	Low efficiency Possibility of overload

Design of ventilation system: worked example

By way of illustration, consider the case of the 250 kW turbocharged diesel engine for which an energy balance is given in Chapter 13, the engine to be coupled to a hydraulic dynamometer.

Ventilation air flow

Assume convection and radiation losses as follows:

engine, 10% of power output	25 kW
exhaust manifold	15 kW
exhaust tailpipe and silencer	15 kW
dynamometer, 5% of power input	12 kW
lights and services	20 kW
forced draught fan	5 kW
Subtotal	92 kW
Less losses from cell by conduction	5 kW
Total	87 kW

Assume air inlet temperature 20°C, Δt of 11°C therefore an outlet temperature 31°C. Then air flow rate, from eq. (5), is:

$$Q_A = \frac{0.84 \times 87}{11} = 6.6 \, \text{m}^3/\text{s}$$

+ induction air, 0.3 m³/s; say 7 m/s in total (= 101 m³/hour per kW engine power output). Assuming cell dimensions 8 × 6 × 4.5 m high, cell volume = 216 m³, this gives 117 air changes per hour.

Table 5.2 suggests a mean duct velocity in the range 15–20 m/s as appropriate, giving a cross-sectional area of 0.37–0.49 m². Heinsohn[2] gives a range of recommended standard duct dimensions of which the most suitable in the present circumstances is 600 × 600 mm, giving a duct velocity of 19.5 m/s and a velocity pressure of 228 Pa. Figure 5.9 shows one possible layout for the ventilation system.

The inlet or forced draught system uses a centrifugal fan and the duct velocity assumed above. For the extraction system, with its simpler layout and smaller pressure losses, an axial flow fan has been chosen. The pressure losses are calculated in Table 5.6 (this process lends itself readily to computer programming) and indicates fan duties as follows:

Forced inlet fan, flow rate 7 m³/s
Static pressure 454 Pa
Extraction fan, flow rate 7 m³/s
Static pressure 112 Pa.

A manufacturer's catalogue offers a fan for the forced draught situation to the following specification:

Centrifugal, backward inclined

Impeller diameter	900 mm
Speed	850 rev/min
Fan static efficiency	65%

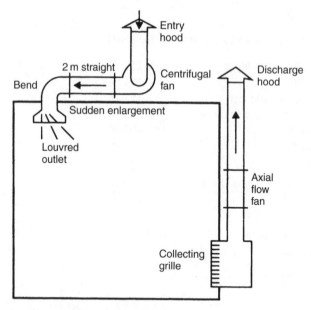

Figure 5.9 *Simplified cell ventilation system layout used in example calculation. Purge air is taken out in combined extract duct situated low in cell*

giving, from eqs (13) and (14)

$$\text{shaft power} = \frac{7.0 \times 454}{0.65} = 4900\,\text{W}$$

$$\text{motor power} = 5\,\text{kW}$$

For the extraction fan, an axial flow unit has the following specification:

Diameter	1 m
Speed	960 rev/min
Velocity pressure	51 Pa
Fan total efficiency	80%

$$\text{shaft power} = \frac{7.0 \times (112 + 51)}{0.80} = 1400\,\text{W}$$

In this case, the manufacturer recommends a motor rated at 2.2 kW.

Ventilation of the control room

This is in general a much less demanding exercise for the test installation designer. Heating loads are moderate, primarily associated with lights and heat generated by

Table 5.6 *Calculation of system pressure losses*

Item	Size (mm)	Area (m²)	Volume flow rate (m³/s)	Velocity (m/s)	Velocity pressure (Pa)	Fitting loss factor K_e	Pressure drop K_e/m	Length (m)	Pressure loss (Pa)	Cumulative loss (Pa)
Forced draught										
Entry hood	800 dia	0.5	7.0	14	118	1.1			129	129
Straight	800 dia	0.5	7.0	14	118		0.02	2	5	134
Bend	800 dia	0.5	7.0	14	118	0.23			27	161
Fan										
Straight	600 × 600	0.36	7.0	19	217		0.025	2	11	172
Bend	600 × 600	0.36	7.0	19	217	0.23			50	222
Sudden enlargement	600 × 600	0.36	7.0	19	217	1.0			217	439
Louvred outlet	1000 × 1000	1.00	7.0	7	29	0.5			15	454
									Total	454
Extraction										
Collecting hood	1000 dia*	0.79	7.0	9	49	1.25			61	61
Straights	1000 dia	0.79	7.0	9	49		0.02	2	2	63
Fan										
Discharge hood	1000 dia	0.79	7.0	9	49	1.0			49	112
									Total	112

* Diameter of branch connection.

electronic apparatus located in the room. Regulations regarding air flow rate per occupant and general conditions of temperature and humidity are laid down in various codes of practice and should be equivalent to those considered appropriate for offices.

Air conditioning

Most people associate this topic with comfort levels under various conditions: sitting, office work, manual work, etc. Levels of air temperature and humidity, as well as air change rates, are laid down by statute (see BS 5720[7] for particulars). Such regulations must be observed with regard to the control room. However, the conditions in an engine test cell are far removed from the normal, and justify special treatment, which follows.

Two properties of the ventilating air entering the cell (and, more particularly, of the induction air entering the engine) are of importance: the temperature and the moisture content. Air conditioning involves four main processes:

- heating the air;
- cooling the air;
- reducing the moisture content (dehumidifying);
- increasing the moisture content (humidifying).

Fundamentals of psychometry

The study of the properties of moist air is known as psychometry. It is treated in many standard texts[1,2] and only a very brief summary will be given here. Air conditioning processes are represented on the psychometric chart (Fig. 5.10).[9]

This relates the following properties of moist air:

- the *moisture content* or *specific humidity*, ω kg moisture/kg dry air. Note that even under fairly extreme conditions (saturated air at 30°C), the moisture content does not exceed 3 per cent by weight
- the *percentage saturation* or *relative humidity*, φ. This is the ratio of the mass of water vapour present to the mass that would be present if the air were saturated at the same conditions of temperature and pressure. The mass of vapour under saturated conditions is very sensitive to temperature:

Temperature °C	10	15	20	25	30
Moisture content ω (kg/kg)	0.0076	0.0106	0.0147	0.0201	0.0273

A consequence of this relationship is the possibility of drying air by cooling. As the temperature is lowered the percentage saturation increases until at the dew point temperature the air is fully saturated and any further cooling results in the deposition of moisture.

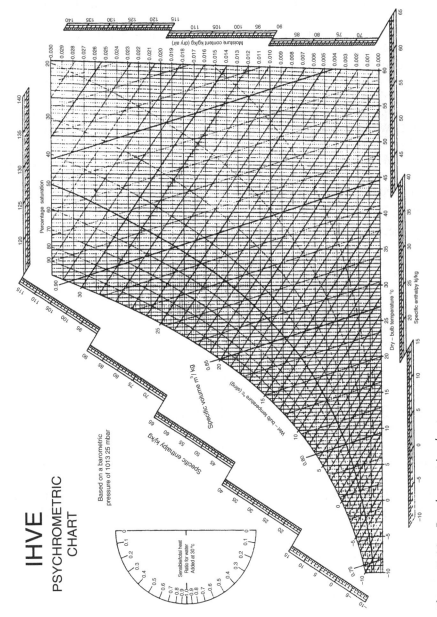

IHVE
PSYCHROMETRIC
CHART

Based on a barometric
pressure of 1013.25 mbar

Percentage saturation

Moisture content kg/kg (dry air)

Specific enthalpy kj/kg

Dry – bulb temperature °c

Specific volume m³/kg

Wet – bulb temperature °c (sling)

Specific enthalpy kj/kg

Sensible/total heat
Ratio for water
Added at 30 °c

Figure 5.10 *Psychometric chart*

- *The wet- and dry-bulb temperatures.* The simplest method of measuring relative humidity is by means of a wet- and dry-bulb thermometer. If unsaturated air flows past a thermometer having a wetted sleeve of cotton around the bulb the temperature registered will be less than the actual temperature of the air, as registered by the dry-bulb thermometer, owing to evaporation from the wetted sleeve. The difference between the wet-bulb and dry-bulb temperatures is a measure of the relative humidity. Under saturated conditions the temperatures are identical, and the depression of the wet-bulb reading increases with increasing dryness. Wet- and dry-bulb temperatures are shown in a psychometric chart.
- *The specific enthalpy of the air*, relative to an arbitrary zero corresponding to dry air at 0°C. We have seen earlier that on this basis the specific enthalpy of dry air

$$h = C_p t_a = 1.01 t_a \text{ kJ/kg} \tag{15a}$$

The specific enthalpy of *moist* air must include both the sensible heat and the *latent heat of evaporation* of the moisture content.

The specific enthalpy of moist air:

$$h = 1.01 t_a + \omega \, (1.86 t_a + 2500) \tag{15b}$$

the last two terms representing the sum of the sensible and latent heats of the moisture.

Taking the example of saturated air at 30°C:

$$h = 1.01 \times 30 + 0.0273 \, (1.86 \times 30 + 2500)$$
$$= 30.3 + 1.5 + 68.3 \text{ kJ/kg}$$

The first two terms represent the sensible heat of air plus moisture, and it is apparent that ignoring the sensible heat of the latter, as is usual in air cooling calculations, introduces no serious error. The third term, however, representing the latent heat of the moisture content, is much larger than the sensible heat terms. This accounts for the heavy cooling load associated with the process of drying air by cooling: condensation of the moisture in the air is accompanied by a massive release of latent heat.

Air conditioning processes

Heating and cooling without deposition of moisture
The expression for dry air:

$$H = \rho Q_A C_p \Delta T \text{ kW} \tag{16}$$

is adequate.

Cooling to reduce moisture content

This is a very energy-intensive process and is best illustrated by a worked example. Increasing moisture content is achieved either by spraying water into the air stream (with a corresponding cooling effect) or by steam injection. With the advent of Legionnaires' disease the latter method, involving steam that is essentially sterile, is favoured.

Calculation of cooling load: worked example. Consider the ventilation system described above. We have assumed an air flow rate $Q_A = 7\,\text{m}^3/\text{s}$, air entering at 20°C. If ambient temperature is 25°C and we are required to reduce this to 20°C then from eq. (16) assuming saturation is not reached,

$$\text{cooling load} = 1.2 \times 7.0 \times 1.01 \times 5 = 42.4\,\text{kW}$$

Now let us assume that the ambient air is 85 per cent saturated, $\varphi = 0.85$ and that we need to reduce the relative humidity to 50 per cent at 20°C.

The psychometric chart shows that the initial conditions correspond to a moisture content:

$$\omega_1 = 0.0171\,\text{kg/kg}$$

The final condition, after cooling and dehumidifying, corresponds to:

$$\omega_2 = 0.0074\,\text{kg/kg}$$

The chart shows that the moisture content ω_2 corresponds to saturation at a temperature of 9.5°C.

From eq. (15b), the corresponding specific enthalpies are:

$$h_1 = 1.01 \times 25 + 0.0171(1.86 \times 25 + 2500)$$

$$= 68.80\,\text{kJ/kg}$$

$$h_2 = 1.01 \times 9.5 + 0.0074(1.86 \times 9.5 + 2500)$$

$$= 28.22\,\text{kJ/kg}$$

The corresponding cooling load:

$$L_C = 1.2 \times 7.0 \times (68.80 - 28.22)$$

$$= 341\,\text{kW}$$

If it is required to warm the air up to the desired inlet temperature of 20°C, the specific enthalpy is increased to:

$$h_3 = 1.01 \times 20 + 0.0074(1.86 \times 20 + 2500)$$

$$= 39\,\text{kJ/kg}$$

The corresponding heating load:

$$L_H = 1.2 \times 7.0 \times (39.00 - 28.22)$$
$$= 90.5\,kW$$

To summarize:

Air flow of $7\,m^3/s$ at 25°C, 85% saturated
Cooling load to reduce temperature to 20°C   42.4 kW
Cooling load to reduce temperature to 9.5°C   341 kW
with dehumidification.
Heating load to restore temperature to 20°C   90.5 kW
50% saturated.

Cooling of air is usually accomplished by heat exchangers fed with chilled water. See Ref. 10 for a description of liquid chilling packages.

This illustrates the fact that any attempt to reduce humidity by cooling as opposed to merely reducing the air temperature without reaching saturation conditions imposes very heavy cooling loads.

Calculation of humidification load: worked example. Let us assume initial conditions, $Q_A = 7\,m^3/s$, temperature 20°C relative humidity $\varphi = 0.3$, and that we are required to increase this to $\varphi = 0.7$ for some experimental purpose.

Relative moisture contents are:

$$\omega_1 = 0.0044\,kg/kg$$
$$\omega_2 = 0.0104\,kg/kg$$

Then rate of addition of moisture

$$= 7 \times 1.2 \times (0.0104 - 0.0044) = 0.050\,kg/s$$

Taking the latent heat of steam as 2500 kJ/kg,

$$\text{heat input} = 2500 \times 0.050 = 125\,kW$$

This is again a very large load, calling for a boiler of at least this capacity.

Effects of humidity: a warning

Electronic equipment is extremely sensitive to moisture. Large temperature changes, when associated with high levels of humidity, can lead to the deposition of moisture on components such as circuit boards, with disastrous results.

This situation can easily arise in hot weather: the plant cools down overnight and dew is deposited on cold surfaces. Some protection may be afforded by continuous air conditioning. The use of chemical driers is possible; the granular substances used are strongly hygroscopic[11] and are capable of achieving very low relative humidities. However, their capacity for absorbing moisture is of course limited and the container should be removed regularly for regeneration by a hot air stream.

The operational answer adopted by many facilities is to leave venerable equipment switched on, in a quiescent state while not in use, or to fit anti-condensation heaters at critical points in cabinets.

Legionnaires' disease

This disease is a severe form of pneumonia and infection is usually the result of inhaling water droplets carrying the causative bacteria (*Legionella pneumophila*). Factors favouring the organism in water systems are the presence of deposits such as rust, algae and sludge, a temperature between 20 and 45°C and the presence of light. Clearly all these conditions can be present in systems involving cooling towers.

Preventative measures include

- treatment of water with scale and corrosion inhibitors to prevent the build-up of possible nutrients for the organisms;
- use of suitable water disinfectant such as chlorine 1–2 ppm or ozone;
- steam humidifiers are preferable to water spray units.

If infection is known to be present flushing, cleaning and hyperchlorination are necessary. If a system has been out of use for some time, heating to about 70–75°C for 1 hour will destroy any organisms present.

This matter should be taken seriously. This is one of the few cases in which the operators of a test facility may be held guilty of endangering life, with consequent ruinous claims for compensation.

Combustion air treatment

The influence of the condition of the combustion air (its pressure, temperature, humidity and purity) on engine performance is discussed in Chapter 12, where it is shown that variations in these factors can have a very substantial effect on performance. In an ideal world engines under test would all be supplied with air at 'standard' conditions. In practice there is a trade-off between the advantages of such standardization and the cost of achieving it.

For routine (non-emissions) production testing variations in the condition of the air supply are not particularly important, but the performance recorded on the test documents with some degree of correlation with other test beds or facilities should be corrected to standard conditions.

The simplest and most widely used method of supplying the combustion air is to allow the engine to take its supply directly from the test cell atmosphere. The great advantage of using cell air for vehicle engines in particular is that rigging of complex 'in-vehicle' air filtration and ducting units is straightforward.

The major disadvantage of drawing the air from within the cell is the uncontrolled variability in temperature and quality arising from air currents and other disturbances in the cell. These can include contamination with exhaust and other fumes and may be aggravated by the use of spot fans.

For research and development testing, and particularly critical exhaust emissions work, it is necessary that, so far as is practicable, the combustion air should be supplied with the minimum of pollution and at constant conditions of temperature, pressure and humidity. While the degree of pollution is a function of the choice of site the other variables may be controlled in the following order of difficulty:

- temperature only;
- temperature and humidity;
- temperature, humidity and pressure.

Centralized combustion air supplies

Centralized combustion air conditioning units can be designed to supply air at 'standard' conditions to a number of cells, and this may be a cost-effective solution providing that all cells require the same standardized conditions of combustion air and if the air is used to condition non-running cell spaces. Feedback from many sites indicates that individual units offer the best operation solution for the majority of research and development test cells.

Specification of operational envelope for humidity and temperature control

To create a realistic operational specification for temperature and humidity control of combustion air requires that the creator has a clear understanding of a psychometric chart (Fig. 5.10), the implications of requiring a large operational envelope plus a clear understanding of the temperature and humidity points actually required by the planned testing regimes is vital. It is recommended that cell users mark on a psychometric chart the operating points prescribed by known (emission homologation, etc.) tests or investigate the specification of available units before producing specifications that may impose unnecessarily high energy demands on the requested unit.

Dedicated combustion air treatment units

It should be noted that standard combustion air treatment units are sold by specialist companies worldwide. The rating of these units matches the common sizes of automotive engines, but due to the different air consumption of gasoline and diesel engines the power rating of the engines supported by one unit will differ for the two engine types as shown in a typical example below:

Model	Flow rate	Suitable for engine of rating below
Unit 1	$800\,m^3/h$	diesel 140 kW, gasoline 200 kW
Unit 2	$1600\,m^3/h$	diesel 280 kW, gasoline 400 kW

Temperature only control, flooded inlet

The simplest form controls air temperature only and supplies an excess of air via a flexible duct terminating in a trumpet-shaped nozzle close to the engine's inlet; this is known as a 'flooded' inlet. The unit should be designed to supply air at a constant volume, calculated by doubling the theoretical maximum air demand of the engine. Because the flow is constant, good temperature control can be achieved; the excess air is absorbed into the cell ventilation inlet flow and compensated for via the ventilation temperature and pressure control.

The normal design of such units contains the following modules:

- inlet filter and fan;
- chiller coils;
- heater matrix;
- insulated duct, via fire damper, terminating in flexible section and bellmouth.

By holding the chiller coils' temperature at a fixed value of say 7°C and heating the air to the required delivery temperature, it is possible to reduce the effect of any humidity change during the day of testing without installing full humidity control.

Switching combustion air units on and off frequently will cause the system control accuracy to decline and therefore it is sensible to use the delivered air in a space heating role when the engine is not running allowing the main ventilation system to be switched off or down to a low extract level. An energy-efficient comfort ventilation regime can be designed by using the combustion air balanced with the purge fan extraction.

Humidity controlled units

Systems for the supply of combustion air of which both temperature and humidity are controlled are expensive, the expense increasing with the range of conditions to be covered and the degree of precision required. They are also energy intensive, particularly when it is necessary to reduce the humidity of the atmospheric air by cooling and condensation. Before introducing such a system, a careful analysis should be made to ensure that the operational envelope proposed is achievable from a standard unit and the services available to it and if not, if it is really justified.

It is sensible to consider some operational restraints on the scheduling of tests having different operational levels of humidity in order to avoid condensation build-up in the system. Corrugated ducting should not be used in systems providing humid air since they tend to collect condensate that has to dry out before good control can be restored.

To indicate more exactly what is involved in full conditioning of combustion air, let us suppose that it is necessary to supply air at standardized conditions to the 250 kW diesel engine for which the energy balance is given in Chapter 13 (Table 13.5).

The calculated air consumption of the engine is 1312 kg/h, corresponding to a volumetric flow rate of about 0.3 m³/s.

The necessary components of an air conditioning unit for attachment to the engine air inlet are shown diagrammatically in Fig. 5.11 and may comprise in succession:

1. air inlet with screen and fan;
2. heater, either electrical or hot water;
3. humidifier, comprising steam or atomized water injector with associated generator;
4. cooling element, with chilled water supply;
5. secondary heating element.

It may be asked why it is necessary to include two heater elements, one before and one after the humidifier. The first element is necessary when it is required to humidify cold dry air, since if steam or water spray is injected into cold air supersaturation will result and moisture will immediately be deposited. On the other hand it is commonly necessary, in order to dry moist air to the desired degree, to cool it to a temperature lower than the desired final temperature and reheat must be supplied downstream of the cooler.

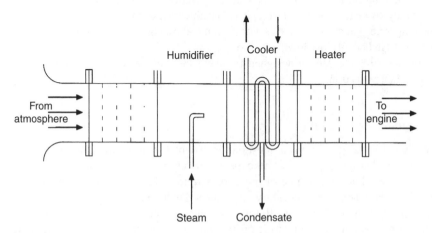

Figure 5.11 *Schematic of typical combustion air system contents*

The internal dimensions of the duct in the present example would be approximately 0.3 × 0.3 m. Assume that the air is to be supplied to the engine at the standard conditions specified in BS 5514 (ISO 3046)

temperature 25°C
relative humidity (r.h.) 30%.

Let us consider two fairly extreme atmospheric conditions and determine the conditioning processes necessary:

hot and humid: 35°C and 80% r.h.
cool to 7°C, cooling load 34 kW
(flow of condensate 0.5 l/min)
reheat to 25°C, heating load 6.5 kW.

It is necessary to cool to 7°C in order to reduce the moisture content, originally 0.030 kg water/kg air, to the required value, 0.006 kg water/kg air; the latter corresponds to saturation at 7°C:

cold and dry: 0°C and 50% r.h.
heat to 25°C, heating load 9 kW
inject steam at rate of 0.1 l/min
heating load 4 kW.

A typical standard unit available for this type of duty has the following technical details:

Unit 2 with delivery of 1600 m³/h at temperature range 15–30°C with option of humidity control 8–20 g/kg:

Total power consumption	64 kVA
C.w. supply at	4–8°C
Pressure	2–4 bar
Flow rate	8 m³/h.

Note that the above equates to ~68 kW, pressure across coils approximately 120 kPa.

Demineralized water for humidifier:

Temperature range	10–30°C
Pressure	2–4 bar
Flow rate	65 dm³/h
Condensate drain	1/2″
Flow rate	80 dm³/h
Compressed air	6–8 bar

Ambient air temperature range (in order for unit to deliver air as per top line) 10–35°C

Humidity	3–30 g/kg

Size of unit: length 2755 mm × width 1050 mm × height 2050 mm.

It will be observed that the energy requirements, particularly for the chiller either internal or from separate chilled water supply, are quite substantial. In the case of large engines and gas turbines any kind of combustion air conditioning is not really practicable and reliance must be placed on correction factors.

Any combustion air supply system must be integrated with the fire alarm and fire extinguishing system and must include the same provisions for isolating individual cells as the main ventilation system.

Important integration points

Combustion air units of all types will require condensate drain lines with free gravity drainage to a building system; back-up or overspill of this fluid can cause significant problems. Those fitted with steam generators will require both condensate and possibility steam 'blow-down' lines piped to safe drainage.

If steam generators are fitted, the condition of feed water will be critical and suitable water treatment plant needs to be installed, particularly where supply water is hard.

Pressure controlled units

A close coupled air supply duct attached to the engine is necessary in special cases, such as anechoic cells, where air intake noise must be eliminated, at sites running legislative tests that are situated at altitudes above those where standard conditions can be achieved and where a precisely controlled conditioned supply is needed.

Since engine air consumption varies more or less directly with speed and that parameters can vary more rapidly than any air supply system can respond, and since it is essential not to impose pressure changes on the engine air, some form of excess air spill or engine bypass strategy has to be adopted to obtain pressure control with any degree of temperature and humidity stability.

For testing at steady state conditions, the pressure control system can operate with a slow and well-damped characteristic; typically having a stabilization time, following an engine speed set-point change, of 30 seconds or more. This is well suited to systems based on regulating the pressure of an intake plenum fitted with a pressure controlling spill valve/flap.

The inlet duct must be of sufficient size to avoid an appreciable pressure drop during engine acceleration and it needs to be attached to the engine inlet in such a way as to provide a good seal without imposing forces on the engine.

It is sensible to encapsulate or connect to the normal engine air inlet filter, if it can sustain the imposed internal pressure imposed by the supply system. If the filter has to be encapsulated it often requires a bulky rig item that has to be suspended from a frame above the engine.

Various strategies have been adopted for fast pressure control, the best of which are able to stabilize and maintain pressure at set point within <2 seconds of an engine step change of 500 r.p.m.

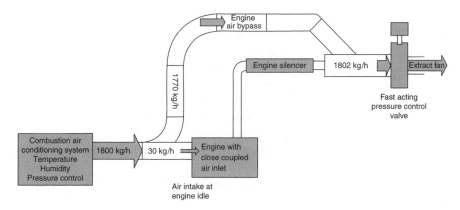

Figure 5.12 *AVL dynamic combustion air pressure control system. Example flows shown at engine idle condition*

Control of inlet pressure only does not simulate true altitude; this requires that both inlet and exhaust are pressurized at the same pressure, a condition that can be simulated using a dynamic pressure control system patented by AVL List GmbH as shown in Fig. 5.12.

Such systems find use in motor sport and other specialist applications, but where the 'ram effect' of the vehicle's forward motion is required, as is the case in F1 and motor cycle engine testing, then considerable energy is required to move the air and the test cell begins to resemble a wind tunnel, which is outside the remit of this book.

Charge (boost) air temperature control

The turbocharging process compresses the combustion air and raises its temperature, the effect of which is discussed in Chapter 12. Charge air cooling frequently has to be accommodated in an engine test cell and there are various strategies that may be used; one is to use the ex-vehicle charge air cooler as a rig item, the other is to use a 'fluid/charge air' cooler as a cell fixture.

Using forced air from a spot fan to cool the vehicle charge air cooler will introduce a great deal of additional heat energy into the cell; it is therefore good practice to use a liquid cooled charge air cooler having very similar flow resistance on the air side to the vehicle unit. A common strategy is to encase the original vehicle air/air intercooler in a special casing and use a water or water spray as the cooling medium. Temperature control is achieved by a flow control valve on the fluid circuit.

Summary

Design of the ventilation system for a test cell is a major undertaking and a careful and thorough analysis of expected heat loads from engine, exhaust system, dynamometer, instrumentation, cooling fans and lights is essential if subsequent difficulties are to be avoided. Once the maximum heat load has been determined it is recommended that a diversity factor is applied so as to arrive at a realistic rating for the cell services.

Any cell using gasoline or other volatile fuels should include a purge system within the ventilation system to remove potentially explosive vapour.

The choice between full flow ventilation and air conditioning has to be made after detailed review of site conditions, particularly building space, and cost.

If a separate air supply for the engine is to be considered, careful thought has to be given to its type and rating in order to keep costs and energy requirements in line with actual test requirements.

The ventilation of the control room should ensure that conditions there meet normal office standards.

Notation

Atmospheric pressure	p_a bar
Atmospheric temperature	t_a °C
Density of air	ρ m^3
Gas constant for air	$R = 287$ J/kgK
Specific heat of air at constant pressure	C_p J/kgK
Mass rate of flow of air	m kg/s
Rate of heat loss, vertical surface	Q_v W^2
Rate of heat loss, horizontal surface	Q_h W^2
Temperature of surface	t_s °C
Rate of heat loss by radiation	Q_r W/m^2
Ventilation air temperature rise	ΔT °C
Total heat load	H_L kW
Ventilation air flow rate	Q_A m^3/s
Fan air flow rate	Q_F m^3/s
Velocity of air	V m/s
Velocity pressure	p_v Pa
Static pressure	p_s Pa
Total pressure	p_t Pa
Pressure loss	Δp_t Pa
Static air power	P_{sF} kW
Total air power	P_{tF} kW
Shaft power	P_A kW
Moisture content	ω kg/kg
Relative humidity	φ
Specific enthalpy of moist air	h kJ/kg
Cooling load	L_C kW

Heating load	L_H kW
Emissivity	ε
Pressure loss coefficient	K_e

References

1. Pita, E.G. (1981) *Air Conditioning Principles and Systems: An Energy Approach*, Wiley, Chichester.
2. Heinsohn, R.J. (1991) *Industrial Ventilation Engineering Principles*, Wiley, Chichester.
3. Bejan, A. (1993) *Heat Transfer*, Wiley, Chichester.
4. Freeston, H.G. (1958) Test bed installations and engine test equipment, *Proc. I. Mech. E.*, **172** (7).
5. TM 8 *Design Notes for Ductwork*, Chartered Institution of Building Services, London.
6. *Industrial Ventilation: A Manual of Recommended Practice* (1982) (17th edn) American Conference of Government Industrial Hygienists: Committee on Industrial Ventilation.
7. BS 5720 *Code of Practice for Mechanical Ventilation and Air Conditioning in Buildings.*
8. BS 848 Part 1 *Fans for General Purposes.*
9. *C.I.B.S. Psychrometric Chart*, Chartered Institution of Building Services, London.
10. BS 7120 *Specification for Rating and Performance of Air to Liquid and Liquid to Liquid Chilling Packages.*
11. BS 2540 *Specification for Granular Desiccant Silica Gel.*

Further reading

BS 599 *Methods of Testing Pumps.*
BS 6339 *Specification for Dimensions of Circular Flanges for General Purpose Industrial Fans.*

6 Test cell cooling water and exhaust gas systems

Introduction

This chapter describes the systems designed to contain cooling water and exhaust gas within a test facility; the largest piped systems normally found within the engine test cell. The thermodynamics of water cooling are considered and the calculation of cooling water requirements is described. A note on water quality follows and the design of test cell cooling water systems is outlined.

Options regarding design of exhaust systems are reviewed and basic recommendations made. Finally, to pull together the various topics considered in chapters of the book relating to energy flows, an energy balance is drawn up for a complete test cell and the process of sizing the various services is considered.

Cooling water supply system: fundamentals

The cooling water system for any heat engine test facility has provided water of suitable quality, temperature and pressure to allow sufficient volume to pass through the equipment in order to have adequate cooling capacity.

The pressure and flow rates have to be sufficiently constant to enable the devices supplied to maintain control. A common fault of badly designed cooling water systems is 'cross talk', where control of one process changes because of sudden supply pressure or temperature changes caused by external events occurring within a shared supply. It is essential for purchasers of water-cooled plant to carefully check the inlet water temperature specified for the required performance, since the higher the cooling water inlet temperature supplied by the factory, the less work the device will be capable of performing before the maximum allowable exit temperature is reached.

Water

Water is the ideal liquid cooling medium. Its specific heat is higher than that of any other liquid, roughly twice that of hydrocarbons. It is of low viscosity, relatively non-corrosive and readily available.

The specific heat of water is usually taken as:

$$C = 4.1868 \, \text{kJ/kg K}$$

Note: this is in fact the value of the 'international steam table calorie' and corresponds to the specific heat at 14°C. The specific heat of water is very slightly higher at each end of the liquid phase range: 4.21 kJ/kg K at 0°C and at 95°C, but these variations may be neglected.

The use of antifreeze (ethylene glycol) as an additive to water permits operation over a wider range of coolant temperatures. A 50 per cent by volume solution of ethylene glycol in water permits operation down to a temperature of −33°C. Such an antifreeze also raises the boiling point of the coolant and a 50 per cent solution will operate at a temperature of 135°C with pressurization of only 1.5 bar.

The specific heat of ethylene glycol is about 2.28 kJ/kg K and, since its density is 1.128 kg/l, the specific heat of a 50 per cent by volume solution is:

$$(0.5 \times 4.1868) + (0.5 \times 2.28 \times 1.128) = 3.38 \, \text{kJ/kg K}$$

or 80 per cent of that of water alone. Thus the circulation rate must be increased by 25 per cent for the same heat transfer rate and temperature rise.

The relation between flow rate, q_w (litres per hour), temperature rise, Δt, and heat transferred to the water is:

$$4.1868 \, q_w \Delta t = 3600H$$
$$q_w \Delta t = 860H$$

where H = heat transfer rate, kW.

(To absorb 1 kW with a temperature rise of 10°C the required flow rate is thus 86 l/h.)

Required flow rates

In the absence of a specific requirement it is good practice, for design purposes, to limit the temperature rise of the cooling medium through the engine water jacket to about 10°C.

In the case of the dynamometer, the flow rate is determined by the maximum permissible cooling water outlet temperature, since it is important to avoid the deposition of scale (temporary hardness) on the internal surfaces of the machine. Eddy current dynamometers, in which the heat to be removed is transferred through the loss plates, are more sensitive in this respect than hydraulic machines, in which heat is generated directly within the cooling water.

Recommended maximum (leaving) temperatures are:

Eddy current machines 60°C
Hydraulic dynamometers 70°C*

provided carbonate hardness of water does not exceed 50 mg CaO/l. For greater hardness values limit temperatures to 50°C.

Approximate cooling loads per kilowatt of engine power output are shown in Table 6.1. Corresponding flow rates and temperature rises are as shown in Table 6.2.

Water quality[1]

At an early stage in planning a new test facility it is essential to ensure that a sufficient supply of water of appropriate quality is made available. Control of water quality, which includes the suppression of bacterial infections, algae and slime, is a complex matter and it is advisable to consult a water treatment expert who is aware of local conditions. If the available water is not of suitable quality then the project must include the provision of water treatment plant.

Table 6.1   *Cooling loads*

	Output (kW/kW)
Automotive gasoline engine, water jacket	0.9
Automotive diesel engine, water jacket	0.7
Medium speed heavy diesel engine	0.4
Oil cooler	0.1
Hydraulic or eddy current dynamometer	0.95

Table 6.2   *Cooling water flow rates*

	In °C	Out °C	l kWh
Automotive gasoline engine	70	80	75
Automotive diesel engine	70	80	60
Medium speed heavy diesel engine	70	80	35
Oil cooler	70	80	5
Hydraulic dynamometer	20	68	20
Eddy current dynamometer	20	60	20

---

* Approaching this outlet temperature, some machines experience flash boiling which can lead to a degrading of control and be heard as a distinctive 'crackling' noise. Prolonged running at these conditions will cause cavitation damage to the working chamber of a dynamometer.

Most dynamometer manufacturers publish tables, prepared by a water chemist, which specify the water quality required for their machines. The following notes are intended for the guidance of non-specialists in the subject.

Solids in water

Circulating water should be as free as possible from solid impurities. If water is to be taken from a river or other natural source it should be strained and filtered before entering the system. Raw surface water usually has significant turbidity caused by minute clay or silt particles which are ionized and may only be removed by specialized treatments (coagulation and flocculation). Other sources of impurities include drainage of dirty surface water into the sump, windblown sand entering cooling towers and casting sand from engine water jackets. Hydraulic dynamometers are sensitive to abrasive particles and accepted figures for the permissible level of suspended solids are in the range 2–5 mg/l. The use of seawater or estuarine water is not to be recommended other than in specially designed marine installations.

Water hardness

The hardness of water is a complex property, while there is a general subjective understanding of the term related to the ease with which soap lather can be created in a water sample; the quality is not easy to measure objectively. Hard water, if its temperature exceeds about 70°C, may deposit calcium carbide 'scale', which can be very destructive to all types of dynamometer and heat exchanger. A scale deposit greatly interferes with heat transfer and commonly breaks off into the water flow when it can jam control valves and block passages. Soft water may have characteristics that cause corrosion, so very soft water is not ideal either.

Essentially, hardness is due to the presence of divalent cations, usually calcium or magnesium, in the water. When a sample of water contains more than 120 mg of these ions per litre, expressed in terms of calcium carbonate, $CaCO_3$, it is generally classified as a hard water.

There are several national scales for expressing hardness, but at present no internationally agreed scale:

American and British: 1° US = 1° UK = 1 mg $CaCO_3$ per kg water = 1 ppm $CaCO_3$
French: 1° F = 10 mg $CaCO_3$ per litre water
German: 1° G = 10 mg $CaCO_3$ per litre water
1° dH = 10 mg CaO per litre water = 1.25° English hardness
(the old British system, 1 Clarke degree = 1 grain per Imperial gallon = 14.25 ppm $CaCO_3$).

Requirements for dynamometers are usually specified as within the range 2–5 Clarke degrees (30 to 70 ppm $CaCO_3$).

Water may be either acid or alkaline/basic. Water molecules, HOH (commonly written H_2O), have the ability to dissociate, or ionize, very slightly. In a perfectly neutral water (neither acidic nor basic) equal concentrations of H+ and OH− are present. The pH value is a measure of the hydrogen ion concentration: its value is important in almost all phases of water treatment, including biological treatments. Acid water has a pH value of less than 7.07 and most dynamometer manufacturers call for a pH value in the range 7–9; the ideal is within the range 8–8.4.

The preparation of a full specification of the chemical and biological properties of a given water supply is a complex matter. Many compounds – phosphates, sulphates, sodium chloride and carbonic anhydride – all contribute to the nature of the water; the anhydrides in particular being a source of dissolved oxygen that may make it aggressively corrosive. This can lead to such problems as the severe roughening of the loss plate passages in eddy current dynamometers, causing failures due to local water starvation and plate distortion (the narrow passages in eddy current dynamometers are particularly liable to blockage which can arise from chemicals used in some water treatment regimes). Water treatment specifications should include the fact that, if used with water brakes, the treated water will be subjected to highly centrifugal regimes and local heating that may cause some degrading of the solution.

Control of water quality also includes the suppression of bacterial infections, algae and slime. BS 4959[2] describes the additives used to prevent corrosion and scale formation, with chemical tests for the control of their concentration and also gives guidance on the maintenance and cleaning of cooling water systems.

A recirculating system should include a small bleed-off to drain, to prevent deterioration of the water by concentration of undesirable elements. A bleed rate of about 1 per cent of system capacity per day should be adequate. If no bleed-off is included, the entire system should be periodically drained, cleaned out and refilled with fresh water.

Finally, consideration should be given at the design stage to the consequences of a power failure. Consider, for example, the consequences of a sudden failure in the water supply to a hydraulic dynamometer absorbing 10 MW at full speed from a marine diesel engine. The system will take some time to bring to rest, during which the brake will be operating on a mixture of air and water vapour, with the possibility of serious overheating. In the case of large engine test facilities some provision for a gravity feed of water in the event of a sudden power failure is advisable.

It is desirable that the supply pressure to hydraulic dynamometers should be stable or the control of the machine will be affected. This implies that the supply pump must be of adequate capacity, with a flat pressure–volume characteristic. Sufficient pressure must be available under maximum demand conditions to meet the specified dynamometer head. The design and installation of a closed (see point 3 below) water supply for a large test installation is a specialist task not to be underestimated. It may require the inclusion of a large number of test and flow balancing valves, together with air bleed points, stand-by pumps and filters with changeover arrangements.

Types of test cell cooling water circuits

Engine coolant control systems and cell cooling water circuits may be classified as follows, with increasing levels of complexity:

1. Direct mains water supplied systems containing service modules and cooling columns that allow heated water to run to waste.
2. Sump or tank stored water systems that are 'open', meaning at some point in the circuit water runs back into the sump via an open pipe. These systems normally incorporate self-regulating water/fluid cooling modules for closed engine cooling systems filled with special coolant/water mix and, if required, for lubricant cooling. They commonly have secondary pumps to circulate water from the sump through evaporative cooling towers when required.
3. Closed pumped circuits with an expansion, pressurization and make-up units in the circuit. Such systems have become the most common as most modern temperature control devices and eddy-current or electrical dynamometers, unlike water-brakes, do not require gravitational discharge. Closed water cooling systems are less prone to environmental problems such as Legionnaires' disease.
4. Chilled water systems (those supplying water below ambient) are almost always closed.

'Open' water cooling circuits

The essential features of these systems are that they store water in a sump lying below floor level from which it is pumped through the various heat exchangers and a cooling tower circulation system. The sump is normally divided into hot and cold areas by a partition weir wall (see Fig. 6.1). Water is circulated from the cool side and drains back into the hot side. When the system temperature reached the control maximum, it is pumped through the cooling tower before draining back into the cool side. A rough rule for deciding sump capacity is that the water should not be turned over more than once per minute. Within the restraints of cost, the largest available volume gives the best results. *Sufficient excess sump capacity, above working level, should be provided to accommodate drain-back from pipework, engines and dynamometers upon system shutdown.*

There is a continuous loss of water due to evaporation plus the small drainage to waste mentioned above and make-up is supplied by way of a float valve fitted to a mains water supply.

To minimize air entrainment the pump suction should be located close to a corner; return flow should be by way of a submerged pipe with air vent.

This is a classic arrangement with thousands of similar systems installed worldwide, but care has to be taken to keep debris such as leaves or flood water 'wash-off' from entering the system via the sump lip or the cooling tower collector. A sensible design feature at sites where freezing conditions are experienced is to use

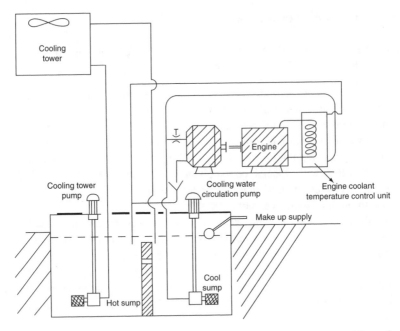

Figure 6.1   *Simple open cooling water system incorporating a partitioned sump*

pumps submerged in the sump so it can be ensured that, when not being used, the majority of pipe work will be empty.

Closed water cooling circuits

Essentially, the system uses one or more pumps to force water through the circuit load where it picks up heat which is then dispersed, usually in the air, via closed 'cooling towers', then the water is returned directly to the pump inlet.

It is vital that air is taken out and kept out of the system and that the whole pipe system be provided with the means of bleeding air out at high points or any trap points in the circuit. To achieve proper circulation, to cope with thermally induced changes of system volume and to make up for any leakage, the closed system has to be fitted with an expansion tank and means of pressurization. Both these requirements can be met by using a form of compressed air/water accumulator connected to a 'make-up' supply of treated water. Balancing water systems is the means by which the required flow, through discrete parts of the circuit having their own particular resistance to flow, is fixed by use of flow regulation valves having test points fitted for setting purposes. The balancing of closed cooling systems can be problematic, particularly if a facility is being brought into commission in several phases meaning that the complete system will have to be rebalanced at each significant change.

None of the devices fitted within a closed and balanced plant water system should have valves that change the flow of the plant water (economizer valves) since that variation will continually unbalance the whole system.

Closed systems are often filled with an ethylene glycol–water mix to cope with freezing weather conditions or have any external pipe work trace heated.

Cooling columns

If special engine coolants are not required, a cooling column is a simple and economical solution commonly used in the USA. It can be portable and located close to the engine under test. This allows the engine outlet temperature to run up to its designed level; at this operating point a thermostatic valve opens, allowing cold water to enter the bottom of the column and hot water to run to waste or the sump from the top. The top of the column is fitted with a standard automotive radiator cap for correct pressurization and use when filling the engine circuit.

Engine coolant and oil temperature control modules

Whether or not the engine under test is fitted with its own thermostat, precise control of coolant temperature is not easily achieved unless the service module used is designed to match the thermal characteristics of the engine with which it is associated; even here it may be difficult to achieve stable temperatures at light load.

The instability of temperature control is increased if the engine is much smaller than that for which the cooling circuit is designed. The capacity of the heat exchanger is the governing factor and it may be advisable, when a wide range of engine powers is to be accommodated, to provide several coolers with a range of capacities.

There are many closed system engine coolant temperature control units on the market, most working on the principle of a closed loop control valve controlling flow of coolant through a heat exchanger and they can be broken down into the following types:

1. mobile pedestal type;
2. special engine pallet-mounted systems;
3. user-specific, wall-mounted systems;
4. complex fixed pedestal type.

Figure 6.2 is an illustration of a typical service module incorporating heat exchangers for jacket coolant and lubricating oil, while Fig. 6.3 shows a simplified schematic of the circuit.

The combined header tank and heat exchanger is a particularly useful feature. This has a filler cap and relief valve and acts in every way as the equivalent of a conventional engine radiator, ensuring that the correct pressure is maintained. If some engines are to be tested without their own coolant pumps the module must

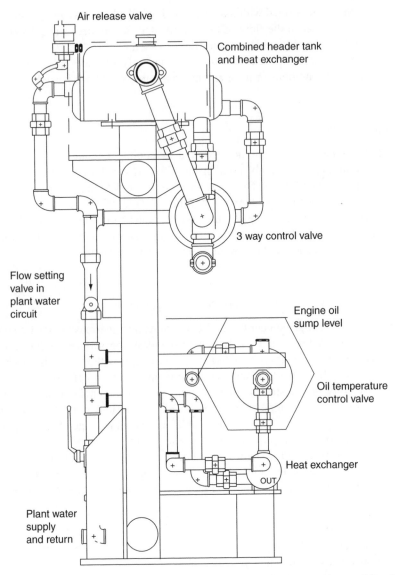

Figure 6.2   *Diagram of movable coolant conditioning unit with combined heat exchanger/header tank*

be fitted with a circulating pump, commonly of the type used in central heating systems. For ease of maintenance, it should be possible to withdraw exchanger tube stacks without major dismantling of the system and a simple means for draining both oil and a coolant circuits should be provided.

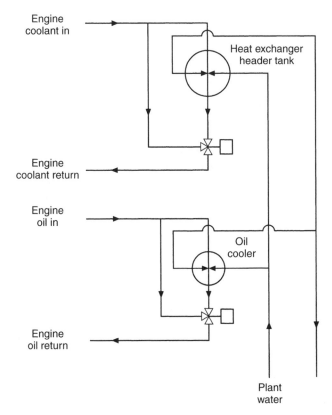

Figure 6.3   *Circuit diagram of unit in Fig. 6.2*

The most usual arrangement is to control the temperature by means of a three-way thermostatically controlled valve in the engine fluid system. The alternative, where temperature is controlled by regulating the primary cooling water flow, will work but gives an inherently lower rate of response to load changes.

The types 2 and 3 listed above are often designed and built by the user, particularly the pallet-mounted systems which may use specific ex-vehicle parts for such items as the header tank and expansion vessel.

Type 4 are the most complex and incorporate a coolant circulation pump, heaters and complex control strategies to deal with low engine loads and transient testing.

None of these devices will operate satisfactorily if not integrated well with the engine and cell pipe work.

Time and distance lag, between a sensor located at the engine (inlet or outlet) and the control valve at the cooler, may be significant and the length and volume of pipe runs between engine and service module should be kept to a minimum. The sum of these phenomena is often referred to as the 'thermal inertia' of the cooling system and can be most easily visualized by considering the speed at which heated or cooled fluid is circulated, detected and diverted within the total engine/cooling system.

To reduce thermal inertia there are two widely used strategies:

1. Reduce the distance and fluid friction head between coolers and engine.
2. Circulate the coolant between engine and coolers with auxiliary pumping (in both cases the interconnecting pipes should be insulated against heat loss/gain).

Strategy 1 is best served by arranging a pallet-mounted cooling module close to the engine.

In cases such as anechoic cells, where the heat exchanger is inevitably remote from the engine, strategy 2 is required to speed up the rate of circulation by an auxiliary pump mounted outside the cell to reduce lag.

Some high-end devices, such as the AVL 553®, provides a continuous circulation of coolant within its own system from which the engine draws off the required flow. However, the quality of control is still dependent on good installation to reduce fluid transit time and heat loss.

Control strategies of heat exchanger systems

Three strategies are commonly adopted for the source of control:

*   independent packaged controllers using electrically operated valves such those produced by Honeywell;
*   indirect control by instruments having internal control 'intelligence' and given set points from a central (engine test) control system;
*   direct control from the engine test controller via auxiliary PID software routines and pneumatically operated valves.

A proportional and integral (P and I) controller will give satisfactory results in most cases, but it is important that thermostatic valves should be correctly sized, since they will not function satisfactorily at flow rates much lower than that for which they are designed. In design terms, the valves have to have 'authority' over the range of flows that the coolers require. A common error, which creates a circuit where valves do not have authority at low demands, is to fit valves that have too higher rated flow rate; it is very difficult to control such a system at low heat levels, since the valve only has to crack open for an excess of cooling to take place and stability is never achieved.

Ensure that the valve is installed with flow in the correct direction!

Flow velocities in cooling water systems

The maximum water velocity in a supply system should not exceed 3 m/s, but may need to reach a minimum of 1.5 m/s in some systems to sweep away deposited matter. Velocities in gravity drains will not normally exceed 0.6 m/s. It is good practice to keep pipe runs as straight as possible and to use bends rather than elbows.

Design of water-to-water heat exchangers

Manufacturers of heat exchangers and devices using cooling water invariably provide simple design procedures for establishing the sizes of flow rates and pressure drop for a given heat exchanger performance. If it should be required to design a heat exchanger from first principles, Ref. 3 gives a detailed design procedure.

Engine thermal shock testing

To accelerate durability testing of engines and engine components, including cylinder head seals, many manufacturers carry out thermal shock tests which may take several forms, but commonly require the sudden exchange of hot coolant within a running engine for cold fluid. The term 'deep thermal shock' is usually reserved for tests having a Δt of around 100°C; such tests are also called, descriptively, 'head cracking' tests. All such sudden changes in fluid temperatures cause significant differential movement between seal faces.

Thermal shock tests are normally carried out in normal (not climatic) test cells and are achieved by having a source of cold fluids that can be switched directly into the engine cooling circuit via three-way valves inserted into the coolant system inlet and outlet pipes. Due to the speed required to chill down the engine and the cyclic nature of the tests it is not usual to use secondary heat exchanges between the chilled fluid and the engine coolant. The energy requirements for such tests clearly depend on the size of engine and the cycle time between hot stabilized running and the chilled engine condition, but they can be very considerable and often require cells with specially adapted fluid services that include a large buffer tank of chilled coolant. Mobile thermal shock systems, complete with chiller and buffer tank, are made and allow any cell fitted with suitable actuating valves in the engine coolant circuit to be used for thermal shock testing. However, their capacity is less than that of a purposely adapted cell.

Commissioning of cooling water circuits

Before dynamometers and any other test cell instrumentation are connected to the building's cooling water system, it must be cleaned and the water treated appropriately. Good practice dictates that during water system commissioning the system is first fully flushed. At each point that is as close as possible to an instrument that has to be supplied with water, the system should be fitted with temporary bypass pipes that allow water to be circulated. If this is not possible or if large heat exchangers have to be flushed then temporary strainers should be put in circuit.

Only when temporary strainers and filters are not picking up debris should instruments be connected.

In locations where water is freely available it is recommended that the first system fill of untreated mains supplied water used for flushing is dumped and the settling tanks cleaned if necessary. Debris will include both magnetic and non-magnetic material injurious to test plant, such as jointing compound, PTFE tape, welding slag and building dust. It is cost-effective to take trouble in getting the coolant up to specification at this stage rather than dealing with the consequences of valve malfunction or medium-term corrosion/erosion problems later.

After flushing, the system should then be filled with clean water that has been treated to balance the hardness, pH level and biocide level to the required specification.

Drawing up the energy balance and sizing the system

In Chapter 2 it is recommended that at an early stage in the design of a new test cell a diagram similar to Fig. 6.4 should be drawn up to show all the flows: air, fuel, water, exhaust gas, electricity and heat into and out of the cell.

The process is best illustrated by an example, and the full-load regime for a 250 kW turbocharged diesel engine driving a hydraulic dynamometer has been chosen. The heat balance for the engine itself is shown diagrammatically in Chapter 13. This shows the results of the engine cooling water calculation, the exhaust energy, the induction air flow and the estimated convection and radiation losses.

The design of the ventilation system for the cell is dealt with in Chapter 5, and the corresponding air and heat flows are shown in Chapter 13.

The cooling water flow for the hydraulic dynamometer may be calculated on the basis of the following assumptions:

- 95 per cent of power absorbed appears in the cooling water;
- cooling water inlet temperature 30°C;
- maximum desirable outlet temperature 70°C.

Then flow to brake:

$$\frac{250 \times 0.95 \times 60}{4.186 \times 40} = 851 \, l/min$$

The electrical power input to the cell for lights, instrumentation and spot fans is assumed to be 20 kW and, finally, the heat losses through the cell walls and ceiling are assumed to be small, 5 kW.

Calculated pipe sizes are as follows:

Fuel, velocity 0.2 m/s, 10.8 mm dia, say 0.5 in
Engine cooling water, 3 m/s, 45 mm dia, say 2 in
Brake cooling water, 3 m/s, 28 mm dia, say 1.5 in

A final point regarding engine cooling water: it may be assumed that this will make use of a service module. Either way, the overall energy flow will be unchanged but in the case where one is not used the temperature rise of the primary cooling water may be substantially higher, leading to a smaller flow rate.

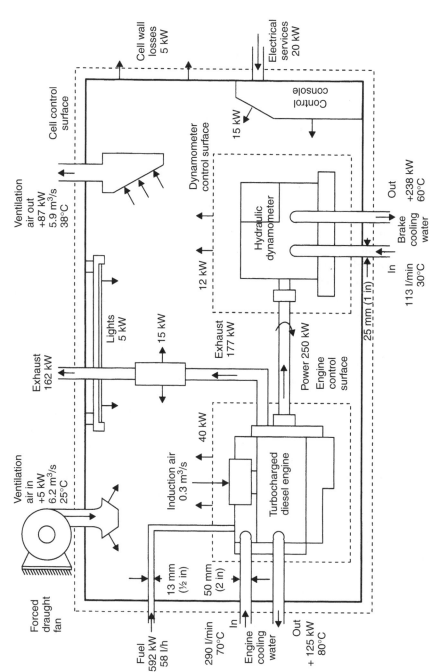

Figure 6.4 *Energy balance and energy flow diagram for example 250 kW cell*

Exhaust gas systems

The layout and design of exhaust systems forming part of full gaseous emission analysis systems is covered in Chapter 16.

It is possible to run into operational safety problems with test cell exhaust systems; there have been accidents because of these systems, some of them fatal.

Particular care should be taken in the detailed design and construction of exhaust ducting containing exhaust gases under positive pressure within building spaces. Within the test cell the most significant risk to operation staff is contact with hot metal exhaust piping. From the research engineer's viewpoint, the ideal exhaust system for an engine under test is one that resembles exactly the system that would be used on the same engine in service; indeed, many modern automotive engines are fitted with emission control systems which require that the actual vehicle exhaust system should be employed. This can impose problems of layout even in cells of large floor area. Similarly, the practical problems of locating the tail pipe in the scavenge duct can be considerable and often calls for modification of the vehicle system or flexible extension to the cell system. If vehicle systems are to be modified, it should be remembered that any change in the length of the primary pipe is particularly undesirable, since this can lead to changes in the exhaust and induction processes as a result of changes in the pattern of exhaust pulses in the system. This can affect the volumetric efficiency and power output of the engine and, in the case of two-stroke engines, it may prove impossible to run the engine at all with a wrongly proportioned exhaust system. An increase in the length of pipe beyond the first chamber is less critical, but care should be taken to limit the back pressure imposed upon the engine, as measured by a manometer at the silencer inlet, to a maximum of perhaps $100 \, mmH_2O$; any greater restriction will probably have a perceptible effect on volumetric efficiency and power output.

Turbocharged engines in particular have complex exhaust systems and run at such high temperatures that large areas of manifold and exhaust pipe can appear incandescent and this can represent a large heat load on the ventilating system. The parts of the exhaust pipe system after the engine, particularly running at low level in the cell should be guarded to avoid burns to staff and vulnerable equipment. The use of bandage-type lagging may be appropriate providing it does not cause overheating of exhaust system components. Unless the system has been specifically designed to run the exhaust pipe in a floor duct and housekeeping is of a high order, it is not generally a good idea.

The practice sometimes adopted of discharging the engine exhaust into the main ventilation extraction duct has several disadvantages:

- The duct should be of stainless steel, to avoid rapid corrosion.
- The air flow must be increased to a level greater than that necessary for basic ventilation, to maintain an acceptable duct temperature. On a cold day this can lead to chilling in the cell.

- Soot deposits and staining condensate in the fan and ducting are unsightly and make maintenance difficult.
- Other difficulties can arise, such as noise, variability in exhaust back pressure, etc.

If 'tail pipe' testers of the type used in vehicle servicing and inspection are to be employed, then easy access to the tail pipe end is necessary.

Test cell exhaust layouts may be classified as follows:

- individual cell, close coupled;
- individual cell, vehicle exhaust system, scavenged duct;
- multiple cells with common scavenged duct;
- specially designed emission cells (see Chapter 16).

Individual cell, close coupled

Such an arrangement is shown schematically in Fig. 6.5a, it may be regarded as the 'standard' arrangement for a general purpose test bed, and is also commonly used for production testing. The exhaust manifold is coupled to a flexible stainless steel pipe, of fairly large diameter to minimize pressure waves, and led by way of a back pressure regulating valve to a pipe system suspended from the cell roof. Condensate, which is highly corrosive, tends to collect in these pipes, which should be laid to a fall with suitable drainage arrangements.

In cells designed to occasionally use smoke or particulate analysis, it is usually a requirement that the sample is taken from the exhaust pipe at particular points; some devices have quite specific requirements regarding size, position and angle of probe insertion and it is desirable, in order to ensure a representative sample, that

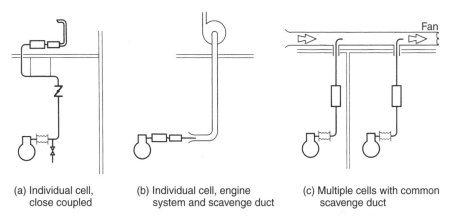

(a) Individual cell, close coupled

(b) Individual cell, engine system and scavenge duct

(c) Multiple cells with common scavenge duct

Figure 6.5 *Exhaust systems*

there should be six diameters length of straight pipe both upstream and downstream of the probe.

It is therefore good practice, when designing the exhaust system, to arrange a straight horizontal run of exhaust pipe; this pipe to be easily replaced by a pipe with specific probe tappings.

Individual cell, engine system scavenged duct

When it is considered necessary to use the vehicle exhaust system, two options are available: one is to take the pipe outside the building through a panel in the cell wall, the other is to use a scavenge air system as shown schematically in Fig. 6.5b. In this case, the tail pipe is simply inserted into a bellmouth through which cell air is drawn; the flow rate should be at least twice the maximum exhaust flow, preferably more. This outflow should be included in the calculations of cell ventilation air flow. The scavenge flow is induced by a fan, usually centrifugal, which must be capable of handling the combined air and exhaust flow at temperatures that may reach 150°C.

Multiple cells common scavenged duct

This arrangement, Fig. 6.5c, may be regarded as standard for large installations. To illustrate the design process for the scavenging system, let us suppose we are setting up an installation of three test cells each like that schematically illustrated in Fig. 6.4. These are to be used for the quality audit testing of the 250 kW turbocharged diesel engine for which maximum exhaust flow rate is shown in the energy balance calculation (see Chapter 13) as 1365 kg/h. To ensure adequate dilution and a sufficiently low temperature in all circumstances, we need to cater for the possibility of all three engines running at full power simultaneously and a scavenge air flow rate of about 10 000 kg/h, say 2.3 m^3/s would be appropriate. Table 5.3 (Chapter 5) indicates a flow velocity in the range 15–20 m/s and hence a duct size in the region of 400 × 300 mm, or 400 mm diameter. As before, the scavenging fan must be suitable for temperatures of at least 150°C.

This arrangement is recommended only for diesel engines. In the case of spark ignition engines there is always the possibility that unburned fuel, say from an engine that is being motored, could accumulate in the ducting and then be ignited by the exhaust from another engine. The possibility may seem remote, but accidents of this kind are by no means unknown.

Note particularly that, in cases (b) and (c), the fan controls must be spark-proof and interlocked with the cell control systems so that engines can only be run when the duct is being evacuated.

Where the engine's exhaust system is not used, the section of exhaust tubing adjacent to the engine must be flexible enough to allow the engine to move on its mountings and a stainless steel bellows section is to be recommended. Exhaust tubing used in this area should be regarded as expendable and the workshop should be equipped to make up replacements and fit transducers.

As a final point, carbon steel silencers that are much oversized for the capacity of the engine will never get really hot and can be rapidly destroyed by corrosion.

Dual use of exhaust gas extraction duct as cell purge systems

Exhaust gas dilution duct fans compliant with European ATEX rules (see Chapter 4) can be integrated into a system design so as to be used as low level hydrocarbon extract/purge systems that run before engine operation, or at a high hydrocarbon gas level alarm. Figure 6.6a shows a possible design for subfloor, close coupled exhaust systems and (b) shows a possible layout for connection to an, above floor, vehicle exhaust system.

Cooling of exhaust gases

There may be an operational requirement to reduce the temperature of exhaust gas exiting the cell. This is usually achieved by using a stainless steel, water-jacketed cooler having several gas tubes that give a low resistance to gas flow. These occasionally used devices are often made specifically for the project in which they are used and there are three important design considerations to keep in mind:

- If water flow is cut off to the exhaust gas cooler, a steam explosion can be caused, therefore primary and secondary safety devices should be fitted.
- High internal stresses can be generated by the differential heating of the cooler elements therefore some expansion of the outer casing should be built into the design.
- Given water jacket pressure will be higher than exhaust gas pressure, tube leakage could fill the engine cylinders unless the system is regularly checked.

Direct water spray has been used in vertical sections of exhaust systems in order to wash and cool exhaust flow into a common evacuated exhaust main.

Exhaust cowls on buildings

When deciding on the position of the exhaust termination outside a building, it is important to consider the possibility of recirculation of exhaust fumes into ventilation inlets and also to avoid the imposition of back pressure through wind flow. Prevention of recirculation requires the careful relative positioning of exhausts and inlet ducts in relation to each other and to the prevailing local (building effected) wind direction.

Figure 6.7a shows a simple termination tube suitable for use on single cell, raw exhaust outlets that incorporates a shroud tube that acts as a plume dilution device and

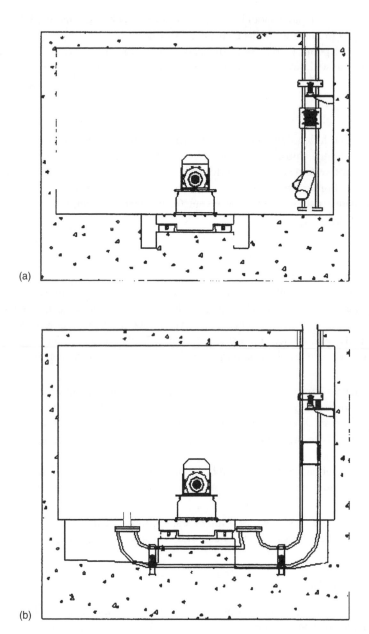

Figure 6.6 *(a) Layout of subfloor exhaust collection duct suitable for left- or right-handed or Vee exhaust configurations and tapping for purge system of subfloor space. (b) Layout of above-floor exhaust dilution duct for collection to vehicle exhaust system and low level purge system*

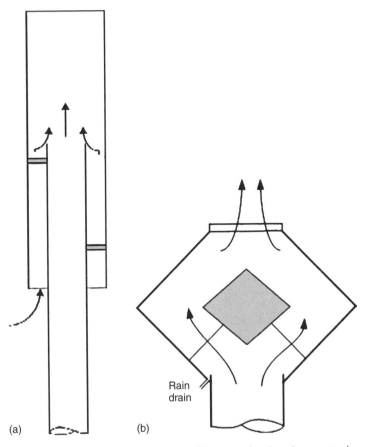

Figure 6.7 *(a) A simple rain excluding dilution tube fitted to a single cell raw exhaust outlet and (b) one design of a building cowl for the termination of a multicell diluted exhaust system*

rain excluder. Figure 6.7b shows a type of rain excluding cowl commonly used on multiple cell diluted exhaust outlets. In both cases, the exit tubes terminate vertically to minimize horizontal noise spread.

Summary

The design of various services concerned with cell cooling water and engine exhaust is outlined. Precautions regarding water are discussed and finally an example of cooling water system design and sizing is given. The possible dangers associated with exhaust systems are emphasized and a number of layouts discussed.

References

1. Vesilind, P.A. Weiner, R.F. and Peirce, J.J. (1988) *Environmental Engineering*, Butterworth-Heinemann, Oxford.
2. BS4959 *Recommendations for Corrosion and Scale Prevention in Engine Cooling Systems.*
3. McAdams, W.H. (1973) *Heat Transmission*, McGraw-Hill, Maidenhead.

7 Fuel and oil storage, supply and treatment

Introduction

Liquid fuels must be stored and transported in conditions which minimize the loss of volatile components, an effect known as 'fuel aging'. The storage and transport of volatile liquid and gaseous fuels worldwide is subject to extensive and ever-developing regulation covering both safety and environmental issues. While much of the legislation quoted in this chapter is British or European in origin, the practices they require are valid in most countries in the world.

In the UK, the Health and Safety Commission (HSC) and the Health and Safety Executive (HSE) are responsible for the regulation of almost all the risks to health and safety arising from work activity, while 410 local authorities (LA) in England, Scotland and Wales have responsibility for the enforcement of health and safety legislation.

The HSE introduced the Dangerous Substances and Explosive Atmospheres Regulations in 2002. Approved Codes of Practice supporting this new legislation require underground storage tanks to be provided with secondary containment or a leak detection system capable of identifying leaks before a hazardous situation can arise. There are European regulations such as the ATEX Directive-94/9/EC, but individual countries within the EEC will then have a level of local regulation, as is the case in different administrative areas in the USA.

Bulk fuel storage and supply systems

In England, if fuel (such as petrol, diesel, vegetable, synthetic or mineral oil) is stored in a container with a storage capacity of over 200 litres (44 gallons) then the owner may need to comply with the Control of Pollution (Oil Storage) (England) Regulations 2001. Similar regulations are in place throughout Europe. Many of the local authorities act as a local Petroleum Licensing Authority. To gain a licence, the following information will usually be required:

1. Site location map to a scale of 1:1250 or 1:2500 indicating all site boundaries.
2. Two copies of site layout to scale 1:100 clearly indicating the present layout of petrol storage depot.

The layout plan must show the following:

- location of storage tanks and tank capacities;
- route taken by road delivery vehicles (this must be a 'drive-through' not a 'cul-de-sac');
- position of fill points and their identification;
- location of pipework including all vent pipes, etc.;
- location of metering pumps, dispensers, etc.;
- all site drainage and its discharge location;
- petrol interceptor location and drainage discharge point;
- all other buildings within the site and their use;
- all neighbouring buildings within 6 m from the boundary;
- position of LPG storage, if applicable;
- location of car wash and drainage if applicable;
- main electrical intake point and distribution board;
- position of all fire fighting appliances.

Very similar restrictions and requirements to those described above for the UK exist in most countries in the world and any engineers taking responsibility, for anything beyond the smallest and simplest test cell, should make themselves familiar with all relevant regulations of this kind, both national and local. Any contractor used to install or modify fuel systems should be selected on the basis of proven competence and, where applicable, relevant licensing to carry out such work.

Both the legislative requirements and the way in which they are interpreted by local officials can vary widely, even in a single country. Where engine test installations are well established there should be a good understanding of the requirements, whereas in localities where such systems are novel, the fire and planning officers may have no experience of the industry and can react with concern; in these cases they may require tactful guidance.

The author refers readers back to the statement made in Chapter 1 concerning the cell being a hazard reduction and containment box.

The vast majority of engine test cells are designed to be classified as zone 2 spaces. The interior of instrument cabinets such as fuel consumption or fuel conditioning devices should be certified by the manufacturer as zone 2.

Figure 7.1 shows a typical arrangement for a fuel oil or gasoline bunded storage tank in accordance with British Standard 799.

The risk of oil being lost during filling or from ancillary pipework is higher than tank rupture; the UK Control of Pollution (Oil Storage) Regulations 2001 recognize this fact and require that tanks have all ancillary equipment such as sight tubes, taps and valves retained within the secondary containment system.

The use of double-skinned tanks is strongly recommended when used in underground installations, as is common in the case for bulk storage of petroleum.

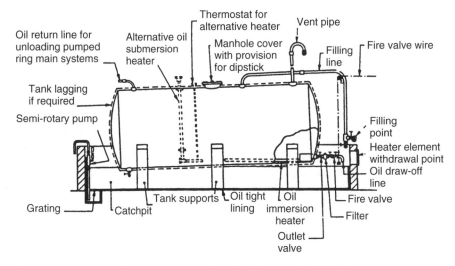

Figure 7.1 *Typical above ground bunded fuel tank with fittings*

Such tanks should also be fitted with an interstitial monitoring device with automatic alarms.

For the storage of fuel in drums or other containers, a petroleum store such as is shown in Fig. 7.2 is to be recommended. This design typically meets the requirements of the local UK authorities, but its location and the volume of fuel allowed to be stored may be subject to site-specific regulation. The lower part forms a bund, not more than 0.6 m deep, capable of containing the total volume of fuel authorized to be kept in the store. Ventilators at high and low level are to be covered by fine wire gauze mesh and protective grilles. Proprietary designs should allow access to a wheeled drum lifter/trolley.

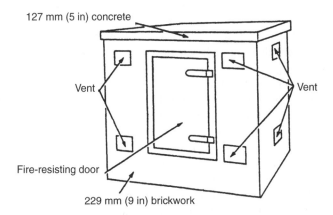

Figure 7.2 *Design of fuel drum store minus statutory warning signs*

Fuel pipes

The expensive consequences of environmental pollution caused by fuel leaks are tending to cause modern best practice to dictate that fuel lines are run above ground, where they can be seen, rather than in underground trenches which was the common practice before the 1990s.

The use of standard (non-galvanized) drawn steel tubing for fuel lines that remain full of liquid fuel is entirely satisfactory, but if they are likely to spend appreciable periods partially drained the use of stainless steel is recommended. The use of threaded fittings is not to be recommended, although provided there is no significant pipe movement, particularly thermally induced movement, and with the use of modern sealants they can be satisfactory. The use of any kind of fibrous 'pipe jointing' should be absolutely forbidden as fibre contamination is difficult to clear.

Preferably, all fuel lines, and certainly all underground lines, should be constructed with orbital welded joints, or with the use of compression fittings approved for the fuels concerned (some fuels such as 'winter diesel' appear to be particularly penetrating). External fuel oil lines should be lagged and must be trace heated if temperatures are likely to fall to a level at which fuel 'waxing' may take place.

Within the test cell, flexible lines may be required in short sections to allow for engine rigging or connection to fixed instrumentation; such tubing and fittings must be specifically made and certified to handle the range of fuels being used; the generally recommended specification is for metal braided, electrically conductive, Teflon hoses.

Storage and treatment of residual fuels

Many large stationary diesel engines and the majority of slow speed marine engines operate on heavy residual fuels that require special treatment before use and must also be heated before delivery to the fuel injection system. In such cases, the test cell fuel supply system is required to incorporate the special features of the fuel supply and treatment systems installed in diesel propelled ships. The bulk storage tanks will require some form of heating coils within it to enable the heavy oil to be pumped into a heated settling tank.

It is always necessary with fuels of this type to remove sludge and water before use in the engine. The problem here is that the density of residual fuels can approach that of water, making separation very difficult. The accepted procedure is to raise the temperature of the oil, thus reducing its density, and then to feed it through a purifier and clarifier in series. These are centrifugal devices, the first of which removes most of the water, while the second completes the cleaning process.

Refs 1 and 2 give detailed recommendations for the design and operation of such systems and Fig. 7.3 shows a schematic arrangement. It is also necessary to provide changeover arrangements in the test cell so that the engine can be started and shut down on a light fuel oil.

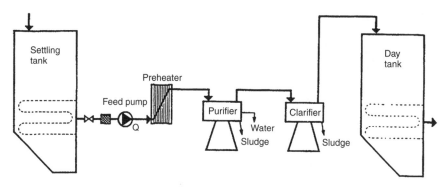

Figure 7.3 *Fuel supply and treatment system for residual fuel oil*

Storage of biofuels

Test facilities involved in the preparation and mixing of biofuels may have special storage problems to consider. Most biodiesel fuels are a mixture of petroleum and vegetable-derived liquids; they are usually designated by the prefix B followed by the percentage of the biomass derived content; thus B20 contains 20 per cent of non-petroleum fuel. Biofuels tend to be hydrophilic, therefore storage tanks and lines can be subject to corrosion from condensation.

Laboratories working with the constituents of biodiesel mixtures will require storage for animal- or vegetable-derived oils in bulk or drums; these may require low-grade heating and stirring in some climatic conditions. Water in the fuel will tend to encourage the growth of microbe colonies in heated fuel tanks which can form soft masses that plug filters.

Ethanol/gasoline mixtures are designated with the prefix E followed by the percentage of ethanol. E5 and E10 (commonly called 'gasohol') are commercially available in various parts of Europe. E20 and E25 are standard fuel mixes in Brazil from where no particular technical problems of storage have been reported.

The storage and handling of 100 per cent ethanol and methanol raises problems of security; the former is highly intoxicating and the latter highly toxic and anyone likely to drink the intoxicant probably lacks the chemical knowledge to distinguish between the two.

Underground fuel lines

Buried mild steel fuel lines, particularly those run under and through concrete floor slabs, should be wrapped with water-repellent bandage (in the UK, 'Denzo tape') and laid on fine gravel within a well-compacted trench. It is good practice to run underground fuel lines in a sealed trench of concrete sections with a load-bearing lid; such ducts should be fitted with hydrocarbon 'sniffers' connected to an alarm system so as to detect fuel leakage before ground contamination takes place. There are certain special requirements for the fuel supply system for an engine test facility that may not apply to systems for other purposes; these include the provision for the use of a number of different fuels including the same quantities of reference fuels. The 'fuel farm' for a modern R&D facility may be provided with a number of storage tanks and several ring mains; these must be clearly labelled at all filling and draw-off points.

Reference fuel drums

Where frequent use is made of special or reference fuels, supplied by the drum, special provision must be made for their transportation from the secure fuel store, such as that shown in Fig. 7.2, and protection while in use. Special drum containers, designed to be transported by fork-lift, are available and these may be parked immediately against an outside wall of the test cell requiring the supply. Connection to the cell system can be achieved by using an automotive 12 V pump system designed to screw into the drum and deliver fuel through a connection point outside the cell fitted with identical fire isolation interlocks as used in the permanently plumbed supplies.

Natural gas, liquefied natural gas, compressed natural gas (NG, LNG, CNG)

Natural gas consists of about 90 per cent methane, which has a boiling point at atmospheric pressure of $-163°C$. Engine test installations requiring natural gas for dual-fuel engines usually draw this from a mains supply at just above atmospheric pressure so high-pressure storage arrangements are not necessary. In the UK, the distribution of LNG to individual commercial users is via a national grid and generally covered by the Gas (Third Party Access) Regulations 2004. Fire hazards are moderate when compared with LPG installations.

Engine test cells using LNG have to comply with gas industry regulations which may include explosion relief panels in the cell and building exhaust system.

Fuel supply to the test cell

Fuel supplies to a cell may be provided either under static head from a day tank, a pumped supply from a reference fuel drum (see above) or by a pressurized reticulation system fed from the central fuel farm (Fig. 7.4).

Day tanks should be fuel specific to prevent cross-contamination, so there may be a requirement for several. Modern safety practice usually, but not exclusively, dictates that gasoline day tanks should be kept on the outside wall of the test facility.

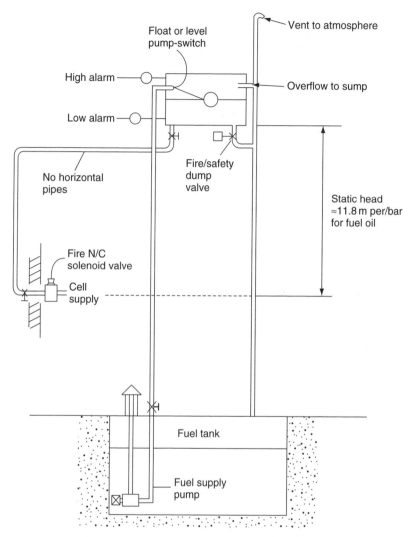

Figure 7.4 *A schematic showing the elements of a typical fuel day tank system*

In all cases, day tanks must be fitted with a dump valve for operation in case of fire; this allows fuel contained in the vulnerable above-ground tanks and pipes to be returned to the fuel farm, or a specific underground dump tank. In addition to having to be vented to atmosphere, day tanks also have to be fitted with a monitored overflow and spill return system in the case of a malfunction of any part of the supply system.

The static head is commonly at 4.5 m or above, but may need to be calculated specifically in order to achieve the 0.5–0.8 bar inlet pressure required by some industrial standard fuel consumption and treatment instruments.

When the tanks are exposed to ambient weather conditions, they should be shielded from direct sunlight and, in the case of diesel fuels, lagged or trace-heated.

If a pumped system is used, it must be remembered that for much of the operating life, the fuel demand from the cells will be below the full rated flow of the pump, therefore the system must be able to operate under bypass, or stall, conditions without cavitation or undue heating.

The use of positive displacement, pneumatically operated pumps, incorporating a rubberized air/fluid diaphragm, have a number of practical advantages: they do not require an electrical supply and they are designed to be able to stall at full (regulated air) pressure, thus maintaining a constant fuel pressure supply to the test cells.

In designing systems to meet this wide range of requirements, some general principles should be borne in mind:

- Pumps handling gasoline should have a positive static suction head to prevent cavitation problems on the suction side.
- Fuel lines (pipes) with low flow resistance where bends rather than sharp elbows are used and no sudden changes in internal cross-section exist.
- Fuel lines should be pressure tested by competent staff prior to filling with fuel.
- Each fuel line penetrating the cell wall should be provided with a normally closed solenoid operated valve interlocked with both the cell control system and the fire protection circuits

In-cell fuel systems

The fuel system in the test cell will vary widely in complexity. In some cases the system may be limited to a single fuel line connected to the engine's fuel pump, but a special purpose test cell may call for the supply of many different fuels all passing through fuel temperature control and fuel consumption measurement devices.

For the occasional test, a simple fuel tank of the type used for outboard engines, capacity not more than 10 litres, may be all that is necessary. These marine devices are certified as safe to use in their designated roles of containing and supplying fuel to an engine; the use of other containers is not safe and would endanger the insurance of the premises in which they are used.

The following points should be considered in the planning of in-cell systems:

- It may be necessary to provide several separate fuel supplies to a cell; a typical provision would be three lines for diesel fuel, 'standard' and 'super' unleaded gasoline, respectively. Problems can arise from carry-over, with consequent danger of 'poisoning' exhaust catalysers. The capacity of the fuel held in the cell system, including such items as filters, needs to be kept to a minimum.

To minimize cross-contamination it is desirable although often difficult to locate the common connection as close to the engine as possible. A common layout provides for an inlet manifold of fuels to be fitted below the in-cell fuel conditioning and measuring systems. Selection of fuel can be achieved via a flexible line fitted with a self-sealing connector from the common system to the desired manifold mounted connector.

- It is good practice to have a cumulative fuel meter in each line for general audit and for contract charging.
- Air entrainment and vapour locking can be a problem in test cell fuel systems. An air eliminating valve should be fitted at the highest point in the system, with an unrestricted vent to atmosphere external to the cell and at a height and in a position that prevents fuel escape.
- Breaking of fuel lines, with consequent spillage, should as far as possible be avoided. It is sensible to mount all control components on a permanent wall-mounted panel or within a special casing, with switching via interlocked and clearly marked valves. The run of the final flexible fuel lines to the engine should not interfere with operator access. A common arrangement is to run the lines via an overhead boom. Self-sealing couplings should be used for engine and other frequently broken connections.
- It is essential to fit oil traps to cell drain connections to avoid the possibility of discharging oil or fuel into the foul water drains.
- Aim for a flow rate during normal operation to be between 0.2 and a maximum recommended fuel line velocity: 1.0 m/s.

Engine fuel pressure control

There are three fuel pressure control problems that may require to be solved when designing or operating an engine test cell:

1. Pressure of fuel supply to the fuel conditioning system or consumption measurement instrument. Devices such as the AVL Fuel Balance require a maximum pressure of 0.8 bar at the instrument inlet; in the case of pumped systems this may require a pressure reducing regulator to be fitted before the instrument.
2. Pressure of fuel supplied to the engine. Typically systems fitted in normal automotive cells are adjustable to give pressures at the engine system inlet of between 0.05 and 4 bar.
3. Pressure of the fuel being returned from the engine. When connected to a fuel conditioning and consumption system, the fuel return line in the test cell may create a greater back pressure than required. In the vehicle, the return line pressure may be between zero and 0.5 bar, whereas cell systems may require over 1.5 bar to force fuel through conditioning equipment and to overcome static head differences between the engine and instruments, therefore a pressure reducing circuit including a pump may be required. Figure 7.5 shows a circuit

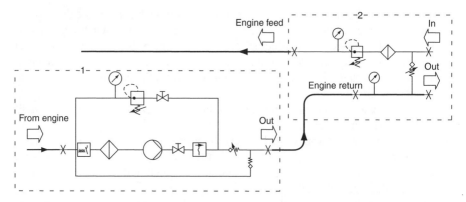

Figure 7.5   *Regulation of engine supply and return pressures within dotted lines (1) is the fuel return pressure control and (2) the fuel feed to the engine from the fuel consumption measurement system (AVL List)*

of a system that allows for the independent regulation of engine supply and return pressures. Care must be exercised in the design of such circuits so as to avoid the fuel pressure being taken below the point at which vapour bubbles form.

Engine fuel temperature control

Clearly all the materials used in the system used in control of fuel temperature, such as heat exchangers, piping and sealing materials, must be checked with the manufacturer as to their suitability for use with all specified fuels.

If fuel supplied to the engine has to be maintained at one standard temperature, as is usually the case in quality audit cells, then a relatively simple control system may provide hot water circulated at controlled temperature through the water-to-fuel heat exchanger. Such an arrangement can give good temperature control despite wide variations in fuel flow rate.

The control of the fluid fuel temperature within the engine circuit is complicated by the fuel rail and spill-back strategy adopted by the engine designer, because of this and variations in the engine rigging pipe work, no one circuit design can be recommended. Commercially available fuel conditioning devices can only specify the temperature control at the unit discharge: the system integrator has to ensure the heat gains and losses within the connecting pipe work do not invalidate experiments.

It is particularly important to minimize the distance between the temperature controlling element and the engine so that, if running is interrupted, the engine receives fuel at the desired temperature with the minimum of delay.

Commercially available units typically have the following basic specifications:

- setting fuel temperature: upto 80°C; providing application system is designed and pressurized to prevent excessive vaporization;
- deviation between set value and actual value at instrument output: <1°C;
- power input: 0.4 kW or 2 kW (with heating);
- cooling load: typically 1.5 kW.

It should be noted that the stability of temperature control below ambient will be dependent on the stability of the chilled water supply which should have a control loop independent of other, larger, cooling loads in any shared system.

A constant fuel temperature over the duration of an experiment is a prerequisite for accurate consumption measurement (see Chapter 12).

As with all systems covered in this book, the complexity and cost of fuel temperature control will depend on the operation range and accuracy specified.

To confirm that temperature change across the fuel measurement circuit has a significant influence on fuel consumption and fuel consumption measurement, especially during low engine power periods Fig. 7.6 should be considered. The graph lines of Fig. 7.6 show that by making a cyclic step change in the set point of a fuel temperature controller of only ±0.2°C within the measurement circuit,

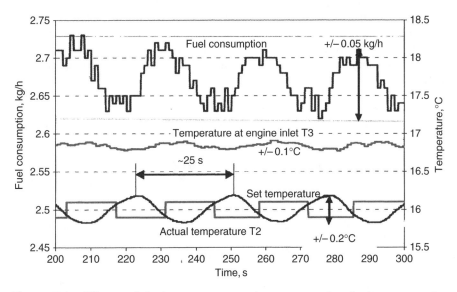

Figure 7.6 *Effects of fuel temperature changes on the fuel consumption measurement values. T2 is the controlled outlet of the fuel conditioning device superimposed on the step demand changes. T3 is the temperature at the engine inlet. The difference between the two is due to system damping (AVL List)*

which produces an oscillation in fuel temperature at the engine inlet of ±0.1°C, then a variation in the measured fuel consumption value of ±0.05 kg/h (top line) is created.

This equals a variation of ±2 per cent of the actual fuel consumption value and shows that instrument manufacturers' claims of fuel consumption accuracy are only valid under conditions of absolutely stable fuel temperatures.

Engine oil cooling systems

Certain special requirements apply to oil cooling units. Unless the cell is operating a 'dry-sump' lubrication system on the engine, the entire lubrication oil temperature control circuit must lie below engine sump level so that there is no risk of flooding the sump. It may be necessary to provide heaters in the circuit for rapid warm-up of the engine; in this case the device must be designed in such a way as to ensure that the skin temperature of the heating element cannot reach temperatures at which oil 'cracking' an occur.

Figure 7.7 shows schematically a separate lubricating oil cooling and conditioning unit.

Where very accurate transient temperature control is necessary, the use of separate pallet-mounted cooling modules located close to the engine may be required; otherwise permanently located oil and cooling water modules offer the best solution for most engine testing. One common position is behind the dynamometer where both units may be fed from the external cooling water system and the engine connection hoses may be run under the dynamometer to a connection point near the shaft end.

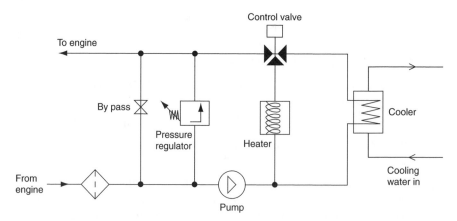

Figure 7.7   *Schematic of oil temperature control unit with sensing points and control connections omitted*

Properties of gasoline and 'shelf life'

Automotive fuels as sold at the retail pump in any country of the world are variable in many details, some of which are appropriate to the geographic location and season and others are due to the original feedstock from which the fuel was refined. Added to the experimental variable of retail fuel properties the storage of gasoline in a partially filled container in conditions where it is exposed to direct sunlight will cause it to degrade rapidly, further increasing the variables in any engine testing. Reference fuels of all types should be stored in dark cool conditions and in containers that are 90–95 per cent full; their shelf life should not be assumed to be more than a few months.

A very important fuel property is the calorific value. Engine development work is often concerned with 'chasing' very small improvements in fuel consumption, differences that can easily be swamped by variations in the calorific value of the fuels. Similarly, comparisons of identical engines manufactured at different sites will be invalid if they are tested on different fuels or fuels that have been allowed to deteriorate at a differential rates.

For example, the LCVs of hexane and benzene, typical constituents of gasoline, are, respectively, 44.8 and 40.2 MJ/kg. A typical value for gasoline would be 43.9 MJ/kg, but this could easily vary by ±2 per cent in different parts of the world, while the presence of alcohol as a constituent can depress the calorific value substantially.

Octane number is the single most important gasoline specification, since it governs the onset of detonation or 'knock' in the engine. This condition, if allowed to continue, will rapidly destroy an engine. Knock limits the power, compression ratio and hence the fuel economy of an engine. Too low an octane number also gives rise to run-on when the engine is switched off.

Three versions of the octane number are used: research octane number (RON), motor octane number (MON) and front end octane number (R100). RON is determined in a specially designed (American) research engine: the Co-operative Fuel Research (CFR) engine. In this engine, the knock susceptibility of the fuel is graded by matching the performance with a mixture of reference fuels, iso-octane with a RON = 100 and n-heptane with a RON = 0.

The RON test conditions are now rather mild where modern engines are concerned and the MON test imposes more severe conditions but also uses the CFR engine. The difference between the RON and the MON for a given fuel is a measure of its 'sensitivity'. This can range from about 2 to 12, depending on the nature of the crude and the distillation process. In the UK, it is usual to specify RON, while in the USA the average of the RON and the (lower) MON is preferred. R100 is the RON determined for the fuel fraction boiling at below 100°C.

Volatility is the next most important property and is a compromise. Low volatility leads to low evaporative losses, better hot start, less vapour lock and less carburettor icing. High volatility leads to better cold starting, faster warm-up and hence better short-trip economy, also to smoother acceleration.

Mixture preparation

Ideally, each cylinder of a multicylinder engine should receive under all operating conditions exactly the same charge of fuel and air, in exactly the same condition, as every other cylinder. This is an unattainable ideal, but a large element in the development programme of any new engine is concerned with approaching it as closely as possible.

A principal variable is the method of introducing the fuel. In order of increasing cost, there are essentially three solutions: carburettor, single (throttle body) injector and individual port injectors. The main factors to be considered are:

* Equal distribution of air quantity, involving detail design of manifold and interaction with flow leaving throttle or carburettor. Particularly difficult at part throttle.
* Equal distribution of fuel. With carburettor or single-point injection there are three aspects to be considered: distribution of air, of fuel droplets and of fuel vapour. The proportion of the latter increases as power is reduced with accompanying increased manifold depression.
* Deposition of liquid fuel on manifold walls.
* Nature of mixture: degree of atomization and of vapour formation.

Properties of diesel fuels

Just as the range of size of the diesel engine is much greater than that of the spark ignition engine, from 1–2 kW to 50 000 kW, the range of fuel quality is correspondingly great (see Chapter 17 for main relevant specifications).

Cetane number is the most important diesel fuel specification. It is an indication of the extent of ignition delay: the higher the cetane number, the shorter the ignition delay, the smoother the combustion and the cleaner the exhaust. Cold starting is also easier the higher the cetane number.

Viscosity covers an extremely wide range. BS 2869 specifies two grades of vehicle engine fuels, class A1 and class A2, having viscosities in the range 1.5–5.0 cSt and 1.5–5.5 cSt, respectively, at 40°C. BS MA 100 deals with fuels for marine engines. It specifies nine grades, class M1, equivalent to class A1, with increasing viscosities up to class M9, which has a viscosity of 130 cSt max at 80°C.

Summary

The bulk storage and handling of automotive fuels has been briefly described and attention drawn to the high degree of safety and environmental legislation that is involved in the design of such systems. A specialist contractor is recommended for any work involving volatile liquid fuels and is required by regulation in many countries when installing gaseous fuel systems.

Liquid fuels are complex mixtures and can be significantly variable; therefore, the need for reference fuels has been discussed as has the possible deterioration of fuels when stored in poor conditions.

The requirement for accurate fuel pressure and temperature control in the engine cell has been demonstrated. For coverage of fuel consumption methods, see Chapter 12; a list of useful publications is to be found below.

References

1. Hughes, J.R. and Swindells, S. (1987) *The Storage and Handling of Petroleum Liquids*, Griffin, London.
2. *Recommendations for Cleaning and Pretreatment of Heavy Fuel Oil*, Alfa Laval, London.

Further reading

BS 3016 *Specification for Pressure Regulators and Automatic Changeover Devices for Liquefied Petroleum Gases.*

BS 4040 *Specification for Leaded Petrol (Gasoline) for Motor Vehicles.*

BS 4250 Part 1 *Specification for Commercial Butane and Propane.*

BS EN589 *Specification for Automotive LPG.*

BS 5355 *Specification for Filling Ratios and Developed Pressures for Liquefiable and Permanent Gases.*

BS 6843 Parts 0 to 3 *Classification of Petroleum Fuels.*

BS EN228 *Specification of Unleaded Petrol (Gasolene) for Motor Vehicles.*

BS 7405 *Guide to Selection and Application of Flowmeters for the Measurement of Fluid Flow in Closed Conduits.*

Dangerous Substances and Explosive Atmospheres Regulations (DSEAR) 2002. BS 2869 Part 2 *Specification for Fuel Oil for Agricultural and Industrial Engines and Burners.*

Installation, Decommissioning and Removal of Underground Storage Tanks: PPG27, UK Environmental Agency.

UK Guidance and rules for the storage of gasoline in portable containers can be found at http://www.hse.gov.uk/lau/lacs/65–9.htm

8 Dynamometers and the measurement of torque

Introduction

The torque produced by a prime mover under test is resisted and measured by the dynamometer to which it is connected. The accuracy with which a dynamometer measures both torque and speed is fundamental to all the other derived measurements made in the test cell.

In this chapter the principles of torque measurement are reviewed and then the types of dynamometer are reviewed in order to assist the purchaser in the selection of the most appropriate machine.

Measurement of torque: trunnion-mounted (cradle) machines

The essential feature of trunnion-mounted or cradled dynamometers is that the power absorbing element of the machine is mounted on bearings coaxial with the machine shaft and the torque is restrained and measured by some kind of transducer acting tangentially at a known radius from the machine axis.

Until the beginning of the present century, the great majority of new and existing dynamometers used this method of torque measurement. In traditional machines the torque measurement was achieved by physically balancing a combination of dead weights and a spring balance against the torque absorbed (Fig. 8.1). As the stiffness of the balance was limited, it was necessary to adjust its position depending on the torque, to ensure that the force measured was accurately tangential.

Modern trunnion-mounted machines, shown diagrammatically in Fig. 8.2, use a force transducer, almost invariably of the strain gauge type, together with an appropriate bridge circuit and amplifier. The strain gauge transducer or 'load cell' has the advantage of being extremely stiff, so that no positional adjustment is necessary, but the disadvantage of a finite fatigue life after a (very large) number of load applications. The backlash and 'stiction'-free mounting of the transducer between carcase and base is absolutely critical.

The trunnion bearings are either a combination of a ball bearing (for axial location) and a roller bearing or hydrostatic type. These bearings operate under unfavourable conditions, with no perceptible angular movement, and the rolling element type is

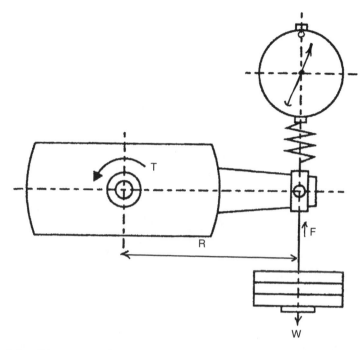

Figure 8.1 *Diagram of Froude type, trunnion-mounted, sluice-gate dynamometer measuring torque with dead weights and spring balance*

consequently prone to brinelling, or local indentation of the races, and to fretting. This is aggravated by vibration that may be transmitted from the engine and periodical inspection and turning of the outer bearing race is recommended in order to avoid poor calibration. A Schenck dynamometer design (Fig. 8.3) replaces the trunnion bearings by two radial flexures, thus eliminating possible friction and wear, but at the expense of the introduction of torsional stiffness, of reduced capacity to withstand axial loads and of possible ambiguity regarding the true centre of rotation, particularly under side loading.

Measurement of torque using in-line shafts or torque flanges

A torque shaft dynamometer is mounted in the drive shaft between engine and brake device. It consists essentially of a flanged torque shaft fitted with strain gauges and designs are available both with slip rings and with RF signal transmission. Figure 8.4 is a brushless torque shaft unit intended for rigid mounting.

More common in automotive testing is the 'disc' type torque transducer, commonly known as a torque flange (Fig. 8.5), which is a device that is bolted directly to the input flange of the brake and transmits data to a static antenna encircling it.

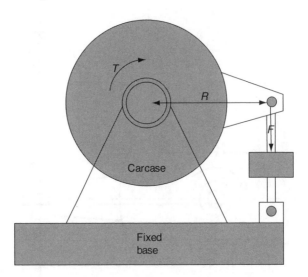

Figure 8.2   *Diagram of trunnion-mounted dynamometer measuring torque with a load cell*

A perceived advantage of the in-line torque measurement arrangement is that it avoids the necessity, discussed below, of applying torque corrections under transient conditions of torque measurement. However, not only are such corrections, using know constants, trivial with modern computer control systems, there are important problems that may reduce the inherent accuracy of this arrangement.

For steady state testing, a well-designed and maintained trunnion machine will give more consistently auditable and accurate torque measurements than the inline systems; the justification for this statement can be listed as follows:

- The in-line torque sensor has to be oversized for the rating of its dynamometer and being oversized the resolution of the signal is lower. The transducer has to be overrated because it has to be capable of dealing with the instantaneous torque peaks of the engine which are not experienced by the load cell of a trunnion-bearing machine.
- The transducer forms part of the drive line and requires very careful installation to avoid the imposition of bending or axial stresses on the torsion sensing element from other components or its own clamping device.
- The in-line device is difficult to protect from temperature fluctuations within and around the drive-line.
- Calibration checking of these devices is not as easy as for a trunnion-mounted machine; it requires a means of locking the dynamometer shaft in addition to the fixing of a calibration arm in a horizontal position without imposing bending stresses.
- Unlike the cradled machine and load cell, it is not possible to verify the measured torque of an in-line device during operation.

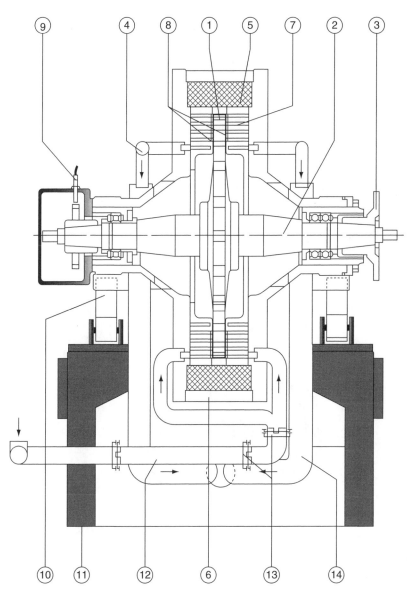

Figure 8.3 *Schenck, dry gap, disc type eddy-current dynamometer 1, rotor; 2, rotor shaft; 3, coupling flange; 4, water outlet with thermostat; 5, excitation coil; 6, dynamometer housing; 7, cooling chamber; 8, air gap; 9, speed pick-up; 10, flexure support; 11, base; 12, water inlet; 13, joint; 14, water outlet pipe*

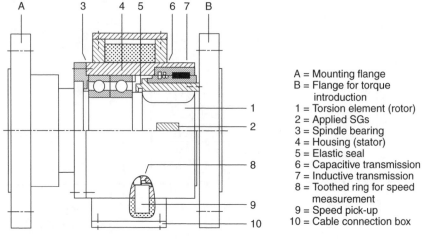

A = Mounting flange
B = Flange for torque
    introduction
1 = Torsion element (rotor)
2 = Applied SGs
3 = Spindle bearing
4 = Housing (stator)
5 = Elastic seal
6 = Capacitive transmission
7 = Inductive transmission
8 = Toothed ring for speed
    measurement
9 = Speed pick-up
10 = Cable connection box

Figure 8.4   *Brushless torque-shaft for mounting in shaft-line between engine and 'brake'*

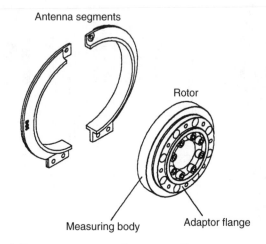

Figure 8.5   *Shaft-line components of a torque flange*

It should be noted that, in the case of modern a.c. dynamometer systems, the tasks of torque measurement and torque control may use different data acquisition paths. In some installations the control of the trunnion-mounted machine may use its own torque calculation and control system, while the test values are taken from an inline transducer such as a torque shaft.

Calibration and the assessment of errors in torque measurement

We have seen that in a conventional dynamometer, torque T is measured as a product of torque arm radius R and transducer force F.

Calibration is invariably performed by means of a *calibration arm*, supplied by the manufacturer, which is bolted to the dynamometer carcase and carries dead weights which apply a load at a certified radius. The manufacturer certifies the distance between the axis of the weight hanger bearing and an axis defined by a line joining the centres of the trunnion bearings (not the axis of the dynamometer, which indeed need not precisely coincide with the axis of the trunnions).

There is no way, apart from building an elaborate fixture, in which the dynamometer user can check the accuracy of this dimension: he is entirely in the hands of the manufacturer. The arm should be stamped with its effective length. For R&D machines of high accuracy the arm should be stamped for the specific machine.

The 'dead weights' should in fact be more correctly termed 'standard masses'. They should be certified by an appropriate standards authority located as near as possible to the geographical location in which they are used. The force they exert on the calibration arm is the product of their mass and the local value of 'g'. This is usually assumed to be 9.81 m/s^2 and constant: in fact this value is only correct at sea level and a latitude of about 47° N. It increases towards the poles and falls towards the equator, with local variations. As an example, a machine calibrated in London, where $g = 9.81$ m/s^2 , will read 0.13 per cent high if recalibrated in Sydney, Australia and 0.09 per cent low if recalibrated in St Petersburg without correcting for the different local values of g.

These are not negligible variations if one is hoping for accuracies better than 1 per cent. The actual process of calibrating a dynamometer with dead weights, if treated rigorously, is not entirely straightforward. We are confronted with the facts that no transducer is perfectly linear in its response, and no linkage is perfectly frictionless. We are then faced with the problem of adjusting the system so as to ensure that the (inevitable) errors are at a minimum throughout the range.

A suitable calibration procedure for a machine using a typical strain-gauge load cell for torque measurement is as follows.

The dynamometer should not be coupled to the engine. After the system has been energized long enough to warm up the load cell output is zeroed with the machine in its normal no-load running condition (cooling water on, etc.) and the calibration arm weight balanced by equal and opposite force. Dead weights are then added to produce approximately the rated maximum torque of the machine. This torque is calculated and the digital indicator set to this value.

The weights are removed, the zero reading noted, and weights are added, preferably in 10 equal increments, the cell readings being noted. The weights are removed in reverse order and the readings again noted.

The procedure described above means that the load cell indicator was set to read zero before any load was applied (it did not necessarily read zero after the weights had been added and removed), while it was adjusted to read the correct maximum torque when the appropriate weights had been added.

We now ask: is this setting of the load cell indicator the one that will minimize errors throughout the range and are the results within the limits of accuracy claimed by the manufacturer?

Let us assume we apply this procedure to a machine having a nominal rating of 600 Nm torque and that we have six equal weights, each calculated to impose a torque of 100 Nm on the calibration arm. Table 8.1 shows the indicated torque readings for both increasing and decreasing loads, together with the calculated torques applied by the weights. The corresponding errors, or the differences between torque applied by the calibration weights and the indicated torque readings are plotted in Figs 8.6 and 8.7.

The machine is claimed to be accurate to within ±0.25 per cent of nominal rating and these limits are shown. It will be clear that the machine meets the claimed limits of accuracy and may be regarded as satisfactorily calibrated.

It is usually assumed, though it is not necessarily the case, that hysteresis effects, manifested as differences between observed torque with rising load and with falling load, are eliminated when the machine is running, due to vibration, and it is a common practice when calibrating to knock the machine carcase lightly with a soft mallet after each load change to achieve the same result.

It is certainly not wise to assume that the ball joints invariably used in the calibration arm and torque transducer links are frictionless. These bearings are designed for working pressures on the projected area of the contact in the range 15 to 20 MN/m^2 and a 'stick slip' coefficient of friction at the ball surface of, at a minimum, 0.1 is to be expected. This clearly affects the effective arm length (in either direction) and must be relaxed by vibration.

Some large dynamometers are fitted with torque multiplication levers, reducing the size of the calibration masses. In increasingly litigious times and ever more stringent health and safety legislation, the frequent handling of multiple 20 or 25 kg

Table 8.1 *Dynamometer calibration (example taken from actual machine)*

Mass (kg)	Applied torque (Nm)	Reading (Nm)	Error (Nm)	Error (% reading)	Error (% full scale)
0	0	0.0	0.0	0.0	0.0
10	100	99.5	−0.5	−0.5	−0.083
30	300	299.0	−1.0	−0.33	−0.167
50	500	500.0	0.0	0.0	0.0
60	600	600.0	0.0	0.0	0.0
40	400	400.5	+0.5	+0.125	+0.083
20	200	200.0	0.0	0.0	0.0
0	0	0.0	0.0	0.0	0.0

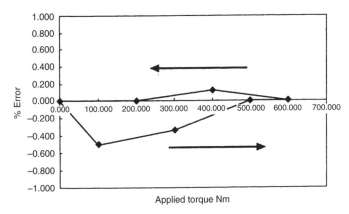

Figure 8.6 *Dynamometer calibration error as percentage of applied torque*

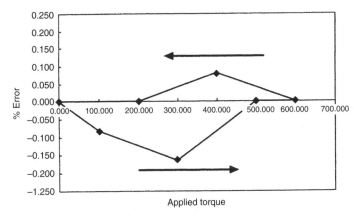

Figure 8.7 *Dynamometer calibration error as percentage of full scale*

weights may not be advisable. It is possible to carry out torque calibration by way of 'master' load cells or proving rings.* These devices have to be mounted in a jig attached to the dynamometer and give an auditable measurement of the force being applied on the target load cell by means of a hydraulic actuator. Such systems produce a more complex 'audit trail' in order to refer the calibration back to national standards.

It is important when calibrating an eddy-current machine that the water pressure in the casing should be at operational level, since pressure in the transfer pipes can give rise to a parasitic torque. Similarly, any disturbance to the run of electrical

---

* A proving ring is a hollow steel alloy ring whose distortion under a rated range of compressive loads is known and measured by means of an internal gauge.

cables to the machine must be avoided once calibration is completed. Finally, it is possible, particularly with electrical dynamometers with forced cooling, to develop small parasitic torques due to air discharged non-radially from the casing. It is an easy matter to check this by running the machine uncoupled under its own power and noting any change in indicated torque.

Experience shows that a high grade dynamometer such as would be used for research work, after careful calibration, may be expected to give a torque indication that does not differ from the absolute value by more than about ±0.1 per cent of the full load torque rating of the machine.

Systematic errors such as inaccuracy of torque arm length or wrong assumptions regarding the value of g will certainly diminish as the torque is reduced, but other errors will be little affected: it is safer to assume a band of uncertainty of constant width. This implies, for example, that a machine rated at 400 Nm torque with an accuracy of ±0.25 per cent will have an error band of ±1 N. At 10 per cent of rated torque, this implies that the true value may lie between 39 and 41 Nm. It is as well to match the size of the dynamometer as closely as possible with the rating of the engine.

All load cells used by reputable dynamometer manufacturers will compensate for changes in temperature, though their rate of response to a change may vary. They will not, however, be able to compensate for internal temperature gradients induced, for example, by air blasts from ventilation fans or radiant heat from exhaust pipes.

The subject of calibration and accuracy of dynamometer torque measurement has been dealt with in some detail, but this is probably the most critical measurement that the test engineer is called upon to make, and one for which a high standard of accuracy is expected but not easily achieved. Calibration and certification of the dynamometer and its associated system should be carried out at the very least once a year, and following any system change or major component replacement.

Torque measurement under accelerating and decelerating conditions

With the increasing interest in transient testing it is essential to be aware of the effect of speed changes on the 'apparent' torque measured by a trunnion-mounted machine.

The basic principle is simple:

Inertia of dynamometer rotor I kg m^2
Rate of increase in speed ω rad/s^2
                                  N rpm/s
Input torque to dynamometer T_1 Nm
Torque registered by dynamometer T_2 Nm

$$T_1 - T_2 = I\omega = \frac{2\pi NI}{60} \text{ Nm}$$
$$= 0.1047 NI \text{ Nm}$$

To illustrate the significance of this correction, a typical eddy-current dynamometer capable of absorbing 150 kW with a maximum torque of 500 Nm has a rotor inertia of 0.11 kg m^2. A d.c. regenerative machine of equivalent rating has a rotational inertia of 0.60 kg m^2.

If these machines are coupled to an engine that is accelerating at the comparatively slow rate of 100 rpm/s the first machine will read the torque low *during the transient phase* by an amount:

$$T_1 - T_2 = 0.1047 \times 100 \times 0.11 = 1.15 \text{ Nm}$$

while the second will read low by 6.3 Nm.

If the engine is decelerating, the machines will read high by the equivalent amount.

Much larger rates of speed change are demanded in some transient test sequences and this can represent a serious variation of torque indication, particularly when using high inertia dynamometers.

With modern computer processing of the data, corrections for these and other electrically induced transient effects can be made with software supplied by test plant manufacturers.

Measurement of rotational speed

Rotational speed of the dynamometer is measured either by a system using a toothed wheel and a pulse sensor within its associated electronics and display or, more recently, by use of an optical encoder system. While the pulse pick-up system is robust and, providing the wheel to transducer gap is correctly set and maintained, reliable, the optical encoders, which use the sensing of very fine lines etched on a small disk, need more care in mounting and operation. Since the commonly used optical encoders transmit over 1000 pulses per revolution, misalignment of its drive may show up as a sinusoidal speed change, therefore they are normally mounted as part of an accurately machined assembly forming part of the machine housing.

It should be remembered that with bidirectional dynamometers and modern electrical machines operating in four quadrants (Fig. 8.8), it is necessary to measure not only speed but also direction of rotation. Encoder systems can use separate tracks of their engraved disks to sense rotational direction. It is extremely important that the operator uses a common and clearly understood convention describing direction of rotation throughout the facility, particularly in laboratories operating reversible prime movers.

As with torque measurement, specialized instrumentation systems may use separate transducers for the measurement of speed or for the control of the dynamometer. In many cases, engine speed is monitored separately and in addition to dynamometer speed. The control system can use these two signals to shut down automatically in the case of a shaft failure. Measurement of crank position during rotation is discussed in Chapter 14.

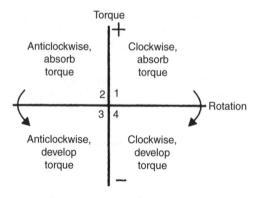

Figure 8.8   *Dynamometer operating quadrants*

Measurement of power, which is the product of torque and speed, raises the important question of sampling time. Engines never run totally steadily and the torque transducer and speed signals invariably fluctuate. An instantaneous reading of speed will not necessarily, or even probably, be identical with a longer-term average. Choice of sampling time and of the number of samples to be averaged is a matter of experimental design and compromise.

Choice of dynamometer

Perhaps the most difficult question facing the engineer setting up a test facility is the choice of the most suitable dynamometer. In this part of the chapter the characteristics, advantages and disadvantages of the various types are discussed and a procedure for arriving at the correct choice is described.

The earliest form of dynamometer, the rope brake dates back to the early years of the last century. An extremely dangerous device, it was nevertheless capable of giving quite accurate measurements of power. Its successor, the Prony brake, also relied on mechanical friction and like the rope brake required cooling by water introduced into the hollow brake drum and removed by a scoop.

Both these devices are only of historical interest. Their successors may be classified according to the means adopted for absorbing the mechanical power of the prime mover driving the dynamometer.

Classification of dynamometers

1. *Hydrokinetic or 'hydraulic' dynamometers (water brakes)*. With the exception of the disc dynamometer, all machines work on similar principles (Fig. 8.9). A shaft carries a cylindrical rotor which revolves in a watertight casing. Toroidal recesses

formed half in the rotor and half in the casing or stator are divided into pockets by radial vanes set at an angle to the axis of the rotor. When the rotor is driven, centrifugal force sets up an intensive toroidal circulation as indicated by the arrows in Fig. 8.9a. The effect is to transfer momentum from rotor to stator and hence to develop a torque resistant to the rotation of the shaft, balanced by an equal and opposite torque reaction on the casing.

A forced vortex of toroidal form is generated as a consequence of this motion, leading to high rates of turbulent shear in the water and the dissipation of power in the form of heat to the water. The centre of the vortex is vented to atmosphere by way of passages in the rotor and the virtue of the design is that power is absorbed with minimal damage to the moving surfaces, either from erosion or from the effects of cavitation.

The machines are of two kinds, depending on the means by which the resisting torque is varied.

1(a) Constant fill machines: *the classical Froude or sluice plate design*, Fig. 8.10. In this machine, torque is varied by inserting or withdrawing pairs of thin sluice plates between rotor and stator, thus controlling the extent of the development of the toroidal vortices.

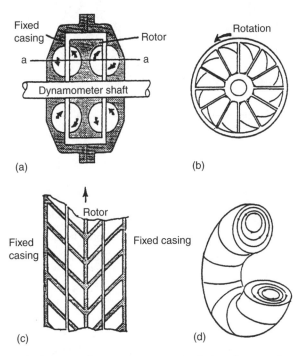

Figure 8.9 *Hydrokinetic dynamometer, principle of operation: (a) section through dynamometer; (b) end view of rotor; (c) development of section a–a of rotor and casing; (d) representation of toroidal vortex*

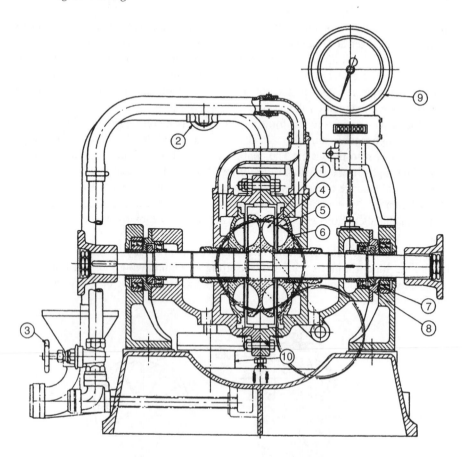

Typical cross-section through casing of Froude dynamometer, type DPX

(1) Rotor	(6) Casing liners
(2) Water outlet valve	(7) Casing trunnion bearing
(3) Water inlet valve	(8) Shaft bearing
(4) Sluice plates for load control	(9) Tachometer
(5) Water inlet holes in vanes	

Figure 8.10   *Froude sluice-plate dynamometer*

1(b) *Variable fill machines*, Fig. 8.11. In these machines, the torque absorbed is varied by adjusting the quantity (mass) of water in circulation within the casing. This is achieved by a valve, usually on the water outlet, associated with control systems of widely varying complexity. The particular advantage of the variable fill machine

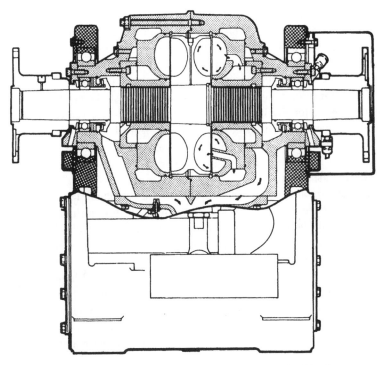

Figure 8.11 *Variable fill hydraulic dynamometer controlled by fast acting outlet valve at bottom of the stator*

is that the torque may be varied much more rapidly than is the case with sluice plate control. Amongst this family of machines are the largest dynamometers ever made with rotors of around 5 m diameter. There are several designs of water control valve and valve actuating mechanisms depending on the range and magnitude of the loads absorbed and the speed of change of load required. For the fastest response, it is necessary to have adequate water available to fill the casing rapidly and it may be necessary to fit both inlet and outlet control valves with an integrated control system.

1(c) *'Bolt-on' variable fill machines.* These machines, available for many years in the USA, operate on the same principle as those described in 1(b) above, but are arranged to bolt directly on to the engine clutch housing or into the truck chassis (see Chapter 4 for more detailed description of use). Machines are available for ratings up to about 1000 kW. In these machines, load is usually controlled by an inlet control valve associated with a throttled outlet. By nature of their simplified design and lower mass, these machines are not capable of the same level of speed holding or torque measurement as the more conventional 1(b) designs.

1(d) *Disc dynamometers.* These machines, not very widely used, consist of one or more flat discs located between flat stator plates, with a fairly small clearance. Power is absorbed by intensive shearing of the water and torque is controlled as in variable fill machines. Disc dynamometers have comparatively poor low speed performance but may be built to run at very high speeds, making them suitable for loading gas turbines. A variation is the perforated disc machine, in which there are holes in the rotor and stators, giving greater power dissipation for a given size of machine.

2. Hydrostatic dynamometers. Not very widely used, these machines consist generally of a combination of a fixed stroke and a variable stroke positive displacement hydraulic pump/motor similar to that found in large off-road vehicle transmissions. The fixed stroke machine forms the dynamometer. An advantage of this arrangement is that, unlike most other, non-electrical machines, it is capable of developing full torque down to zero speed and is also capable of acting as a source of power to 'motor' the engine under test.

3. Electrical motor-based dynamometers. The common feature of all these machines is that the power absorbed is transformed into electrical energy, which is 'exported' from the machine via its associated 'drive' circuitry. The energy loss within both the motor and its drive in the form of heat is transferred to a cooling medium, which may be water or is more commonly forced air flow.

All motor-based dynamometers have associated with them large drive cabinets that produce heat and noise. The various sections of these cabinets contain high voltage/power devices and complex electronics; they have to be housed in suitable conditions which have a clean and non-condensing atmosphere with sufficient space for access and cooling. When planning a facility layout, the designer should remember that these large and heavy cabinets have to be positioned after the building work has been completed. The position of the drives should normally be within 15 m of the dynamometer, but this should be minimized so far as is practical to reduce the high cost of the connecting power cables.

3(a) Direct current (d.c.) dynamometers. These machines consist essentially of a trunnion-mounted d.c. motor generator. Control is almost universally by means of a thyristor based a.c.→d.c.→a.c. converter.

These machines have a long pedigree in the USA, are robust, easily controlled, and capable of motoring and starting as well as of absorbing power. Disadvantages include limited maximum speed and high inertia, which can present problems of torsional vibration (see Chapter 9) and limited rates of speed change. Because they contain a commutator, the maintenance of d.c. machines may be higher than those based on a.c. squirrel cage motors.

3(b) Asynchronous or alternating current (a.c.) dynamometers. These asynchronous machines consist essentially of an induction motor with squirrel cage rotor, the speed of which is controlled by varying the supply frequency. The modern power control stage of the control will invariably be based upon insulated gate bipolar transistor (IGBT) technology.

The squirrel cage rotor machines have a lower rotational inertia than d.c. machines of the same power and are therefore capable of better transient performance. Being based on an asynchronous motor they have proved very robust in service requiring low maintenance.

However, it is misleading to think that any motor's mechanical design may be used without adaptation as a dynamometer. During the first decade of their wide industrial use, it was discovered that several different dynamometer/motor designs suffered from bearing failures caused by an electrical arcing effect within the rolling elements; this was due to the fact that, in their dynamometer role, a potential difference developed between the rotor and the stator (ground). Ceramic bearing elements and other design features are now used to prevent such damage occurring.

3(c) *Synchronous, permanent magnet dynamometers.* The units represent the latest generation of dynamometer development and while using the same drive technology as the asynchronous dynamometers are capable of higher dynamic performance because of their inherently lower rotational inertia. It is this generation of machine that will provide the high dynamic test tools required by engine and vehicle system simulation in the test cell.

Acceleration rates of 160 000 rpm/s and air-gap torque rise times of less than 1 ms have been achieved, which makes it possible to use these machines as engine simulators where the full dynamic fluctuation speed and torque characteristic of the engine is required for drive-line component testing.

3(d) *Eddy-current dynamometers*, Fig. 8.3. These machines make use of the principle of electromagnetic induction to develop torque and dissipate power. A toothed rotor of high-permeability steel rotates, with a fine clearance, between water-cooled steel loss plates. A magnetic field parallel to the machine axis is generated by two annular coils and motion of the rotor gives rise to changes in the distribution of magnetic flux in the loss plates. This in turn gives rise to circulating eddy currents and the dissipation of power in the form of electrical resistive losses. Energy is transferred in the form of heat to cooling water circulating through passages in the loss plates, while some cooling is achieved by the radial flow of air in the gaps between rotor and plates.

Power is controlled by varying the current supplied to the annular exciting coils and rapid load changes are possible. Eddy-current machines are simple and robust, the control system is simple and they are capable of developing substantial braking torque at quite low speeds. Unlike a.c. or d.c. dynamometers, however, they are unable to develop motoring torque.

There are two common forms of machine both having air circulating in the gap between rotor and loss (cooling) plates, hence 'dry gap':

1. Dry gap machines fitted with one or more tooth disk rotors. These machines have lower inertia than the drum machines and a very large installed user base, particularly in Europe. However, the inherent design features of their loss plates place certain operational restrictions on their use. It is absolutely critical to maintain the required water flow through the machines at all times; even a very short loss of cooling will cause the loss plates to distort leading to the rotor/plate gap closing with disastrous results. These machines must be fitted with flow detection devices interlocked with the cell control system; pressure switches should not be used since in a closed water system it is possible to have pressure without flow.

2. Dry gap machines fitted with a drum rotor. These machines usually have a higher inertia than the equivalent disc machine, but may be less sensitive to cooling water conditions.

Although no longer so widely used, an alternative form of eddy-current machine is also available. This employs a simple disc or drum design of rotor in which eddy currents are induced and the heat developed is transferred to water circulated through the gaps between rotor and stator. These 'wet gap' machines are liable to corrosion if left static for any length of time, have higher inertia, and have a high level of minimum torque, arising from drag of the cooling water in the gap.

4. *Friction dynamometers,* Fig. 8.12. These machines, in direct line of succession from the original rope brake, consist essentially of water-cooled, multidisc friction brakes. They are useful for low-speed applications, for example for measuring the power output of a large, off-road vehicle transmission at the wheels, and have the advantage, shared with the hydrostatic dynamometer, of developing full torque down to zero speed.

5. *Air brake dynamometers.* These devices, of which the Walker fan brake was the best-known example, are now largely obsolete. They consisted of a simple arrangement of radially adjustable paddles that imposed a torque that could be approximately estimated. They survive mainly for use in the field testing of helicopter engines, where high accuracy is not required and the noise is no disadvantage.

Hybrid and tandem dynamometers

For completeness, mention should be made of both a combined design that is occasionally adopted for cost reasons and the use of two dynamometers in line for special test configurations.

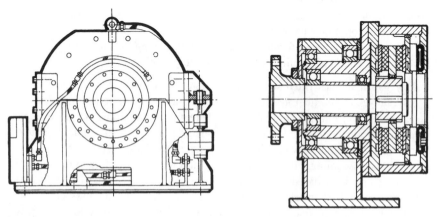

Figure 8.12   *Water-cooled friction brake used as a dynamometer*

The d.c. or a.c. electrical dynamometer is capable of generating a motoring torque almost equal to its braking torque. However, the motoring torque required in engine testing seldom exceeds 30 per cent of the engine power output. Since, for equal power absorption, a.c. and d.c. machines are more expensive than other types, it is sometimes worth running an electrical dynamometer in tandem with, for example, a variable fill hydraulic machine. Control of these hybrid machines is a more complex matter and the need to provide duplicate services, both electrical power and cooling water, is a further disadvantage. The solution may, however, on occasion be cost-effective.

Tandem machines are used when the torque/speed envelope of the prime mover cannot be covered by a standard dynamometer, usually this is found in gas turbine testing when the rotational speed is too high for a machine fitted with a rotor capable of absorbing full rated torque. The first machine in line has to have a shaft system capable of transmitting the combine torques.

Tandem machines are also used when the prime mover is producing power through two contrarotating shafts as with some aero and military applications; in these cases the first machine in line is of a special design with a hollow rotor shaft to allow the housing of a quill shaft connecting the second machine.

One, two or four quadrant?

Figure 8.8 illustrates diagrammatically the four 'quadrants' in which a dynamometer may be required to operate. Most engine testing takes place in the first quadrant, the engine running anticlockwise when viewed on the flywheel end. On occasions it is necessary for a test installation using a unidirectional water brake to accept engines running in either direction; one solution is to fit the dynamometer with couplings at both ends mounted on a turntable. Large and some 'medium speed' marine engines are usually reversible.

All types of dynamometer are naturally able to run in the first (or second) quadrant. Hydraulic dynamometers are usually designed for one direction of rotation, though they may be run in reverse at low fill state without damage. When designed specifically for bidirectional rotation they may be larger than a single-direction machine of equivalent power and torque control may not be as precise as that of the unidirectional designs. The torque measuring system must of course operate in both directions. Eddy-current machines are inherently reversible.

When it is required to operate in the third and fourth quadrants (i.e. for the dynamometer to produce power as well as to absorb it) the choice is effectively limited to d.c. or a.c. machines, or to the hydrostatic or hybrid machine. These machines are generally reversible and therefore operate in all four quadrants.

There is an increasing requirement for four-quadrant operation as a result of the growth in transient testing, with its call for very rapid load changes and even for torque reversals.

If mechanical losses in the engine are to be measured by 'motoring', a four-quadrant machine is obviously required.

Table 8.2 *Operating quadrants of dynamometer designs*

Type of machine	Quadrant
Hydraulic sluice plate	1 or 2
Variable fill hydraulic	1 or 2
'Bolt on' variable fill hydraulic	1 or 2
Disc type hydraulic	1 and 2
Hydrostatic	1, 2, 3, 4
d.c. electrical	1, 2, 3, 4
a.c. electrical	1, 2, 3, 4
Eddy current	1 and 2
Friction brake	1 and 2
Air brake	1 and 2
Hybrid	1, 2, 3, 4

A useful feature of such a machine is its ability also to start the engine. Table 8.2 summarizes the performance of machines in this respect.

Matching engine and dynamometer characteristics

The different types of dynamometer have significantly different torque–speed and power–speed curves, and this can affect the choice made for a given application. Figure 8.13 shows the performance curves of a typical hydraulic dynamometer. The different elements of the performance envelope are as follows:

- Dynamometer full (or sluice plates wide open). Torque increases with square of speed, no torque at rest.
- Performance limited by maximum permitted shaft torque.
- Performance limited by maximum permitted power, which is a function of cooling water throughput and its maximum permitted temperature rise.
- Maximum permitted speed.
- Minimum torque corresponding to minimum permitted water flow.

Figure 8.14 shows the considerably different performance envelope of an electrical machine, made up of the following elements:

- Constant torque corresponding to maximum current and excitation.
- Performance limited by maximum permitted power output of machine.
- Maximum permitted speed.

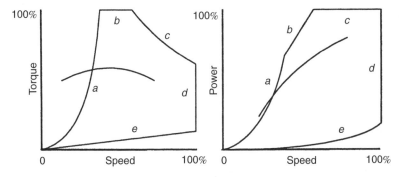

Figure 8.13 *Engine torque curves plotted on hydraulic dynamometer torque curves*

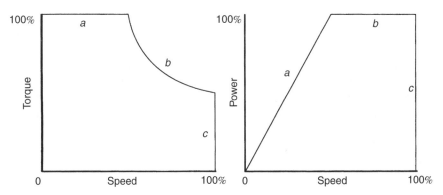

Figure 8.14 *Performance curve shapes of electrical motor-based dynamometers*

Since these are 'four-quadrant' machines, power absorbed can be reduced to zero and there is no minimum torque curve.

Figure 8.15 shows the performance curves for an eddy-current machine, which lie between those of the previous two machines:

- Low speed torque corresponding to maximum permitted excitation.
- Performance limited by maximum permitted shaft torque.
- Performance limited by maximum permitted power, which is a function of cooling water throughput and its maximum permitted temperature rise.
- Maximum permitted speed.
- Minimum torque corresponding to residual magnetization, windage and friction.

In choosing a dynamometer for an engine or range of engines, it is essential to superimpose the maximum torque– and power–speed curves on to the dynamometer

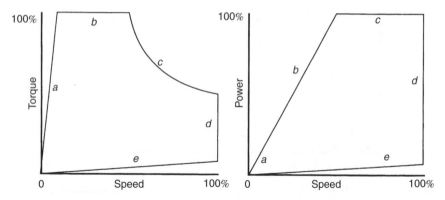

Figure 8.15   *Performance curve shapes of an eddy-current dynamometer*

envelope. See the example in Fig. 8.13 which demonstrates a typical problem: the hydraulic machine is incapable of developing sufficient torque at the bottom end of the speed range.

For best accuracy, it is desirable to choose the smallest machine that will cope with the largest engine to be tested. Hydraulic dynamometers are generally able to deal with a moderate degree of overload and overspeed, but it is undesirable to run electrical machines beyond their rated limits: this can lead to damage to commutators, overheating and distortion of eddy-current loss plates.

Careful attention must also be given to the arrangements for coupling engine and dynamometer, see Chapter 9.

Engine starting and cranking

Starting an engine when it is connected to a dynamometer may present the cell designer and operator with a number of problems, and is a factor to be borne in mind when selecting the dynamometer. If the engine is fitted with a starter motor, the cell system must provide the high current d.c. supply and associated switching; in the absence of an engine mounted starter a complete system to start and crank the engine must be available which compromises neither the torsional characteristics (see Chapter 9) nor the torque measurement accuracy.

Engine cranking, no starter motor

The cell cranking system must be capable of accelerating the engine to its normal starting speed and, in most cases, of disengaging when the engine fires. A four-quadrant dynamometer, suitably controlled, will be capable of starting the engine directly. The power available from any four-quadrant machine will always be greater than that required, therefore excessive starting torque must be avoided by an alarm system otherwise an engine locked by seizure or fluid in a cylinder may cause damage to the drive line.

The preferred method of providing other types of dynamometer with a starting system is to mount an electric motor at the non-engine end of the dynamometer shaft, driving through an over-running or remotely engaged clutch, and generally through a speed-reducing belt drive. The clutch half containing the mechanism should be on the input side, otherwise it will be affected by the torsional vibrations usually experienced by dynamometer shafts. The motor may be mounted above, below or alongside the dynamometer to save cell length.

The sizing of the motor must take into account the maximum break-away torque expected, usually estimated as twice the average cranking torque, while the normal running speed of the motor should correspond to the desired cranking speed. The choice of motor and associated starter must take into account the maximum number of starts per hour that may be required, both in normal use and when dealing with a faulty engine. The running regime of the motor is demanding, involving repeated bursts at overload, with the intervening time at rest, and an independent cooling fan may be necessary.

Some modern diesel engines, when 'green',* require cranking at more than the normal starting speed, sometimes as high as 1200 rev/min, in order to prime the fuel system. In such cases a two-speed or fully variable speed starter motor may be necessary.

The system must be designed to impose the minimum parasitic torque when disengaged, since this torque will not be sensed by the dynamometer measuring system.

In some cases, to avoid this source of inaccuracy, the motor may be mounted directly on the dynamometer carcase and permanently coupled to the dynamometer shaft by a belt drive. This imposes an additional load on the trunnion bearings, which may lead to brinelling, and it also increases the effective moment of inertia of the dynamometer. However, it has the advantage that motoring and starting torque may be measured by the dynamometer system.

An alternative solution is to use a standard vehicle engine starter motor in conjunction with a gear ring carried by a 'dummy flywheel' carried on a shaft with separate bearings incorporated in the drive line, but this may have the disadvantage of complicating the torsional behaviour of the system.

Engine-mounted starter systems

If the engine is fitted with its own starter motor on arrival at the test stand, all that must be provided is the necessary 12 or 24 V supply. The traditional approach has been to locate a suitable battery as close as possible to the starter motor, with a suitable battery charger supply. This system is not ideal, as the battery needs to be

---

* A green engine is one that has never been run. The rubbing surfaces may be dry, the fuel system may need priming, and there is always the possibility that it, or its control system, is faulty and incapable of starting.

in a suitably ventilated box, to avoid the risk of accidental shorting, and will take up valuable space. Special transformer/rectifier units designed to replace batteries for this duty are on the market. They will include an 'electrical services box' to provide power in addition for ignition systems and diesel glow plugs. In large integrated systems there may be a bus bar system for the d.c. supplies.

The engine starter will be presented with a situation not encountered in normal service: it will be required to accelerate the whole dynamometer system in addition to the engine while a 'green' engine may exhibit a very high breakaway torque and require prolonged cranking at high speed to prime the fuel system before it fires.

Non-electrical starting systems

Diesel engines larger than the automotive range are usually started by means of compressed air, admitted to the cylinders by way of starting valves. In some cases it is necessary to move the crankshaft to the correct starting position, either by barring or using an engine-mounted inching motor. The test facility should include a compressor and a receiver of capacity at least as large as that recommended for the engine in service.

Compressed air or hydraulic motors are sometimes used instead of electric motors to provide cranking power but have no obvious advantages over a d.c. electric motor, apart from a marginally reduced fire risk in the case of compressed air, provided the supply is shut off automatically in the case of fire.

In Chapter 9, attention is drawn to the possibility of overloading flexible couplings in the drive line during the starting process, and particularly when the engine first fires. This should not be overlooked.

Choice of dynamometer

Table 8.3 lists the various types of dynamometer and indicates their applicability for various classes of engine being tested in steady or mild transient states.

In most cases, several choices are available and it will be necessary to consider the special features of each type of dynamometer and to evaluate the relative importance of these in the particular case. These features are listed in Table 8.3 and other special factors are considered later.

Some additional considerations

The final choice of dynamometer for a given application may be influenced by some of the following factors:

1. The speed of response required by the test sequences being run: steady state, transient, dynamic or high dynamic. This will determine the technology and probably the number of quadrants of operation required.

Table 8.3 *Dynamometers: advantages and disadvantage of available types*

Dynamometer type	Advantages	Disadvantages
Froude sluice plate	Obsolete, but many cheap and reconditioned models in use worldwide, robust	Slow response to change in load. Manual control not easy to automate
Variable fill water brakes	Capable of medium speed load change, automated control, robust and tolerant of overload. Available for largest prime-movers	'Open' water system required. Can suffer from cavitation or corrosion damage
'Bolt-on' variable fill water brakes	Cheap and simple installation. Up to 1000 kW	Lower accuracy of measurement and control than fixed machines
Disc type hydraulic	Suitable for high speeds	Poor low speed performance
Hydrostatic	For special applications, provides four quadrant performance	Mechanically complex, noisy and expensive. System contains large volumes of high pressure oil
d.c electrical motor	Mature technology. Four quadrant performance	High inertia, commutator may be fire and maintenance risk
asychronous motor (a.c.)	Lower inertia than d.c. Four quadrant performance	Expensive. Large drive cabinet needs suitable housing
Permanent magnet motor	Lowest inertia, most dynamic four quadrant performance. Small size in cell	Expensive. Large drive cabinet needs suitable housing
Eddy current	Low inertia (disk type air gap). Well adapted to computer control. Mechanically simple	Vulnerable to poor cooling supply. Not suitable for sustained rapid changes in power (thermal cycling)

(continued)

Table 8.3   *(cont.)*

Dynamometer type	Advantages	Disadvantages
Friction brake	Special purpose applications for very high torques at low speed	Limited speed range
Air brake	Cheap. Very little support services needed	Noisy. Limited control accuracy
Hybrid	Possible cost advantage over sole electrical machine	Complexity of construction and control

2. Load factor. If the machine is to spend long periods out of use, the possibilities of corrosion must be considered, particularly in the case of hydraulic or wet gap eddy-current machines. Can the machine be drained readily? Should the use of corrosion inhibitors be considered?
3. Overloads. If it may be necessary to consider occasional overloading of the machine a hydraulic machine may be preferable, in view of its greater tolerance of such conditions. Check that the torque measuring system has adequate capacity.
4. Large and frequent changes in load. This can give rise to problems with eddy-current machines, due to expansion and contraction with possible distortion of the loss plates.
5. Wide range of engine sizes to be tested. It may be difficult to achieve good control and adequate accuracy when testing the smallest engines, while the minimum dynamometer torque may also be inconveniently high.
6. How are engines to be started? If a non-motoring dynamometer is favoured it may be necessary to fit a separate starter to the dynamometer shaft. This represents an additional maintenance commitment and may increase inertia.
7. Is there an adequate supply of cooling water of satisfactory quality? Hard water will result in blocked cooling passages and some water treatments can give rise to corrosion. This may be a good reason for choosing d.c. or a.c. dynamometers, despite extra cost.
8. Is the pressure of the water supply subject to sudden variations? Sudden pressure changes or regular pulsations will affect the stability of control of hydraulic dynamometers. Eddy-current and indirectly cooled machines are unaffected providing inlet flow does not fall below emergency trip levels.
9. Is the electrical supply voltage liable to vary as the result of other loads on the same circuit? With the exception of air brakes and manually controlled hydraulic machines, all dynamometers are affected by electrical interference and voltage changes.

10. Is it proposed to use a shaft docking system for coupling engine and dynamometer? Are there any special features or heavy overhung or axial loads associated with the coupling system? Such features should be discussed with the dynamometer manufacturer before making a decision. Some machines, notably the Schenck flexure plate mounting system, are not suited to taking axial loads.

The supplier of any new dynamometer should offer an acceptance test and basic training in operation, calibration and safety of the new machine. A careful check should be made on the level of technical support, including availability of calibration services, spares and local service facilities, offered by the manufacturer.

9 Coupling the engine to the dynamometer

Introduction

The selection of suitable couplings and shaft for the connection of the engine to the dynamometer is by no means a simple matter. Incorrect choice or faulty system design may give rise to a number of problems:

- torsional oscillations;
- vibration of engine or dynamometer;
- whirling of coupling shaft;
- damage to engine or dynamometer bearings;
- excessive wear of shaft line components;
- catastrophic failure of coupling shafts;
- engine starting problems.

This whole subject, the coupling of engine and dynamometer, can give rise to more trouble than any other routine aspect of engine testing, and a clear understanding of the many factors involved is desirable. The following chapter covers all the main factors, but in some areas more extensive analysis may be necessary and appropriate references are given.

The nature of the problem

The special feature of the problem is that it must be considered afresh each time an engine not previously encountered is installed. It must also be recognized that unsatisfactory torsional behaviour is associated with the whole system – engine, coupling shaft and dynamometer – rather than with the individual components, all of which may be quite satisfactory in themselves.

Problems arise partly because the dynamometer is seldom equivalent dynamically to the system driven by the engine in service. This is particularly the case with vehicle engines. In the case of a vehicle with rear axle drive, the driveline consists of a clutch, which may itself act as a torsional damper, followed by a gearbox, the characteristics of which are low inertia and some damping capacity. This is followed

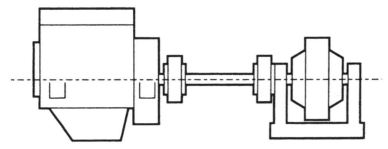

Figure 9.1 *Simple form of dynamometer/engine drive line*

by a drive shaft and differential, itself having appreciable damping, two half shafts and two wheels, both with substantial damping capacity and running at much slower speed than the engine, thus reducing their effective inertia.

When coupled to a dynamometer this system, Fig. 9.1, with its built-in damping and moderate inertia, is replaced by a single drive shaft connected to a single rotating mass, the dynamometer, running at the same speed as the engine. The clutch may or may not be retained.

Particular care is necessary where the moment of inertia of the dynamometer is more than about twice that of the engine. A further consideration that must be taken seriously concerns the effect of the difference between the engine mounting arrangements in the vehicle and on the testbed. This can lead to vibrations of the whole engine that can have a disastrous effect on the drive shaft.

Overhung mass on engine and dynamometer bearings

Care must be taken when designing and assembling a shaft system that the loads imposed by the mass and unbalanced forces do not exceed the overhung weight limits of the engine bearing at one end and the dynamometer at the other. Steel adaptor plates required to adapt the bolt holes of the shaft to the dynamometer flange or engine flywheel can increase the load on bearings significantly and compromise the operation of the system. Dynamometer manufacturers produce tables showing the maximum permissible mass at a given distance from the coupling face of their machines; the equivalent details for most engines is more difficult to obtain, but the danger of overload should be kept in mind by all concerned.

Background reading

The mathematics of the subject is complex and not readily accessible. Den Hartog[1] gives what is possibly the clearest exposition of fundamentals. Ker Wilson's classical treatment in five volumes[2] is probably still the best source of comprehensive information; his abbreviated version[3] is sufficient for most purposes. Mechanical

Engineering Publications have published a useful practical handbook[4] while Lloyd's Register[5] gives rules for the design of marine drives that are also useful in the present context. A listing of the notation used is to be found at the end of this chapter.

Torsional oscillations and critical speeds

In its simplest form, the engine–dynamometer system may be regarded as equivalent to two rotating masses connected by a flexible shaft, Fig. 9.2. Such a system has an inherent tendency to develop torsional oscillations. The two masses can vibrate 180° out of phase about a node located at some point along the shaft between them. The oscillatory movement is superimposed on any steady rotation of the shaft. The resonant or critical frequency of torsional oscillation of this system is given by:

$$n_c = \frac{60}{2\pi}\sqrt{\frac{C_c(I_e + I_b)}{I_e I_b}} \tag{1}$$

If an undamped system of this kind is subjected to an exciting torque of constant amplitude T_{ex} and frequency n, the relation between the amplitude of the induced oscillation θ and the ratio n/n_c is as shown in Fig. 9.3.

At low frequencies, the combined amplitude of the two masses is equal to the static deflection of the shaft under the influence of the exciting torque, $\theta_0 = T_{ex}/C_s$. As the frequency increases, the amplitude rises and at $n = n_c$ it becomes theoretically infinite: the shaft may fracture or non-linearities and internal damping may prevent actual failure. With further increases in frequency the amplitude falls and at $n - \sqrt{2n_c}$ it is down to the level of the static deflection. Amplitude continues to fall with increasing frequency.

The shaft connecting engine and dynamometer must be designed with a suitable stiffness C_s to ensure that the critical frequency lies outside the normal operating

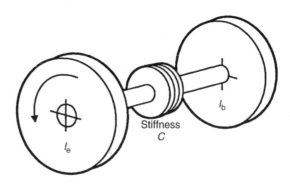

Figure 9.2   *Two mass system (compare with Fig. 9.1)*

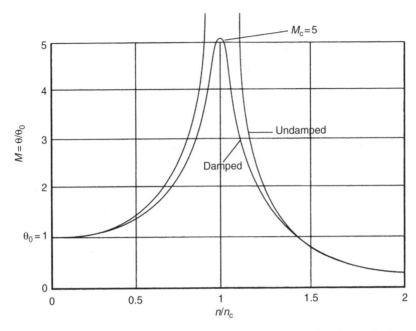

Figure 9.3 *Relationship between frequency ratio, amplitude and dynamic amplifier M*

range of the engine, and also with a suitable degree of damping to ensure that the unit may be run through the critical speed without the development of a dangerous level of torsional oscillation. Figure 9.3 also shows the behaviour of a damped system. The ratio θ/θ_0 is known as the dynamic magnifier M. Of particular importance is the value of the dynamic magnifier at the critical frequency, M_c. The curve of Fig. 9.3 corresponds to a value $M_c = 5$.

Torsional oscillations are excited by the variations in engine torque associated with the pressure cycles in the individual cylinders (also, though usually of less importance, by the variations associated with the movement of the reciprocating components).

Figure 9.4 shows the variation in the case of a typical single cylinder four-stroke diesel engine. It is well known that any periodic curve of this kind may be synthesized from a series of *harmonic components*, each a pure sine wave of a different amplitude having a frequency corresponding to a multiple or submultiple of the engine speed and Fig. 9.4 shows the first six components.

The *order* of the harmonic defines this multiple. Thus a component of order $N_0 = 1/2$ occupies two revolutions of the engine, $N_0 = 1$ one revolution and so on. In the case of a four cylinder four-stroke engine, there are two firing strokes per revolution of the crankshaft and the turning moment curve of Fig. 9.4 is repeated at intervals of 180°. In a multicylinder engine, the harmonic components of a given

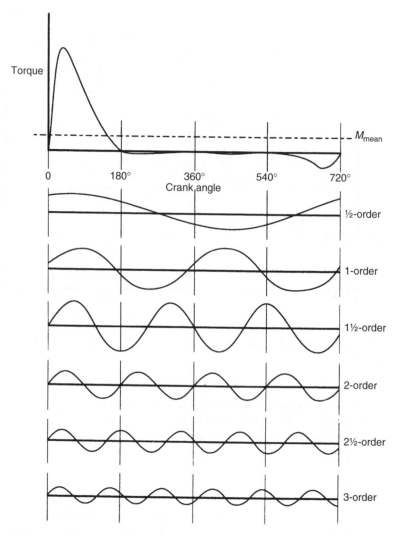

Figure 9.4 *Harmonic components of turning moment, single cylinder four-stoke gasoline engine*

order for the individual cylinders are combined by representing each component by a vector, in the manner illustrated in Chapter 3, for the inertia forces. A complete treatment of this process is beyond the scope of this book, but the most significant results may be summarized as follows.

The first major critical speed for a multicylinder, in-line engine is of order:

$$N_0 = N_{CYL}/2 \quad \text{for a four-stroke engine} \tag{2a}$$

$$N_0 = N_{\text{CYL}} \quad \text{for a two-stroke engine} \tag{2b}$$

Thus, in the case of a four cylinder, four-stroke engine the major critical speeds are of order 2, 4, 6, etc. In the case of a six cylinder engine, they are of order 3, 6, 9, etc.

The distinction between a major and a minor critical speed is that in the case of an engine having an infinitely rigid crankshaft it is only at the major critical speeds that torsional oscillations can be induced. This, however, by no means implies that in large engines having a large number of cylinders, the minor critical speeds may be ignored.

At the major critical speeds the exciting torques T_{ex} of all the individual cylinders in one line act in phase and are thus additive (special rules apply governing the calculation of the combined excitation torques for Vee engines).

The first harmonic is generally of most significance in the excitation of torsional oscillations, and for engines of moderate size, such as passenger vehicle engines, it is generally sufficient to calculate the critical frequency from eq. (1), then to calculate the corresponding engine speed from:

$$N_c = n_c/N_0 \tag{3}$$

The stiffness of the connecting shaft between engine and dynamometer should be chosen so that this speed does not lie within the range in which the engine is required to develop power.

In the case of large multicylinder engines, the 'wind-up' of the crankshaft as a result of torsional oscillations can be very significant and the two-mass approximation is inadequate; in particular, the critical speed may be overestimated by as much as 20 per cent and more elaborate analysis is necessary. The subject is dealt with in several different ways in the literature; perhaps the easiest to follow is that of Den Hartog.[1] The starting point is the value of the mean turning moment developed by the cylinder, M_{mean} (Fig. 9.4). Values are given for a so-called 'p factor', by which M_{mean} is multiplied to give the amplitude of the various harmonic excitation forces. Table 9.1, reproduced from Den Hartog, shows typical figures for a four-stroke medium speed diesel engine.

Exciting torque:

$$T_{\text{ex}} = p. \, M_{\text{mean}} \tag{4}$$

Table 9.1 p factors

Order	$\frac{1}{2}$	1	$1\frac{1}{2}$	2	$2\frac{1}{2}$	3...	8
p factor	2.16	2.32	2.23	1.91	1.57	1.28	0.08

The relation between M_{mean} and imep (indicated **mean effective pressure**) is given by:

for a four-stroke engine $$M_{mean} = p_i \cdot \frac{B^2 S}{16} \cdot 10^{-4} \qquad (5a)$$

for a two-stroke engine $$M_{mean} = p_i \cdot \frac{B^2 S}{8} \cdot 10^{-4} \qquad (5b)$$

Lloyd's *Rulebook*,[5] the main source of data on this subject, expresses the amplitude of the harmonic components rather differently, in terms of a 'component of tangential effort', T_m. This is a pressure that is assumed to act upon the piston at the crank radius $S/2$. Then exciting torque per cylinder:

$$T_{ex} = T_m \frac{\pi B^2}{4} \frac{S}{2} \times 10^{-4} \qquad (6)$$

Lloyd's give curves of T_m in terms of the indicated mean effective pressure p_i and it may be shown that the values so obtained agree closely with those derived from Table 9.1.

The amplitude of the vibratory torque T_v induced in the connecting shaft by the vector sum of the exciting torques for all the cylinders, $\sum T_{ex}$ is given by:

$$T_v = \frac{\sum T_{ex} M_c}{(1 + I_e/I_b)} \qquad (7)$$

The complete analysis of the torsional behaviour of a multicylinder engine is a substantial task, though computer programmes are available which reduce the effort required. As a typical example, Fig. 9.5 shows the 'normal elastic curves' for the first and second modes of torsional oscillation of a 16 cylinder Vee engine coupled to a hydraulic dynamometer. These curves show the amplitude of the torsional oscillations of the various components, relative to that at the dynamometer which is taken as unity. The natural frequencies are respectively $n_c = 4820$ c.p.m. and $n_c = 6320$ c.p.m. The curves form the basis for further calculations of the energy input giving rise to the oscillation. In the case of the engine under consideration, these showed a very severe fourth-order oscillation, $N_0 = 4$, in the first mode. (For an engine having eight cylinders in line the first major critical speed, from eq. (2a), is of order $N_0 = 4$.) The engine speed corresponding to the critical frequency of torsional oscillation is given by:

$$N_c = n_c/N_0 \qquad (8)$$

giving, in the present case, $N_c = 1205$ rev/min, well within the operating speed range of the engine. Further calculations showed a large input of oscillatory energy at $N_0 = 4\frac{1}{2}$, a minor critical speed, in the second mode, corresponding to a critical engine speed of $6320/4\frac{1}{4} = 1404$ rev/min, again within the operating range. Several

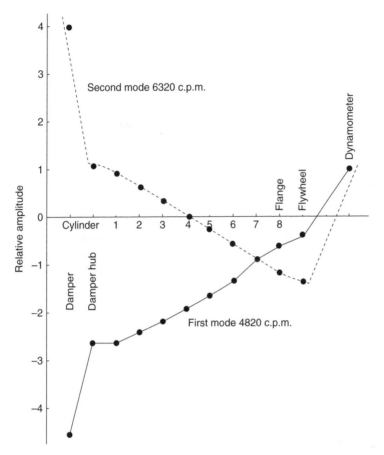

Figure 9.5 *Normal elastic curves for a particular 16 cylinder V engine coupled to a hydraulic dynamometer (taken from an actual investigation)*

failures of the shaft connecting engine and dynamometer occurred before a safe solution was arrived at.

This example illustrates the need for caution and for full investigation in setting up large engines on the test bed.

It is not always possible to avoid running close to or at critical speeds and this situation is usually dealt with by the provision of torsional vibration dampers, in which the energy fed into the system by the exciting forces is absorbed by viscous shearing. Such dampers are commonly fitted at the non-flywheel end of engine crankshafts. In some cases it may also be necessary to consider their use as a component of engine test cell drive lines, when they are located either as close as possible to the engine flywheel, or at the dynamometer. The damper must be 'tuned' to be most effective at the critical frequency and the selection of a suitable damper involves equating the energy fed into the system per cycle with the energy absorbed by viscous shear in the damper. This leads to an estimate of the magnitude of the oscillatory stresses at the critical speed. For a clear treatment of the theory, see Den Hartog.[1]

Points to remember:

- As a general rule, it is good practice to avoid running the engine under power at speeds between 0.8 and 1.2 times critical speed. If it is necessary to take the engine through the critical speed, this should be done off load and as quickly as possible. With high inertia dynamometers the transient vibratory torque may well exceed the mechanical capacity of the drive line and the margin of safety of the drive line components may need to be increased.
- Problems frequently arise when the inertia of the dynamometer much exceeds that of the engine: a detailed torsional analysis is desirable when this factor exceeds 2. This situation usually occurs when it is found necessary to run an engine of much smaller output than the rated capacity of the dynamometer.
- the simple 'two mass' approximation of the engine-dynamometer system is inadequate for large engines and may lead to overestimation of the critical speed.

Design of coupling shafts

The maximum shear stress induced in a shaft, diameter D, by a torque T Nm is given by:

$$\tau = \frac{16T}{\pi D^3} \, \text{Pa} \tag{9a}$$

In the case of a tubular shaft, bore diameter d, this becomes:

$$\tau = \frac{16TD}{\pi(D^4 - d^4)} \, \text{Pa} \tag{9b}$$

For steels, the shear yield stress is usually taken as equal to 0.75 × yield stress in tension. A typical choice of material would be a nickel–chromium–molybdenum steel, to specification BS 817M40 (previously En 24) heat-treated to the 'T' condition.

The various stress levels for this steel are roughly as follows:

ultimate tensile strength	not less than 850 MPa (55 t.s.i.)
0.1% proof stress in tension	550 MPa
ultimate shear strength	500 MPa
0.1% proof stress in shear	300 MPa
shear fatigue limit in reversed stress	±200 MPa

It is clear that the permissible level of stress in the shaft will be a small fraction of the ultimate tensile strength of the material.

The choice of designed stress level at the maximum rated steady torque is influenced by two principal factors.

Stress concentrations, keyways and keyless hub connection

For a full treatment of the very important subject of stress concentration see Ref. 6. There are two particularly critical locations to be considered:

- At a shoulder fillet, such as is provided at the junction with an integral flange. For a ratio fillet radius/shaft diameter = 0.1 the stress concentration factor is about 1.4, falling to 1.2 for $r/d = 0.2$.
- At the semicircular end of a typical rectangular keyway, the stress concentration factor reaches a maximum of about 2.5× nominal shear stress at an angle of about 50° from the shaft axis. The authors have seen a number of failures at this location and angle.

Cyclic stresses associated with torsional oscillations is an important consideration and as, even in the most carefully designed installation involving an internal combustion engine, some torsional oscillation will be present, it is wise to select a conservative value for the nominal (steady state) shear stress in the shaft.

In view of the stress concentration inherent in shaft keyways and the backlash present that can develop in splined hubs, the use of keyless hub connection systems of the type produced by the Ringfeder Corporation are now widely used. These devices are supported by comprehensive design documentation; however, the actual installation process must be exactly followed for the design performance to be ensured.

Stress concentration factors apply to the cyclic stresses rather than to the steady state stresses. Figure 9.6 shows diagrammatically the Goodman diagram for a steel having the properties specified above. This diagram indicates approximately the

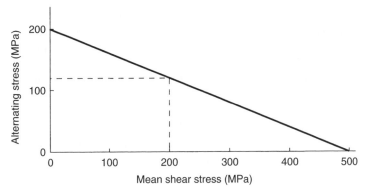

Figure 9.6 *Goodman diagram, steel shaft in shear*

relation between the steady shear stress and the permissible oscillatory stress. The example shown indicates that, for a steady torsional stress of 200 MPa, the accompanying oscillatory stress (calculated after taking into account any stress concentration factors) should not exceed ±120 MPa. In the absence of detailed design data, it is good practice to design shafts for use in engine test beds very conservatively, since the consequences of shaft failure can be so serious. A shear stress calculated in accordance with eq. (9) of about 100 MPa for a steel with the properties listed should be safe under all but the most unfavourable conditions. To put this in perspective, a shaft 100 mm diameter designed on this basis would imply a torque of 19 600 Nm, or a power of 3100 kW at 1500 rev/min.

The torsional stiffness of a solid shaft of diameter D and length L is given by:

$$C_s = \frac{\pi D^4 G}{32L} \tag{10a}$$

for a tubular shaft, bore d:

$$C_s = \frac{\pi (D^4 - d^4)}{32L} \tag{10b}$$

Shaft whirl

The coupling shaft is usually supported at each end by a universal joint or flexible coupling. Such a shaft will 'whirl' at a rotational speed N_w (also at certain higher speeds in the ratio $2^2 N_w$, $3^2 N_w$, etc.).

The whirling speed of a solid shaft of length L is given by:

$$N_w = \frac{30\pi}{L^2} \sqrt{\frac{E\pi D^4}{64 W_s}} \tag{11}$$

It is desirable to limit the maximum engine speed to about $0.8 N_w$. *When using rubber flexible couplings it is essential to allow for the radial flexibility of these couplings, since this can drastically reduce the whirling speed.* It is sometimes the practice to fit self-aligning rigid steady bearings at the centre of flexible couplings in high-speed applications, but these are liable to give fretting problems and are not universally favoured.

As is well known, the whirling speed of a shaft is identical with its natural frequency of transverse oscillation. To allow for the effect of transverse coupling flexibility the simplest procedure is to calculate the transverse critical frequency of the shaft plus two half couplings from the equation:

$$N_t = \frac{30}{\pi} \sqrt{\frac{k}{W}} \tag{12a}$$

where W = mass of shaft + half couplings and k = combined radial stiffness of the two couplings.

Then whirling speed N taking this effect into account will be given by:

$$\left(\frac{1}{N}\right)^2 = \left(\frac{1}{N_w}\right)^2 + \left(\frac{1}{N_t}\right)^2 \tag{12b}$$

Couplings

The choice of the appropriate coupling for a given application is not easy: the majority of drive line problems probably have their origin in an incorrect choice of components for the drive line, and are often cured by changes in this region. A complete discussion would much exceed the scope of this book, but the reader concerned with drive line design should obtain a copy of Ref. 4, which gives a comprehensive treatment together with a valuable procedure for selecting the best type of coupling for a given application. A very brief summary of the main types of coupling follows.

Quill shaft with integral flanges and rigid couplings

This type of connection is best suited to the situation where a driven machine is permanently coupled to the source of power, when it can prove to be a simple and reliable solution. It is not well suited to test bed use, since it is intolerant of relative vibration and misalignment.

Quill shaft with toothed or gear type couplings

Gear couplings are very suitable for high powers and speeds, and can deal with relative vibration and some degree of misalignment, but this must be very carefully controlled to avoid problems of wear and lubrication. Lubrication is particularly important as once local tooth to tooth seizure takes place deterioration may be rapid and catastrophic. Such shafts are inherently stiff in torsion.

Conventional 'cardan shaft' with universal joints

These shafts are readily available from a number of suppliers, and are probably the preferred solution in the majority of cases. However, standard automotive type shafts can give trouble when run at speeds in excess of those encountered in vehicle applications. A correct 'built-in' degree of misalignment is necessary to avoid fretting of the needle rollers.

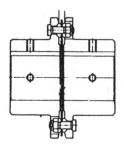

Figure 9.7   *Multiple steel disc type flexible coupling*

Multiple membrane couplings

These couplings, Fig. 9.7, are stiff in torsion but tolerant of a moderate degree of misalignment and relative axial displacement. They can be used for very high speeds.

Elastomeric element couplings

There is a vast number of different designs on the market and selection is not easy. Ref. 8 may be helpful. The great advantage of these couplings is that their torsional stiffness may be varied widely by changing the elastic elements and problems associated with torsional vibrations and critical speeds dealt with (see the next section).

Damping: the role of the flexible coupling

The earlier discussion leads to two main conclusions: the engine–dynamometer system is susceptible to torsional oscillations and the internal combustion engine is a powerful source of forces calculated to excite such oscillations. The magnitude of these undesirable disturbances in any given system is a function of the damping capacities of the various elements: the shaft, the couplings, the dynamometer and the engine itself.

The couplings are the only element of the system the damping capacity of which may readily be changed, and in many cases, for example with engines of automotive size, the damping capacity of the remainder of the system may be neglected, at least in an elementary treatment of the problem, such as will be given here.

The dynamic magnifier M (Fig. 9.3) has already been mentioned as a measure of the susceptibility of the engine–dynamometer system to torsional oscillation. Now referring to Fig. 9.1, let us assume that there are two identical flexible couplings, of stiffness C_c, one at each end of the shaft, and that these are the only sources of damping. Figure 9.8 shows a typical torsionally resilient coupling in which torque is transmitted by way of a number of shaped rubber blocks or bushes which provide torsional flexibility, damping and a capacity to take up misalignment. The torsional characteristics of such a coupling are shown in Fig. 9.9.

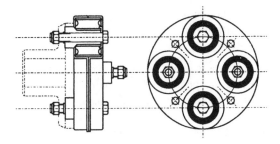

Figure 9.8 *Rubber bush type torsionally resilient coupling*

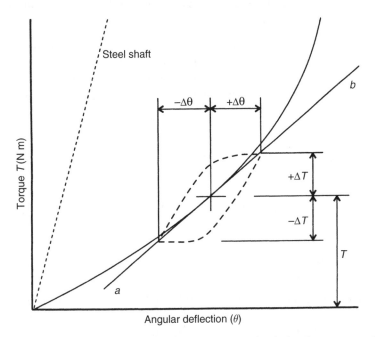

Figure 9.9 *Dynamic torsional characteristic of multiple bush type coupling*

These differ in three important respects from those of, say, a steel shaft in torsion:

1. The coupling does not obey Hooke's law: the stiffness or coupling rate $C_c = \Delta T/\Delta \theta$ increases with torque. This is partly an inherent property of the rubber and partly a consequence of the way it is constrained.
2. The shape of the torque–deflection curve is not independent of frequency. Dynamic torsional characteristics are usually given for a cyclic frequency of 10 Hz. If the load is applied slowly the stiffness is found to be substantially less. The following values of the ratio dynamic stiffness (at 10 Hz) to static stiffness of natural rubber of varying hardness are taken from Ref. 4.

Shore (IHRD) hardness	40	50	60	70
$\dfrac{\text{Dynamic stiffness}}{\text{Static stiffness}}$	1.5	1.8	2.1	2.4

Since the value of C_c varies with the deflection, manufacturers usually quote a single figure which corresponds to the slope of the tangent *ab* to the torque–deflection curve at the rated torque, typically one third of the maximum permitted torque.

3. If a cyclic torque $\pm\Delta T$, such as that corresponding to a torsional vibration, is superimposed on a steady torque T, Fig. 9.9, the deflection follows a path similar to that shown dotted. It is this feature, the hysteresis loop, which results in the dissipation of energy, by an amount ΔW proportional to the area of the loop that is responsible for the damping characteristics of the coupling.

Damping energy dissipated in this way appears as heat in the rubber and can, under adverse circumstances, lead to overheating and rapid destruction of the elements. The appearance of rubber dust inside coupling guards is a warning sign.

The damping capacity of a component such as a rubber coupling is described by the *damping energy ratio*:

$$\psi = \frac{\Delta W}{W}$$

This may be regarded as the ratio of the energy dissipated by hysteresis in a single cycle to the elastic energy corresponding to the wind-up of the coupling at mean deflection:

$$W = \frac{1}{2}T\theta = \frac{1}{2}T^2/C_c$$

The damping energy ratio is a property of the rubber. Some typical values are given in Table 9.2. The dynamic magnifier is a function of the damping energy ratio: as would be expected a high damping energy ratio corresponds to a low dynamic magnifier. Some authorities give the relation:

$$M = 2\pi/\psi$$

Table 9.2 *Damping energy ratio ψ*

Shore (IHRD) hardness	50/55	60/65	70/75	75/80
Natural rubber	0.45	0.52	0.70	0.90
Neoprene		0.79		
Styrene-butadiene (SBR)		0.90		

Table 9.3 *Dynamic magnifier M*

Shore (IHRD) hardness	50/55	60/65	70/75	75/80
Natural rubber	10.5	8.6	5.2	2.7
Neoprene		4.0		
Styrene-butadiene (SBR)		2.7		

However, it is pointed out in Ref. 2 that for damping energy ratios typical of rubber the exact relation:

$$\psi = \left(1 - e^{-2\pi/M}\right)$$

is preferable. This leads to values of M shown in Table 9.3, which correspond to the values of ψ given in Table 9.2.

It should be noted that when several components, e.g. two identical rubber couplings, are used in series the dynamic magnifier of the combination is given by:

$$\left(\frac{1}{M}\right)^2 = \left(\frac{1}{M_1}\right)^2 + \left(\frac{1}{M_2}\right)^2 + \left(\frac{1}{M_3}\right)^2 + \cdots \tag{13}$$

(this is an empirical rule, recommended in Ref. 5).

An example of drive shaft design

The application of these principles is best illustrated by a worked example. Figure 9.1 represents an engine coupled by way of twin multiple-bush type rubber couplings and an intermediate steel shaft to an eddy current dynamometer, with dynamometer starting.

Engine specification is as follows:

Four cylinder four-stroke gasoline engine
Swept volume 2.0 litre, bore 86 mm, stroke 86 mm

Maximum torque	110 Nm at 4000 rev/min
Maximum speed	6000 rev/min
Maximum power output	65 kW
Maximum bmep	10.5 bar
Moments of inertia	$I_e = 0.25\,\text{kgm}^2$
.	$I_d = 0.30\,\text{kgm}^2$

Table 9.4 indicates a service factor of 4.8, giving a design torque of $110 \times 4.8 = 530\,\text{Nm}$.

Table 9.4 *Service factors for dynamometer/engine combinations*

Dynamometer type	Number of cylinders									
	Diesel					Gasolene				
	1/2	3/4/5	6	8	10+	1/2	3/4/5	6	8	10+
Hydraulic	4.5	4.0	3.7	3.3	3.0	3.7	3.3	3.0	2.7	2.4
Hyd. + dyno. start	6.0	5.0	4.3	3.7	3.0	5.2	4.3	3.6	3.1	2.4
Eddy current (EC)	5.0	4.5	4.0	3.5	3.0	4.2	3.8	3.3	2.9	2.4
EC + dyno. start	6.5	5.5	4.5	4.0	3.0	5.7	4.8	3.8	3.4	2.4
d.c. + dyno. start	8.0	6.5	5.0	4.0	3.0	7.2	5.8	4.3	3.4	2.4

It is proposed to connect the two couplings by a steel shaft of the following dimensions:

Diameter $D = 40\,\text{mm}$
Length $L = 500\,\text{mm}$
Modulus of rigidity $G = 80 \times 10^9\,\text{Pa}$

From eq. (9a), torsional stress $\tau = 42\,\text{Mpa}$, very conservative.
From eq. (10a)

$$C_s = \frac{\pi \times 0.04^4 \times 80 \times 10^9}{32 \times 0.5} = 40\,200\,\text{Nm/rad}$$

Consider first the case when rigid couplings are employed:

$$n_c = \frac{60}{2\pi} \sqrt{\frac{40\,200 \times 0.55}{0.25 \times 0.30}} = 5185\,\text{c.p.m.}$$

For a four cylinder, four-stroke engine, we have seen that the first major critical occurs at order $N_0 = 2$, corresponding to an engine speed of 2592 rev/min. This falls right in the middle of the engine speed range and is clearly unacceptable. This is a typical result to be expected if an attempt is made to dispense with flexible couplings.

The resonant speed needs to be reduced and it is a common practice to arrange for this to lie between either the cranking and idling speeds or between the idling and minimum full load speeds. In the present case these latter speeds are 500 and

1000 rev/min, respectively. This suggests a critical speed N_c of 750 rev/min and a corresponding resonant frequency $n_c = 1500$ cycles/min.

This calls for a reduction in the torsional stiffness in the ratio:

$$\left(\frac{1500}{5185}\right)^2$$

i.e. to 3364 Nm/rad.

The combined torsional stiffness of several elements in series is given by:

$$\frac{1}{C} = \frac{1}{C_1} + \frac{1}{C_2} + \frac{1}{C_3} + \cdots \tag{14}$$

This equation indicates that the desired stiffness could be achieved by the use of two flexible couplings each of stiffness 7480 Nm/rad. A manufacturer's catalogue shows a multi-bush coupling having the following characteristics:

Maximum torque	814 Nm (adequate)
Rated torque	170 Nm
Maximum continuous vibratory torque	±136 Nm
Shore (IHRD) hardness	50/55
Dynamic torsional stiffness	8400 Nm/rad

Substituting this value in eq. (14) indicates a combined stiffness of 3800 Nm/rad. Substituting in eq. (1) gives $n_c = 1573$, corresponding to an engine speed of 786 rev/min, which is acceptable.

It remains to check on the probable amplitude of any torsional oscillation at the critical speed. Under no-load conditions, the imep of the engine is likely to be in the region of 2 bar, indicating, from eq. (5a), a mean turning moment $M_{mean} = 8$ Nm.

From Table 9.1, p factor $= 1.91$, giving $T_{ex} = 15$ Nm per cylinder:

$$\sum T_{ex} = 4 \times 15 = 60 \text{ Nm}$$

Table 9.3 indicates a dynamic magnifier $M = 10.5$, the combined dynamic magnifier from eq. (13) $= 7.4$.

The corresponding value of the vibratory torque, from eq. (7), is then:

$$T_v = \frac{60 \times 7.4}{(1 + 0.25/0.30)} = \pm 242 \text{ Nm}$$

This is in fact outside the coupling continuous rating of ±136 Nm, but multiple bush couplings are tolerant of brief periods of quite severe overload and this solution should be acceptable provided the engine is run fairly quickly through the critical speed. An alternative would be to choose a coupling of similar stiffness using SBR bushes of 60/65 hardness. Table 9.3 shows that the dynamic magnifier is reduced from 10.5 to 2.7, with a corresponding reduction in T_v.

If in place of an eddy current dynamometer we were to employ a d.c. machine, the inertia I_b would be of the order of 1 kgm², four times greater.

This has two adverse effects:

1. Service factor, from Table 9.4 increased from 4.8 to 5.8;
2. The denominator in eq. (7) is reduced from $(1 + 0.25/0.30) = 1.83$ to $(1 + 0.25/1.0) = 1.25$, corresponding to an increase in the vibratory torque for a given exciting torque of nearly 50 per cent.

This is a general rule: the greater the inertia of the dynamometer the more severe the torsional stresses generated by a given exciting torque.

An application of eq. (1) shows that for the same critical frequency the combined stiffness must be increased from 3364 Nm/rad to 5400 Nm/rad. We can meet this requirement by changing the bushes from Shore Hardness 50/55 to Shore Hardness 60/65, increasing the dynamic torsional stiffness of each coupling from 8400 Nm/rad to 14000 Nm/rad (in general, the usual range of hardness numbers, from 50/55 to 75/80, corresponds to a stiffness range of about 3:1, a useful degree of flexibility for the designer).

Eq. (1) shows that with this revised coupling stiffness n_c changes from 1573 cycles/min to 1614 cycles/min, and this should be acceptable. The oscillatory torque generated at the critical speed is increased by the two factors mentioned above, but reduced to some extent by the lower dynamic magnifier for the harder rubber, $M = 8.6$ against $M = 10.5$. As before, prolonged running at the critical speed should be avoided.

For completeness, we should check the whirling speed from eq. (11). The mass of the shaft per unit length is: $W_s = 9.80$ kg/m.

$$N_w = \frac{30\pi}{0.50^2} \sqrt{\frac{200 \times 10^9 \times \pi \times 0.04^4}{64 \times 9.80}} = 19\,100 \text{ r.p.m.}$$

The mass of the shaft + half couplings is found to be 12 kg and the combined radial stiffness 33.6 MN/m. From eq. (12a):

$$N_t = \frac{30}{\pi} \sqrt{\frac{33.6 \times 10^6}{12}} = 16\,000 \text{ r.p.m.}$$

then from eq. (12b), whirling speed = 12 300 rev/min, which is satisfactory.

Note, however, that, if shaft length were increased from 500 to 750 mm, whirling speed would be reduced to about 7300 rev/min, which is barely acceptable. This is a common problem, usually dealt with by the use of tubular shafts, which have much greater transverse stiffness for a given mass.

There is no safe alternative, when confronted with an engine of which the characteristics differ significantly from any run previously on a given test bed, to following through this design procedure.

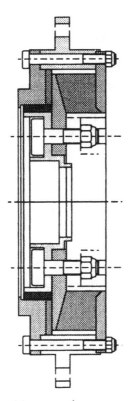

Figure 9.10 *Annular type rubber coupling*

An alternative solution

The above worked example makes use of two multiple-bush type rubber couplings with a solid intermediate shaft. An alternative is to make use of a conventional propeller shaft with two universal joints, as used in road vehicles, with the addition of a coupling incorporating an annular rubber element in shear to give the necessary torsional flexibility. These couplings, Fig. 9.10, are generally softer than the multiple bush type for a given torque capacity, but are less tolerant of operation near a critical speed. If it is decided to use a conventional universal joint shaft, the supplier should be informed of the maximum speed at which it is to run. This will probably be much higher than is usual in the vehicle and may call for tighter than usual limits on balance of the shaft.

Shock loading of couplings due to cranking, irregular running and torque reversal

Systems for starting and cranking engines are described in Chapter 8, where it is emphasized that during engine starting severe transient torques can arise. These have been known to result in the failure of flexible couplings of apparently adequate torque capacity. The maximum torque that can be necessary to get a green engine over t.d.c. or that can be generated at first firing should be estimated and checked against maximum coupling capacity.

Irregular running or imbalance between the powers generated by individual cylinders can give rise to exciting torque harmonics of lower order than expected in a multicylinder engine and should be borne in mind as a possible source of rough running. Finally, there is the possibility of momentary torque reversal when the engine comes to rest on shutdown.

However, the most serious problems associated with the starting process arise when the engine first fires. Particularly when, as is common practice, the engine is motored to prime the fuel injection pump, the first firing impulses can give rise to severe shocks. Annular type rubber couplings, Fig. 9.10, can fail by shearing under these conditions. In some cases, it is necessary to fit a torque limiter or slipping clutch to deal with this problem.

Axial shock loading

Engine test systems that incorporate automatic shaft docking systems have to provide for the axial loads on both the engine and dynamometer imposed by such a system. In some high volume production facilities, an intermediate pedestal bearing isolates the dynamometer from both the axial loads of normal docking operation and for cases when the docking splines jam 'nose to nose'; in these cases the docking control system should be programmed to back off the engine, spin the dynamometer and retry the docking.

Selection of coupling torque capacity

Initial selection is based on the maximum rated torque with consideration given to the type of engine and number of cylinders, dynamometer characteristics and inertia. Table 9.4, reproduced by courtesy of Twiflex Ltd, shows recommended service factors for a range of engine and dynamometer combinations. The rated torque multiplied by the service factor must not exceed the permitted maximum torque of the coupling.

Other manufacturers may adopt different rating conventions, but Table 9.4 gives valuable guidance as to the degree of severity of operation for different situations. Thus, for example, a single cylinder diesel engine coupled to a d.c. machine with dynamometer start calls for a margin of capacity three times as great as an eight cylinder gasoline engine coupled to a hydraulic dynamometer.

Figure 9.11 shows the approximate range of torsional stiffness associated with three types of flexible coupling: the annular type as illustrated in Fig. 9.10, the multiple bush design of Fig. 9.8 and a development of the multiple bush design which permits a greater degree of misalignment and makes use of double-ended bushed links between the two halves of the coupling. The stiffness figures of Fig. 9.11 refer to a single coupling.

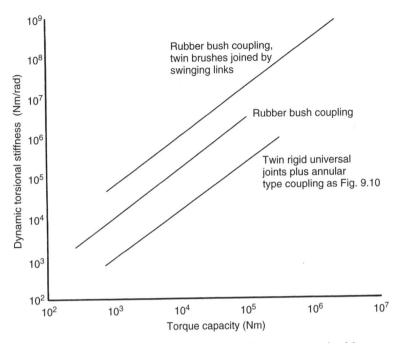

Figure 9.11 *Ranges of torsional stiffness for different types of rubber coupling*

The role of the engine clutch

Vehicle engines are invariably fitted with a clutch and this may or may not be retained on the test bed. The advantage of retaining the clutch is that it acts as a torque limiter under shock or torsional vibration conditions. The disadvantages are that it may creep, particularly when torsional vibration is present, leading to ambiguities in power measurement, while it is usually necessary, when the clutch is retained, to provide an outboard bearing. Clutch disc springs may have limited life under test bed conditions.

Balancing of drive line components

This is a matter which is often not taken sufficiently seriously and can lead to a range of troubles, including damage (which can be very puzzling) to bearings, unsatisfactory performance of such items as torque transducers, transmitted vibration to unexpected locations and serious drive line failures. Particular care should be taken in the choice of couplings for torque shaft dynamometers: couplings such as the multiple disc type, Fig. 9.7, cannot be relied upon to centre these devices sufficiently accurately.

It has sometimes been found necessary to carry out in situ balancing of a composite engine drive line where the sum of the out of balance forces in a particular radial relationship causes unacceptable vibration; specialist companies with mobile plant exist to provide this service.

Conventional universal joint type cardan shafts are often required to run at higher speeds in test bed applications than is usual in vehicles; when ordering, the maximum speed should be specified and, possibly, a more precise level of balancing than the standard specified.

BS 5265, Mechanical balancing of rotating bodies,[9] gives a valuable discussion of the subject and specifies 11 Balance Quality Grades. Drive line components should generally be balanced to Grade G 6.3, or, for high speeds, to grade G 2.5. The standard gives a logarithmic plot of the permissible displacement of the centre of gravity of the rotating assembly from the geometrical axis against speed. To give an idea of the magnitudes involved, G 6.3 corresponds to a displacement of 0.06 mm at 1000 rev/min, falling to 0.01 mm at 5000 rev/min.

Alignment of engine and dynamometer

This is a fairly complex and quite important matter. For a full treatment and description of alignment techniques, see Ref. 4. Differential thermal growth and the movement of the engine on its flexible mountings when on load should be taken into account and if possible the mean position should be determined. The laser-based alignment systems now available greatly reduce the effort and skill required to achieve satisfactory levels of accuracy. In particular, they are able to bridge large gaps between flanges without any compensation for droop and deflection of arms carrying dial indicators, a considerable problem with conventional alignment methods.

There are essentially three types of shaft having different alignment requirements to be considered:

1. Rubber bush and flexible disc couplings should be aligned as accurately as possible as any misalignment encourages heating of the elements and fatigue damage.
2. Gear type couplings require a minimum misalignment of about 0.04° to encourage the maintenance of an adequate lubricant film between the teeth.
3. Most manufacturers of universal joint propeller shafts recommend a small degree of misalignment to prevent brinelling of the universal joint needle rollers. Note that it is essential, in order to avoid induced torsional oscillations, that the two yokes of the intermediate shaft joints should lie in the same plane.

Distance between end flanges can be critical, as incorrect positioning can lead to the imposition of axial loads on bearings of engine or dynamometer.

Guarding of coupling shafts

Not only is the failure of a high speed coupling shaft potentially dangerous, as very large amounts of energy can be released, it is quite common. The ideal shaft-guard will contain the debris resulting from a failure of any part of the drive line and prevent accidental or casual contact with rotating parts, while being sufficiently light and adaptable not to interfere with engine rigging and alignment. A guard system that is very inconvenient to use will eventually be misused or neglected.

A really substantial guard, preferably a steel tube not less than 6 mm thick, split and hinged in the horizontal plane for shaft access, is an essential precaution. The hinged 'lid' should be interlocked with the emergency stop circuit to prevent engine cranking or running while it is open. Many designs include shaft restraint devices loosely fitting around the tubular portion, made of wood or a non-metallic composite and intended to prevent a failing shaft from whirling; these should not be so close to the shaft as to be touched by it during normal starting or running otherwise they will be the cause of failure rather than a prevention of damage.

Engine to dynamometer coupling: summary of design procedure

1. Establish speed range and torque characteristic of engine to be tested. Is it proposed to run the engine on load throughout this range?
2. Make a preliminary selection of a suitable drive shaft. Check that maximum permitted speed is adequate. Check drive shaft stresses and specify material. Look into possible stress raisers.
3. Check manufacturer's recommendations regarding load factor and other limitations.
4. Establish rotational inertias of engine and dynamometer and stiffness of proposed shaft and coupling assembly. Make a preliminary calculation of torsional critical speed from eq. (1). (In the case of large multicylinder engines consider making a complete analysis of torsional behaviour.)
5. Modify specification of shaft components as necessary to position torsional critical speeds suitably. If necessary, consider use of viscous torsional dampers.
6. Calculate vibratory torques at critical speeds and check against capacity of shaft components. If necessary specify 'no go' areas for speed and load.
7. Check whirling speeds.
8. Specify alignment requirements.
9. Design shaft guards.

Flywheels

No treatment of the engine/dynamometer drive line would be complete without mention of flywheels which may form a discrete part of the shaft system. A flywheel is a device that stores kinetic energy. The energy stored may be precisely calculated and is instantly available. The storage capacity is directly related to the mass moment of inertia which is calculated by:

$$I = k \times M \times R^2$$

where:

I = moment of inertia (kgm^2)
k = inertia constant (dependent upon shape)
m = mass (kg)
R = radius of flywheel mass

In the case of a flywheel taking the form of a uniform disk, which is the common form found within dynamometer cells and chassis dynamometer designs:

$$I = \frac{1}{2}MR^2$$

The engine or vehicle test engineer would normally expect to deal with flywheels in two roles:

1. As part of the test object, as in the common case of an engine flywheel where it forms part of the engine/dynamometer shaft system and contributes to the system's inertial masses taken into account during a torsional analysis.
2. As part of the test equipment where one or more flywheels may be used to provide actual inertia that would, in 'real life', be that of the vehicle or some part of the engine driven system.

No mention of flywheels should be made without consideration of the safety of the application. The uncontrolled discharge of energy from any storage device is hazardous. The classic case of a flywheel failing by bursting is now exceptionally rare and invariably due to incompetent modification rather than the nineteenth century problems of poor materials, poor design or overspeeding.

In the case of engine flywheels, the potential danger in the test cell is the shaft system attached to it. This may be quite different in mass and fixing detail from its final application connection, and can cause overload leading to failure. Cases are on record where shock loading caused by connecting shafts touching the guard system due to excessive engine movement has created shock loads that have led to the cast engine flywheel fracturing, with severe consequential damage.

The most common hazard of test rig mounted flywheels is caused by bearing or clutch failure where consequential damage is exacerbated by the considerable energy available to fracture connected devices or because of the time that the flywheel and connected devices will rotate before the stored energy is dissipated and movement is stopped.

It is vital that flywheels are guarded in such a manner as to prevent absolutely accidental entrainment of clothing or cables, etc.

A common and easy to comprehend use of flywheels is as part of a vehicle brake testing rig. In these devices, flywheels supply the energy that has to be absorbed and dissipated by the brake system under test. The rig motor is only used to accelerate the chosen flywheel combinations up to the rotational speed required to simulate the vehicle axle speed at the chosen vehicle speed. Flywheel brake rigs have been made up to the size that can provide the same kinetic energy as fully loaded high speed trains. Flywheels are also used on rigs used to test automatic automotive gearboxes.

Test rig flywheel sets need to be rigidly and securely mounted and balanced to the highest practical standard. Multiples of flywheels forming a common system that can be engaged in different combinations and in any radial relationship require particular care in the design of both their base frame and individual bearing supports. Such systems can produce virtually infinite combinations of shaft balance and require each individual mass to be as well balanced and aligned on as rigid a base as possible.

Simulation of inertia* versus iron inertia

Modern a.c. dynamometer systems and control software have significantly replaced the use of flywheels in chassis and engine dynamometer systems in the automotive industry. Any perceived shortcoming in the speed of response or accuracy of the simulation is usually considered to be of less concern than the mechanical simplicity of the electric dynamometer system and the reduction in required cell space.

Finally, it should be remembered that, unless engine rig flywheels are able to be engaged through a clutch then the engine starting/cranking system will have to be capable of accelerating engine, dynamometer and flywheel mass up to engine start speed.

Notation

Frequency of torsional oscillation	n cycles/min
Critical frequency of torsional oscillation	n_c cycles/min
Stiffness of coupling shaft	C_s Nm/rad
Rotational inertia of engine	I_e kg m^2
Rotational inertia of dynamometer	I_b kg m^2
Amplitude of exciting torque	T_x Nm
Amplitude of torsional oscillation	θ rad
Static deflection of shaft	θ_0 rad
Dynamic magnifier	M
Dynamic magnifier at critical frequency	M_c
Order of harmonic component	N_o

---

* Some readers may object to the phrase 'simulation of inertia' since one is simulating the effects rather than the attribute, but the concept has wide industrial acceptance.

Number of cylinders	N_{cyl}
Mean turning moment	M_{mean} Nm
Indicated mean effective pressure	p_i bar
Cylinder bore	B mm
Stroke	S mm
Component of tangential effort	T_m Nm
Amplitude of vibratory torque	T_v Nm
Engine speed corresponding to n_c	N_c rev/min
Maximum shear stress in shaft	τ N/m^2
Whirling speed of shaft	N_W rev/min
Transverse critical frequency	N_t cycles/min
Dynamic torsional stiffness of coupling	C_c Nm/rad
Damping energy ratio	ψ
Modulus of elasticity of shaft material	E Pa
Modulus of rigidity of shaft material	G Pa
(for steel, $E = 200 \times 10^9$ Pa, $G = 80 \times 10^9$ Pa)	

References

1. Den Hartog, J.P. (1956) *Mechanical Vibrations*, McGraw-Hill, Maidenhead, UK.
2. Ker Wilson, W. (1963) *Practical Solution to Torsional Vibration Problems* (5 vols), Chapman and Hall, London.
3. Ker Wilson, W. (1959) *Vibration Engineering*, Griffin, London.
4. Neale, M.J., Needham, P. and Horrell, R. (1998) *Couplings and Shaft Alignment*, Mechanical Engineering Publications, London.
5. *Rulebook*, Chapter 8. Shaft vibration and alignment, Lloyd's Register of Shipping, London.
6. Pilkey, W.D. (1997) *Peterson's Stress Concentration Factors*, Wiley, New York.
7. Young, W.C. (1989) *Roark's Formulas for Stress and Strain*, McGraw-Hill, New York.
8. BS 6613 *Methods for Specifying Characteristics of Resilient Shaft Couplings*.
9. BS 5265 Parts 2 and 3 *Mechanical Balancing of Rotating Bodies*.

Further reading

BS 4675 Parts 1 and 2 *Mechanical Vibration in Rotating Machinery*.
BS 6861 Part 1 *Method for Determination of Permissible Residual Unbalance*.
BS 6716 *Guide to Properties and Types of Rubber*.
Nestorides, E.J. (1958) *A Handbook of Torsional Vibration*, Cambridge University Press, Cambridge.

10 Electrical design considerations

Introduction

The electrical system of a test facility provides the power, nerves and operating logic that control the test piece, the test instrumentation and modules of building services. The power distribution to, and integration of, these many parts falls significantly within the remit of the electrical engineer, whose drawings will be used as the primary documentation in system commissioning or any subsequent fault finding tasks. Integration of test cell modules equipment built to American standards in Europe and the rest of the world and vice versa can create detailed interface problems not appreciated by specialists at either end of the project unless they have experience in dealing with each other's practices.

The theme of system integration is nowhere more pertinent than within the role of the electrical designer.

Note: In this text and in context with electrical engineering, the UK English terms 'earth' and 'earthing' have been generally used in preference to the US English terms 'ground' and 'grounding' with which they are, in the context of this book, interchangeable.

The electrical engineer's design role

The electrical engineer's prime guidance are the operational and functional specifications (see Chapter 1) from which he will develop the final detailed functional specification, system schematics, power distribution and alarm logic matrices.

In the design process much of the mechanical and civil plant has to be identified before the electrical engineer can start to calculate the electrical power required and produce an electrical distribution scheme, complete with the control cabinets and interconnecting cable ways; all of which may have implications for the other specialist designers. System integration is therefore an iterative process with the electrical engineer as a key team member.

Engine test facilities are, by the nature of their component parts and their role, particularly vulnerable to signal distortion due to various forms of electrical noise. To avoid signal corruption and cell downtime due to instrumentation errors, special and detailed attention must be made to the standard of electrical installation within a test facility. The electrical designer and installation supervisor must be able to take a holistic approach and be aware of the need to make an electrically integrated and mutually compatible system. Facilities that are developed on an ad hoc basis often fall foul of unforeseen signal interference or control logic errors within interacting subsystems.

General characteristics of the electrical installation

Perhaps more than any other aspect of test cell design and construction the electrical installation is subject to regulations, most of which have statutory force. It is essential that any engineer responsible for the design or construction of a test cell in the UK should be familiar with BS 7671 *Requirements for Electrical Installations*. There are also many British Standards specifying individual features of an electrical system; other countries will have their own national standards. In the USA the relevant electrical standards may be accessed through the ANSI website.

The rate of change of regulations, particularly European regulation, will outpace that of any general text book (Table 10.1). (BS EN) 60204 is of particular relevance to test cell design since it includes general rules concerning safety interlocking and the shutting down of rotating plant as in section 9.2.2 covering shutting down of engine/dynamometer combinations (see Chapter 11, section Safety systems).

While these regulations cover most aspects of the electrical installation in test cells, there are several particular features that are a consequence of the special conditions associated with the test cell environment which are not explicitly covered. These will be included within this chapter, as will the special electrical design features required if four quadrant electrical dynamometers (see Chapter 8) are being included within the facility.

Table 10.1   *Electrical installation regulations*

IEC 60204–1/97 EN 60204–1/97	Safety of machinery – electrical equipment of machines
IEC 1010–1 EN 61010–1/A2	Safety requirement for electrical equipment for measurement, control and laboratory use
EN 50178/98	Electronic equipment of power installation
IEC 61800–3/96	Adjustable speed electrical power drive systems
EN 61800–3/96	EMC product standard
EN 61000–6–4	Electromagnetic influence – emission
EN 61000–6–2	Electromagnetic influence – immunity
EN 61010–1	Safety requirements of laboratory (test cell) equipment
IEC 61010–1	designed for measurement and control

Physical environment

The physical environment inside a test cell can vary between the extremes used to test the engine or vehicle and the test cell designer must ensure that the test plant is rated such that it can withstand the full range of operating conditions specified.

The control room and services spaces will, at worst, experience the range of ambient conditions met at the geographical location, which can exceed the 'normal' conditions expected by some instrument manufacturers. If air conditioning of control room or services spaces is installed in atmospheres of high humidity, the electrical designer must consider protection against condensation within control cabinets and dynamometer drive cabinets.

The guidelines for acceptable ambient conditions for most electrical plant are as follows:

- The ambient temperature of the control cabinet should be between +5°C and +35°C.
- The ambient temperature for a control room printer is restricted to +35°C and for the PC to +38°C.
- The ambient temperature of the test cell should be between +5°C and +45°C.
- The air must be free of abnormal amounts of dust, acids, corroding gases and salt (use filter).
- The relative humidity at +40°C must not be more than 50 and 90 per cent at +20°C (use heating).

The rating of electrical devices, such as motors and enclosures, must be correctly defined at the design stage.

In some vehicle and engine cells, particularly large diesel and engine rebuild facilities, it is common practice for a high pressure heated water washer to be used in which case the test bed enclosures should have a rating of at least IP×5 (see Table 10.2 below).

Electrical signal and measurement interference

The protection methodology against signal interference in engine test cells has changed significantly over the period since the mid-1990s because of the arrival of electromagnetic interference in the radio frequencies as the dominating source of signal corruption.

The pulse width modulating drive technology based on fast switching *insulated bipolar gate transistors* (IGBT) devices used with a.c. dynamometers has reduced the *total harmonic distortion* (THD) experienced in power supplies from that approaching the 30 per cent that was produced by a d.c. thyristor controlled drive to <5 per cent THD.

Table 10.2 *Ingress protection rating details*

IP54 = IP Letter Code ——————— IP			
1st Digit ————————————— 5			
2nd Digit ——————————————— 4			

1st Digit	Protection from solid objects	2nd Digit	Protection from moisture
0	Non-protected	0	Non-protected
1	Protected against solid objects greater than 50 mm	1	Protected against dripping water
2	Protected against solid objects greater than 12 mm	2	Protected against dripping water when tilted up to 15°
3	Protected against solid objects greater than 2.5 mmØ	3	Protected against spraying water
4	Protected against solid objects greater than 1.0 mmØ	4	Protected against splashing water
5	Dust protected	5	Protected against water jets
6	Dust tight	6	Protected against heavy seas
		7	Protected against the effects of immersion
		8	Protected against submersion

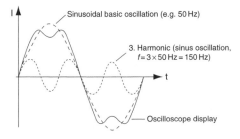 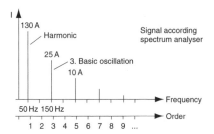

Figure 10.1 *Harmonic distortion*

$$\text{THD\%} = \sqrt{\sum_{2}^{k} \left(\frac{H_x}{H_f}\right)^{-2}}$$

where H_x is the amplitude of any harmonic order and H_f is the amplitude of the fundamental harmonic (1st order). Most sensitive devices should be unaffected by a THD of less than 8 per cent.

The harmonic distortion produced by thyristor drives associated d.c. machines is load-dependent and causes a voltage drop at the power supply (see Fig. 10.1).

While IGBT technology has reduced THD that caused problems in facilities fitted with d.c. dynamometers, it has joined other digital devices in introducing disturbance in the frequency range of 150 to 30 MHz.

IBGT drive systems produce a common mode, load-independent disturbance that causes unpredictable flow through the facility earth system.

This has meant that standard practices concerning signal cable protection and provision of 'clean earth' connections that are separate from 'protective earths' are tending to change in order to fight the new enemy: electromagnetic interference (EMI). The connection of shielded cables between devices shown in Fig. 10.2 was the

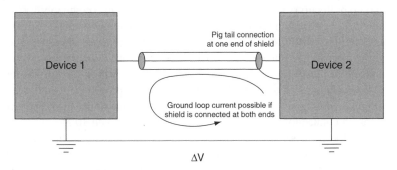

Figure 10.2 *Recommended method of connecting shielded cable between devices to prevent ground loop current distortion of signals*

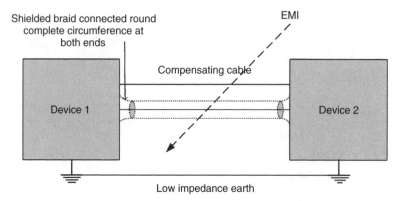

Figure 10.3 *Recommended method of connecting shielded cable between devices to prevent EMI and ground loop corruption of signals*

standard method aimed at preventing earth loop distortion of signals in the shielded cable. *Earth loops* are caused by different earth potential values occurring across a measuring or signal circuit which induces compensating currents.

This single end connection of the cable shield shown in Fig. 10.2 offers no protection from EMI which requires connection at both ends (at both devices). This requires that the ground loops are defeated by different measures. At the level of shielded cable connection the countermeasure is to run a compensation lead of high surface area in parallel with the shield as shown in Fig. 10.3. This method also reduces the vulnerability of the connection to external magnetic flux fields.

Earthing system design

The earthing systems used in many industrial installations are primarily designed as a human protection measure. However, electromagnetic compatibility (EMC) requirements increasingly require high frequency equipotential bonding, achieved by continuous linking of all ground potentials. This modern practice of earthing devices has significantly changed layouts in which there was once a single protective earth (PE) connection, plus when deemed necessary, a 'clean earth' physically distant for the PE. A key feature of modern EMC practice is the provision of multiple earth connections to a common earthed grid of the lowest possible impedance. The physical details of such connection practice are shown in Figs 10.4 and 10.5.

To ensure the most satisfactory functioning of these EMC systems, the earthing system needs to be incorporated into a new building design and the specification of the electrical installation. Ideally the building should be constructed with a ground mat made of welded steel embedded in the concrete floor and a circumferential earth strap.

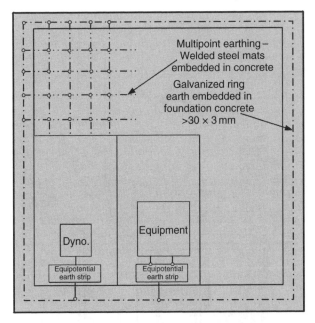

Figure 10.4 *EMC equipotential earthing grid incorporated in new building*

The layout of cabling

Transducer signals are usually 'conditioned' as near to the transducer as possible; nevertheless, the resultant conditioned signals are commonly in the range 0 to 10 V d.c. or 0 to 5 mA, very small when compared with the voltage differences and current flows that may be present in power cables in the immediate vicinity of the signal lines. Cable separation and layout is of vital importance. Practices that create chaotic jumbles of cables beneath trench covers and 'spare' cable lengths looped or coiled in the base of distribution boxes are simply not acceptable in modern laboratories.

Most engine test facilities will contain the following types of wiring which have distinctly different roles and which must be prevented from interfering with each other:

- power cables, mains supply ranging from high power wiring for dynamometers through three-phase and single-phase distribution for services and instruments to low power supplies for special transducers, also high current d.c. supply for starter systems;
- control cables for inductive loads, relays, etc.;
- signal cables:
  - digital control with resistive load;
  - ethernet, RS232, RS422, IEEE1394;

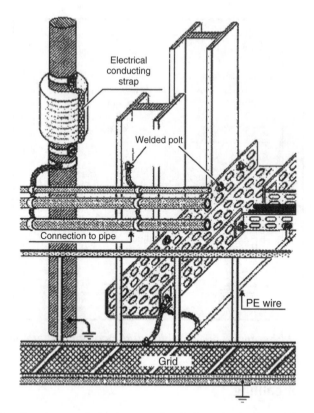

Figure 10.5   *Continuous linked network of all ground potentials*

  - bus systems, such as CAN;
  - 24 Vd.c. supplies;
- measuring cables associated with transducers and instrumentation transmitting analogue signals.

In the following paragraphs, the common causes of signal interference are identified and practices recommended to avoid the problems described.

Inductive interference

Inductive interference is caused by the magnetic flux generated by electrical currents inducing voltages in nearby conductors.

  Countermeasures include:

- Do not run power cables close to control or signal cables (also see Capacitive interference).

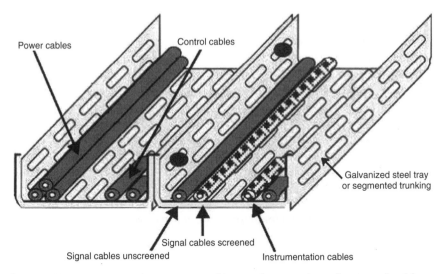

Power cables

Control cables

Galvanized steel tray
or segmented trunking

Signal cables screened

Signal cables unscreened

Instrumentation cables

Figure 10.6 *Segmented trays or trunking and separation of types of cable*

- Use either segmented trunking or different cable tray sections as shown in Fig. 10.3.
- Use twisted pair cables for connection of devices requiring supply and return connection (30 twists per metre will reduce interference voltage by a factor of 25).
- Use shielded signal cables and connect the shield to earth at both ends.

Figure 10.6 below shows a suggested layout of cables running in open cable trays. The same relationship and metal division between cable types may be achieved using segmented trunking, trays or ladders. It is important that the segments of such metallic support systems are connected together as part of the earth bonding system; it is not acceptable to rely on metal to metal contact of the segments.

Note that cables of different types crossing through metal segments at 90° to the main run do not normally cause problems.

Capacitive interference

Capacitive interference can occur if signal cables with different voltage levels are run closely together. It can also be caused by power cables running close to signal lines.

Countermeasures include:

- separate signal cables with differing voltage levels;
- do not run signal cables close to power cables;
- if possible use shielded cable for vulnerable signal lines.

Electromagnetic interference

Electromagnetic interference can induce both currents and voltages in signal cables. It may be caused by a number of 'noise' transmitters ranging from spark plugs to mobile communication devices.

RF noise produced by motor drives based on IBGT devices is inherent in the technology with inverter frequencies of between 3 and 4 kHz and is due to the steep leading and failing edge of the, typically 500 V, pulses that have a voltage rise of around 2 kV/μsec. There may be some scope in varying the pattern of frequency disturbances produced at a particular drive/site combination by adjustment of the pulsing frequency; the supplier would need to be consulted.

Countermeasures for a.c. drive-induced noise include:

- Choice and layout of both motor/drive cabinet and drive cabinet to supply cables is very important:
  - Motor cable should have the three multicore power conductors in a symmetrical arrangement within common braided and foil screens. The bonding of the screen at the termination points must contact 360° of the braid and be of low impedance.
  - The power cables should contain symmetrically arranged conductors of low inductivity within a concentric PE conductor.
  - Single core per phase cables should be laid as shown in Fig. 10.7.
- Signal cables should be screened.
- Signal cables should be encased in metal trunking or laid within metal cable tray.

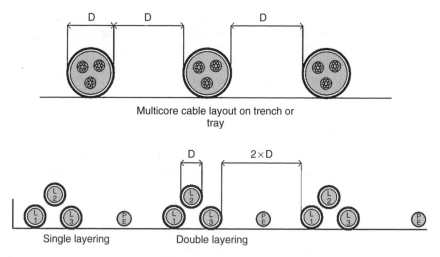

Multicore cable layout on trench or tray

Single layering   Double layering

Figure 10.7 *Recommended spacing and layout of power cables*

The layout of power cables of both multicore and single core is shown in Fig. 10.7 following recommended practice.

Conductive coupling interference

Conductive coupling interference may occur when there is a supply voltage difference between a number of control or measuring devices. It is usually caused by long supply line length or inappropriate distribution layout.

Countermeasures include:

- Keep supply cables short and of sufficient conductor size to minimize voltage drop.
- Avoid common return lines for different control or measurement devices by running separate supply lines for each device.

The same problem may occur when two devices are fed from different power supplies.

To minimize interference in analogue signal cables, an equipotential bonding cable or strap should connect the two devices running as close as possible to the signal cable. It is recommended that the bonding cable resistance be <1/10 (one tenth) of the cable screen resistance.

Integration of a.c. dynamometer systems

Careful consideration should be made when integrating an a.c drive system into an engine test facility. Where possible it is most advisable to provide a dedicated electrical supply to an a.c drive or number of a.c. drives.

There needs to be a clear understanding of the status of the existing electrical supply network, and the work involved in providing a new supply for an a.c dynamometer system. It is important to calculate the correct rating for a new supply transformer. This is a highly specialized subject and the details may change depending on the design of the supply system and a.c. devices, but the general rule is that the mains short circuit apparent power (S_{SC}) needs to be at least 20 times greater than the nominal apparent power of one dynamometer (S_N). Hence,

$$S_{SC}/S_N > 20.$$

In a sample calculation based on a single 220 kW dynamometer, the following sizing of the supply transformer may be found:

$P_N = 220\,\text{kW}$ nominal power, $\cos \varphi = 1$, overload 25%
$S_N = P_N \times \cos \varphi = 220\,\text{kVA}$
$S_{SC} = S_N \times 20 = 4400\,\text{kVA}$
Standard transformer (ST) $\sim U_{sc} = 6\%$

$ST = S_{SC} \times 0.06 = 264 \, kVA$ $\Rightarrow$ Transformer $= 315 \, kVA$
$ST = S_N \times 1.25 = 275 \, kVA$ $\Rightarrow$ Transformer $= 315 \, kVA$.

In the case of multiple dynamometer installation, some manufacturers in some conditions will allow a reduction in the S_{SC}/S_N ratio; this can reduce the cost of the primary supply.

Shown below is the basic calculation of power supply transformer in the case of $3 \times 220 \, kW$ dynamometers:

$P_N = 220 \, kW$ nominal power, $\cos \varphi = 1$, overload 25 per cent, diversity 0.7
$S_N = 3 \times P_N \times \cos \varphi = 660 \, kVA$
$S_{SC1} = S_N \times 20 = 13\,200 \, kVA$ in the case of no reduction for multiple machines
$S_{SC2} = S_N \times 10 = 6600 \, kVA$ in the case of ratio reduction to minimum
Standard transformer $\sim U_{sc} = 6\%$
$ST = S_{SC1} \times 0.06 = 792 \, kVA$ $\Rightarrow$ Transformer $= 800 \, kVA$
$ST = S_{SC2} \times 0.06 = 396 \, kVA$ $\Rightarrow$ Transformer $= 400 \, kVA$
$ST = S_N \times 1.25 \times 0.7 = 577 \, kVA$ $\Rightarrow$ Transformer $= 630 \, kVA$.

The ratio reduction and diversity (0.7) will reduce the transformer size from 800 kVA to 630 kVA.

Supply interconnection of disturbing and sensitive devices

An infinite variation of transformer and connection systems is possible, but Fig. 10.8 shows the range from poor (a) to a recommended layout (c).

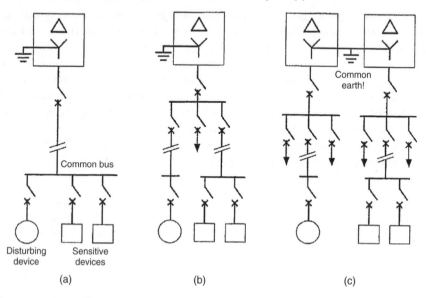

Figure 10.8 *Different power connection layouts ranging from poor (a) to recommended (c)*

The worst connection scheme is shown in Fig. 10.8a, where both the sensitive and a disturbing device, such as a PWM drive, are closely connected on a common local bus. Figure 10.8b shows an improved version of (a) where the common bus is local to the transformer rather than the devices. The best layout is based on Fig. 10.8c, where the devices are fed from separate transformers that share a common earth connection.

Power utility representatives drawn in projects involving a.c. dynamometer or chassis dynamometer installations may not at first appreciate that, for the majority of their working life, these machines are not operating as motors in the usual sense but are working as engine/vehicle power absorbers, or in electrical terms they are exporting, rather than importing, electrical power. It may be necessary to use the expertise of the equipment suppliers at the planning stage if upgraded power systems are being provided.

Electrical power supply specification

Most suppliers of major plant will include the power supply conditions their plant will accept; in the case of a.c. dynamometers this may require a dedicated isolating transformer to provide both isolation and the required supply voltage.

The EEC Directive (89/336/EEC) gives the acceptable limits of high frequency distortion on mains supply; the categories EN 55011 QP A2* and A2 are applicable to supplies for a.c. automotive dynamometers.

A typical mains power specification in the UK is shown in Table 10.3.

Fire stopping of cable penetrations in cell structure

Where cables or cable trunking break through the test cell walls, roof or floor they must pass through a physical 'fire block'. There are some designs of cable wall-box that allow the fire block to be disassembled and extra cables added; none of these are particularly easy to use after a year or more in service and, if used, it is advisable to build in a number of spare cables. Since most cells have a fire rating of at least one hour it is important not to compromise this by casual misuse, poor maintenance or inappropriate cable ways through the structure. A hole in the control room wall loosely stuffed with bags of intumescent material (fire barrier) is not adequate protection for cells fitted with gas-based fire suppression systems, particularly CO_2, because the pressure wave created by gas release will be capable of blowing out such stuffing unless it is correctly retained by a wall plate on either side.

Plastic 'soil pipes' cast into the floor are a convenient way of carrying signal cables between test cell and control room. These ducts should be laid to a fall in the direction of the cell and should have a raised lip to prevent drainage of liquid into them. Spare cables should be laid during installation. These pipes can be

Table 10.3   *Typical UK electrical supply specification*

Voltage	230/240 VAC ± 10%
Frequency	0.99 to 1.01 of nominal frequency 50 or 60 Hz continuously 0.98 to 1.02 of nominal frequency 50 or 60 Hz short time
Harmonics distortion	Harmonics distortion is not to exceed 10% of total r.m.s. voltage between the live conductors for the sum of the 2nd to 5th harmonic, or 12% max of total r.m.s. voltage between the live conductors for the sum of the 6th to 30th harmonic
Voltage interruption	Supply must not be interrupted or at zero voltage for more than 3 ms at any time in the supply cycle and there should be more than 1 second between successive interruptions
Voltage dips	Voltage dips must not exceed 20% of the peak voltage or the supply for more than one cycle and there should be more than 1 second between successive dips

'capped' by foam or filled with dried casting sand to create a noise, fire and vapour barrier.

Large power cables may enter a cell via concrete trenches cast under the wall and filled with dense dry sand below a floor plate. This method gives good sound insulation, a fire barrier and it is relatively easy to add cables at a later date.

Electrical cabinet ventilation

Instrument errors caused by heat are particularly difficult to trace, as the instrument will probably be calibrated when cold; so they should be eliminated at the design and installation phases. Many instrumentation packages produce quite appreciable quantities of heat and if mounted low down within the confined space of a standard 19-inch rack cabinet they may raise the temperature of apparatus mounted above them over the generally specified maximum of 40°C. Control and instrument cabinets should be well ventilated and it may be necessary to supplement individual ventilation fans by extraction fans high in the cabinet. Cabinet ventilation systems should have filtered intakes that are regularly changed or cleaned.

Special attention should be given to heat insulation and ventilation when instruments are carried on an overhead boom. Thermal 'tell-tale' strips installed in cabinets may be considered as a good maintenance or fault finding device.

European safety standards and CE marking

The European Community has, since 1985, been developing regulations to achieve technical harmonization and standards to permit free movement of goods within the community. There are currently four directives of particular interest to the builders and operators of engine test facilities (Table 10.4).

The use of the CE Mark (abbreviation for Conformité Europeen) implies that the manufacturer has complied with all directives issued by the EEC that are applicable to the product to which the mark is attached.

An engine test cell must be considered as the sum of many parts. Some of these parts will be items under test that may not meet the requirements of the relevant directives. Some parts will be standard electrical products that are able to carry their individual CE marks, while other equipment may range from unique electronic modules to assemblies of products from various manufacturers. The situation is further complicated by the way in which electronic devices may be interconnected.

If standard and tested looms join units belonging to a family of products, then the sum of the parts may comply with the relevant directive. If the interconnecting loom is unique to the particular plant, the sum of the CE-marked parts may not meet the strict requirements of the directive.

It is therefore not sensible for a specification for an engine test facility to include an unqualified global requirement that the facility be CE marked. Some products are specifically excluded from the regulations, while others are covered by their own rules; for example the directive 72/245/EEC covers radio interference from spark-ignition vehicle engines. Experimental and prototype engines may well fail to comply with this directive: an example of the impossibility of making any unqualified commitment to comply in all respects at all times with bureaucratic requirements drawn up by legislators unaware of our industry.

There are three levels of CE marking compliance that can be considered:

- All individual control and measuring instruments should individually comply with the relevant directives and bear a CE mark.
- 'Standardized' test bed configurations that consist exclusively of compliant instru-mentation, are configured in a documented configuration, installed to assembly

Table 10.4 *EC directives for CE marking relevant to engine test facilities*

Directive	Reference	Optional	Mandatory
Electromagnetic compatibility	89/336/EEC	1 January 1996	
Machinery	98/37EC	1 January 1993	1 January 1995
Low voltage	73/23/EWG plus 93/68/EEC	1 January 1995	1 January 1997
Pressure equipment		November 1999	May 2002

instructions/codes and have been subjected to a detailed and documented risk analysis may be CE marked. This requires a test bed equipment supplier to define and document such a package which would allow the whole 'cell' to be CE marked.

- Project specific test cells. As stated above, the CE marking of the complete hybrid cell at best will require a great deal of work in documentation and at worst may be impossible, particularly if required retrospectively for cells containing instrumentation of different generations and manufacturers.

The reader is advised to consult specific health and safety literature, or that produced by trade associations, if in doubt regarding the way in which these directives should be treated.

Note on 'as built' electrical documentation (drawing) standards

As stated at the start of this chapter, the electrical schematics of a test facility will be used, more than other documentation, by technical staff having to modify and maintain a test facility. It is therefore vital that such drawings reflect the 'as built' state of the facility at the time of acceptance, and that they are arranged in a logical manner. There are many national and company standards covering the layout of electrical schematics and the symbols used. Some are significantly different from others, which can cause problems for system integrators and maintenance staff. Documentation standards should be stated in the functional specification (Chapter 1).

Safety interaction matrix

A key document of any integrated system is some form of safety or alarm interaction matrix; it is the documentary proof that a comprehensive risk analysis has been carried out. To be most effective, the base document should be verified between the system integrator and user group since it is the latter that have to decide, within a framework of safety rules, upon the secondary reactions triggered in the facility by a primary event. The control logic of the building management system (BMS) has to be integrated with that of the test control system; if the contractual responsibly for the two systems is split then the task of producing an integrated safety matrix needs to be allocated and sponsored.

Summary

To enable a well-integrated system to be created, the electrical engineer will need to have both a complete operational specification and a mechanically completed functional specification. The final function and alarm matrix which defines how the many component parts and subsystems must interact will be created as part of the electrical design role.

It has been pointed out that electrical engineering regulation and practices around the world differ and while electrical engineers may use the same words the local practices may be significantly different, making integration of international projects fraught with traps for the unwary.

Notation

Ampere, unit of electric current	A
Bayonet cap	BC
Ambient temperature correction factor	C_a
European Committee for Electro technical Standardization	CENELEC
Cable grouping correction factor	C_g
Thermal insulation correction factor	C_i
Correction factor for the conductor operating temperature	C_t
Combined neutral and earth	CNE
Power factor (sinusoidal systems)	$\cos 0$
Circuit protective conductor	CPC
Cross-sectional area	c.s.a.
Overall cable diameter	D_e
Electrical Contractors Association	ECA
Earthed equipotential bonding and automatic disconnection	EEBAD
Earth leakage circuit breaker	ELCB
Extra-low voltage	ELV
Electromagnetic compatibility	EMC
Electromotive force	e.m.f.
Electromagnetic interference	EMI
Edison screw	ES
Frequency	f
Functional extra-low voltage	FELV
Guidance note	GN
High breaking capacity (fuse)	HBC
High rupturing capacity (fuse)	HRC
Hertz, unit of frequency	Hz
Symbol for electric current	I
Operating current (fuse or circuit breaker)	I_2
Current to operate protective device	I_a
Design current	I_b
Fault current	I_d

International Electrotechnical Commission	IEC
Institution of Electrical Engineers	IEE
Current setting of protective device	I_n
Tabulated current	I_t
Current carrying capacity	I_z
Earthing system	IT
Kilo, one thousand times	k
Kilovolt (1000 V)	kV
Lines of three-phase system	L1, L2, L3
Metre	m
Milli, one thousandth part of	m
Meg or mega, one million times	M
Milli-ampere	mA
Miniature circuit breaker	MCB
Moulded case circuit breaker	MCCB
Maximum demand	MD
Mineral-insulated	m.i.
National Inspection Council for Electrical Installation Contracting	NICEIC
Potential difference	p.d.
Protective extra-low voltage	PELV
Protective earth	PE
Combined protective earth and neutral	PEN
Protective multiple earthing	PME
Programmable logic controller	PLC
Prospective short-circuit current	PSC
Polyvinyl chloride	p.v.c.
Resistance (electrical)	R
The total resistance of the earth electrode and the protective conductor connecting it to exposed conductive parts	R_a
Resistance of the human body	R_p
Residual current circuit breaker	RCCB
Residual current device	RCD
Root-mean-square (effective value)	r.m.s.
Second, unit of time	s
Conductor cross-sectional area	S
Separated extra-low voltage	SELV
Time	t
Earthing system (L1, L2, L3, PEN)	TN-C
Earthing system (L1, L2, L3, N, PE)	TN-C-S
Earthing system	TN-S
Earthing system	TT

Symbol for voltage (alternative for V)	U
Alternating voltage	U_{ac}
Direct voltage	U_{dc}
Phase voltage	U_o
Volt, unit of e.m.f. or p.d.	V
Watt, unit of power	W
Reactance	X
Impedance (electrical)	Z
Earth loop impedance external to installation	Z_e
Earth fault loop impedance	Z_s
Phase angle	$^\circ$

Further reading

Electromagnetic Compatibility and Functional Safety. The Institution of Engineering and Technology, www.theiet.org/factfiles.

Goedbloed, J.J. (1992) *Electromagnetic Compatibility*, ISBN: 0-13-249293-8, Prentice Hall, New Jersey.

Middleton, J. *The Engineer's EMC Workbook*, ISBN: 0-9504941-3-5.

Williams, T. (1992) *EMC for Product Designers*, ISBN: 0-7506-1264-9, Newnes, Oxford.

11 Test cell control and data acquisition

Introduction

This chapter deals with the test cell as an integrated working system and the practical aspects of how that system is controlled, collects data and functions safely.

The forms of transducers used in engine test cells are described. Once again the theme of system integration is important within this chapter as the switching on, normal, abnormal operation and shutting down of each piece of apparatus has to be considered for its possible effect on, or interaction with, any other apparatus and as part of the whole facility. The subject of data handling and post processing is covered in Chapter 19.

Safety systems

Emergency stop function

The emergency stop (EM stop) function is a hard-wired system, normally quite separate from the alarm monitoring system embedded into the cell control system. In general practice it is a shutdown system initiated by humans rather then computers, although electrical links are often made to safety critical systems such as the fire alarm system.

Most emergency stop systems may be visualized as a chain of switches in series all of which need to be closed before engine ignition or some other significant process can be initiated. When the chain is opened, normally by use of manually operated press buttons, the process is shut down and the operated button is latched open until manually reset. This feature allows staff working on any part of the test cell complex to use the EM stop as a self-protection device by pressing an EM stop button and hanging a 'Do not activate notice' bearing name and time on it.

The traditional function of the emergency stop circuit is to shut off electrical power from all devices directly associated with the running of the engine. In this model, dynamometers are usually switched to a 'de-energized' condition, fuel supply isolation systems are activated, ventilation fans are shut down, but services shared with other cells are not interrupted and fire dampers are not closed.

Modern practice for cells fitted with electrical dynamometers is for the emergency stop function to shut down the engine and bring the shaft rotation to a stop via a fast regenerative stop before cutting off power to the dynamometer. BS EN 60204 recognizes the following categories:

- *Uncontrolled stop (removal of power): Category 0.* This is commonly used in some engine cells fitted with manual control systems and two quadrant dynamometers.
- *Controlled shutdown (fast stop with power removed at rotation stop or after timed period): Category 1.* This is the normal procedure for engine cells fitted with four quadrant dynamometers where, under some circumstances, shutting down the engine may not stop rotation.
- *Controlled stop (fast stop with power retained at rotation stop): Category 2.* Normally operated as a 'fast stop' rather than an EM stop in test cell practice (see below).

The control desk may be fitted with the following 'emergency action' buttons:

1. large red emergency stop button (latched) shutting down engine and cell systems;
2. fast stop button (hinged cover to prevent accidental operation) operating only on engine and dynamometer systems to stop rotation;
3. fire suppression release button (wall-mounted and possibly break-glass protected).

It should be clear to all readers that operational training is required to be able to distinguish between situations requiring use of emergency control devices.

It is important to use the *safety interaction action matrix,* introduced in Chapter 10, to record what equipment will be switched to what state in the event of any of the many alarm or shutdown states being monitored.

Everything possible should be done to avoid spurious or casual operation of the EM stop system or any other system that initiates cutting off power to cell systems. Careful consideration has to be given before allowing the shutting down of plant that requires long stabilization times, such as humidity control for combustion air or oil temperature control heaters. It is also important not to exceed the numbers of starts per hour of motors driving large fans, etc., that would cause overheating of windings and the operation of thermal protection devices.

Operational and safety instrumentation

This term covers the basic instruments and controls necessary for the safe running of the engine and cell equipment; unlike the EM stop the shutdown parameters are monitored by the control system. Operational instrumentation safety systems can be either software operated, or hard wired,* with duplicate signals going to any associated computer, and includes devices associated with the cell shutdown procedure, such as fire dampers. The following indications would normally be included under this heading:

- engine speed (shaft line speed too high);
- oil pressure (after start-up, pressure too low);
- facility cooling water flow and compressed air pressure;
- signals associated with fire suppression system being available;
- status of cell ventilation, purge system and fire dampers;
- overspeed trips on both engine and dynamometer to deal with shaft failures;
- cell door and shaft guard interlocks;
- in-cell beacon/alarm automatically triggered by starting sequence.

Computerized monitoring of alarm signals

Primary alarms should refer to limits which, if exceeded, would endanger the facility personnel and/or equipment. These are additional to alarms that are specific to the engine (which are often lower than the absolute limits of the facility or instrumentation).

During a test, the condition of both engine and facility may be monitored continuously by the test cell control computer so that the appropriate preprogrammed action may be triggered should one or more parameters fall outside a preset limit.

Alarms are of two kinds: those safeguarding the equipment and those safeguarding the integrity of the test. The first type will unconditionally cause protective actions to be taken.

Most engine test computer software make available four levels of alarm for each data channel:

1. High level, warning
2. High level, test shutdown
3. Low level, warning
4. Low level, test shutdown.

Such comprehensive warning arrangements are likely to be necessary on only a few channels; in most cases they will either be 'switched off' or assigned values outside the operating range. Some software suites allow stage-specific alarms to be set; this can be useful when, for example, particularly close control of fuel temperature is required during a consumption measurement, the warning limits can be closed only for the critical period.

---

* The term 'hard wired', as used in this book, means that the instrument or other device receives and sends its status signal via its own unique cable rather than by way of a computer (software-dependent) signal or a multiplexing system. Increasing use of PLC-based building management systems is making the concept of hardwiring of service status signals more difficult.

Most engine test software contains a rolling buffer that allows examination of the values of all monitored channels for a short time before any alarm shutdown. Known as 'historic' or 'dying seconds' logging, this buffer will also record the order in which the alarm channels have triggered, information of help in tracing the location of the prime cause of a failure.

Security of supply

The possible consequences of a failure of the electricity supply, or of such factors as undetected fuse failures, must be carefully thought out. Particular dangers can arise if the power is restored unexpectedly after a mains supply failure.

It is usual to design safety systems as normally de-energized (energize for normal operation), but all aspects of cell operation must be considered. In many cases the cell control logic should require that equipment, such as solenoid-operated fuel valves at the cell entry, be reset after mains failure via an orderly, operator initiated, restart procedure.

The installation of an uninterrupted power supply (UPS) for critical control and data acquisition devices is to be recommended. This relatively inexpensive piece of equipment need only provide power for the few minutes required to save or transfer computer data and bring instrumentation to a safe state.

Building management systems (BMS) and services status displays

Test cell function is usually totally dependent on the operation of a number of building and site services, remote and hidden from the operator. A failure of one of these services will directly or indirectly (through alarm state caused by secondary effects) result in a cell shutdown. It is strongly recommended that indication of the status of each building service is available to the operator in the cell or, in the case of multicell installations, at some common point. The common and time-wasting alternative is for operators to tour the facility in order to check the status of individual pieces of plant at start-up, a fault condition and shutdown. A centralized control and indication panel is of great value; however, such system status displays must indicate the actual rather than the requested status.*

A modern PLC-controlled BMS can have an interactive mimic display showing not only the running state of services but also the status of fire dampers, valves and the temperature and pressure of fluids. These PLC systems may have the services

* The Three Mile Island nuclear power station accident, initiated by a failure of the main feed water pumps, was greatly exacerbated by incorrect actions taken by operations staff whose only indications showed a critical valve as being closed, but in fact this only represented that the signal to close the valve was sent, in fact it was open. A 'positive feedback' lamp in the control room indicating the true position of the valve had been eliminated in the original construction to save time and money.

system start, stop and alarm state logic programmed into them but the cell designers must carefully consider whether each piece of plant's primary control should come from the cell controller or the BMS; in the case of compressed air supply the choice is clear but the case of fuel supply to the cell it may not be.

Test cell start-up and shutdown procedures

Failure of any test cell service, or failure to bring the services on stream in the correct order, can have serious consequences and it is essential to devise a logical system for start-up and shutdown, with all the operator procedures, the necessary alarms and automatic shutdowns. In all but highly automated production tests, a visual inspection of the unit under test must be made before each and every test run. Long-established test procedures can give rise to complacency and lead to accidents.

Checks before start-up

It will be assumed that routine calibration procedures have been followed, and that instrument calibrations are correct and valid; then the following type of checks should be made or status confirmed:

- No 'work in progress' or maintenance labels are attached to any system switches.
- Engine/dynamometer alignment within set limits and each shaft bolt has been tightened to correct and recorded torque.
- Shaft-guard is in place and centred so that no contact with shaft is possible (if appropriate, it is a good practice to rock the engine on its mounts to see that the rigging system, including exhaust tubing, is secure and flexes correctly).
- All loose tools, bolts, etc. are removed from the test bed.
- Engine support system is tightened down.
- Fuel system is connected and leak proof.
- Engine oil is at the correct level and oil pressure alarm hard wired into control system.
- All fluid services such as dynamometer water are on.
- Fire system is primed.
- Ventilation system is available and switched on to purge cell of flammable vapour prior to start-up of engine.
- Check that cell access doors remote from the control desk have warnings set to deter casual entry during test.

Engine start process is initiated.

Checks immediately after start-up

- Oil pressure is above trip setting. In case of manually operated cells the alarm over-ride is released when pressure is achieved.
- If stable idle is possible and access around the engine is safe, a quick in-cell inspection should be made. Particularly look for engine fuel or oil leaks, cables or pipes chafing or being blown against the exhaust system and listen for abnormal noises.
- In the case of a run scheduled for prolonged unmanned running, test the emergency shutdown system by operating any one of the stations.
- Restart and run test.

Checks immediately after shutdown

- Allow cooling period with services and ventilation left on.
- Shut off fuel system as dictated by site regulations (draining down of day-tanks, etc.).
- Carry out data saving, transmission or back-up procedures.

The test sequence: modes of control

It should be recognized that a test programme for an engine coupled to a dynamometer is, first and foremost, a sequence of desired values of engine torque and speed.

This sequence is achieved by manipulating only two controls: the engine speed ('throttle') and the dynamometer torque setting. (It is convenient to refer to the throttle, even when it is an electric signal direct to the engine control unit or, in the case of a diesel engine actuation of the fuel pump rack setting.)

The dynamometer torque is set according to the design of dynamometer used (see Chapter 8 for details).

For any given setting of the throttle, the engine has its own inherent torque-speed characteristic and similarly each dynamometer has its own torque-speed curve for a given control setting. The interaction of these two characteristics determines the inherent stability of the engine dynamometer combination.

Note on mode nomenclature

In this book, the naming convention of modes of control used in the UK and widely throughout the English-speaking world has been adopted, the alternative nomenclature and word order used in some major German-speaking organizations is shown in Table 11.1.

Table 11.1 *Control modes nomenclatures*

English (first term refers to engine)	German (first term may refer to dynamometer)
Speed	n
Position	Alpha
Torque	Torque

The engine or throttle control may be manipulated in three different ways:

- to maintain a constant throttle opening (position mode);
- to maintain a constant speed (speed mode);
- to maintain a constant torque (torque mode).

The dynamometer control may be manipulated:

- to maintain a constant control setting (position mode);
- to maintain a constant speed (speed mode);
- to maintain a constant torque (torque mode);
- to reproduce a particular torque–speed characteristic (power law mode).

Various combined modes, usually described in terms of engine mode/dynamometer mode, are possible.

Position/position mode

This describes the classic manually operated engine test. The throttle is set to a fixed position, the dynamometer control similarly set, and the system settles down, hopefully in a stable state. There is no feedback: this is an open loop system. Figure 11.1a shows a typical combination in which the engine has a fairly flat torque–speed characteristic at fixed throttle opening, while the dynamometer torque rises rapidly with speed, typical of most but not all water-brake machines. The two characteristics meet at an angle approaching 90°, and operation is quite stable.

Certain designs of variable fill dynamometer with simple outlet valve control, see Chapter 8, may become unstable at light loads, this can lead to hunting, or to the engine running away. The two characteristics meet at an acute angle, Fig. 11.1b. A friction brake, Fig. 11.1c operating at a given control setting, develops a torque almost independent of speed and is clearly unsuitable for loading an engine in position mode.

Position/position mode is useful in system fault finding in that the inherent stability of the dynamometer, independent of control system influence may be checked.

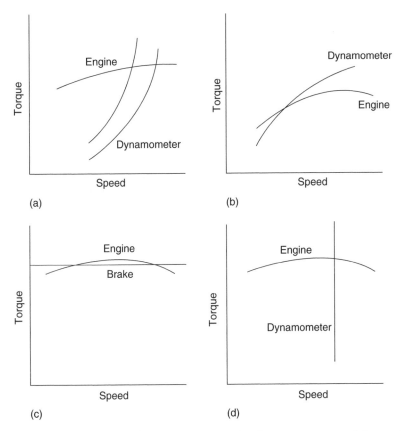

Figure 11.1 *Control modes, engines and dynamometers: (a) position mode stable hydraulic dynamometer; (b) position mode unstable hydraulic dynamometer; (c) position mode, friction brake; (d) position/speed mode; (e) position/torque, governed engine; (f) speed/torque mode; (g) torque/speed mode*

Position and power law mode

This is a variation on position mode, in which the dynamometer controller is manipulated to give a torque–speed characteristic of the form

$$\text{Brake torque} = \text{constant} \times \text{speed}^n$$

When $n = 2$, this approximates to the torque characteristic of a marine propeller and the mode is thus useful when testing marine engines. It is also a safe mode, tending to prevent the engine from running away if the throttle is opened by an operator in the cell.

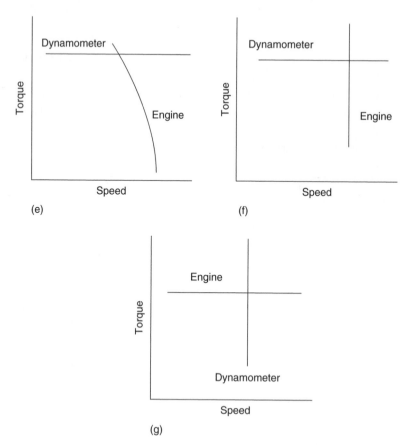

Figure 11.1   *(Cont.)*

Position and speed mode

In this mode, the throttle position continues to be set manually, but the dynamometer is equipped with an automatic controller which adjusts the torque absorbed by the machine to maintain the engine speed constant whatever the throttle position and power output, Fig. 11.1d. This is a very stable mode, and is generally used for plotting engine torque–speed curves at full and part throttle opening.

Position and torque mode (governed engines)

Governed engines have a built-in torque–speed characteristic, usually slightly 'drooping' (speed falling as torque increases). They are therefore not suited for coupling to a dynamometer in speed mode. They can, however, be run with a dynamometer in torque mode. In this mode the automatic controller on the dynamometer adjusts the torque absorbed by the machine to a desired value, Fig. 11.1e. Control is quite stable. Care must be taken not to set the dynamometer controller to a torque that may stall the engine.

Speed and torque mode

This is a useful mode for running in a new engine, when it is essential not to apply too much load. As the internal friction of the engine decreases, it tends to develop more power and since the torque is held constant by the dynamometer the tendency is for the speed to increase. This is sensed by the engine speed controller, which acts to close the throttle, Fig. 11.1f.

Torque and speed mode

This can be useful as an approximate simulation of the performance of a vehicle engine when climbing a hill. The dynamometer holds the speed constant while the engine controller progressively opens the throttle to increase torque at the set speed, Fig. 11.1g.

Precautions in modes of the four-quadrant dynamometer

Where the dynamometer is also capable of generating torque, it will be necessary to take precautions in applying some of the above modes of control. This applies when the dynamometer is in speed mode, Figs 11.1d, g. When running in either of these modes, the four-quadrant dynamometer may be controlled by logic that will maintain the set speed even if the engine is being 'motored'. This could clearly be dangerous should the engine have failed in some way and has ceased to deliver power.

Throttle actuation

Except in the cases when an engine is controlled directly with a marine type of cable system and lever at the control station, engines fitted with a speed control lever require the test cell to be equipped with some form of actuation that can be remotely and precisely operated, under manual or automatic control from the control room.

The actuators may be based on rotary servomotors, printed circuit motors or linear actuators, in all cases fitted with some form of positional feedback to allow closed loop control. Most actuators will be fitted with some form of stroke adjustment to allow for 0 (shut)–100 per cent (WOT) positions to be determined for different engines and linkage configurations and some form of force limiter device. Typical specification ranges of proprietary devices are:

Constant force applied at Bowden cable connection	100–400 N
Maximum stroke length	100–160 mm
Cable connector speed of travel	0.5–1.5 m/s
Positional repeat accuracy	±0.05 mm

It must be remembered that the forces applied by the actuator to the engine lever will depend on the mechanical linkage within the engine rigging.

It should be clear from the specification of positional repeat accuracy (above) that cable and linkage used must be as free from backlash as is possible if good closed loop control is to be achieved. An important safety feature of a throttle actuator is that it should automatically move the engine control lever to the 'stop' position in the event of a power failure.

Many modern engines do not have a speed control lever, but instead require an electrical signal proportional to speed requested (fly-by-wire). In the test cell this can be achieved using one of the following strategies:

1. As part of the engine rigging where part or all of a vehicle accelerator pedal mechanism is connected via cable or lever to a throttle actuator.
2. The electronic signal required is supplied via the test cell control computer as a discrete 'analogue out' signal.

Problems in achieving control

An engine test system is an assembly of an engine, a dynamometer and various actuators and peripherals. Each of these has its own control characteristics and, in many cases, its own controller. These controllers have not been designed as a group and it is not surprising that the combined control characteristics are often very far from ideal. The problem is compounded when the entire system is subjected to overall computerized control. At the simplest level, a dynamometer with a control loop intended to produce a particular torque–speed characteristic can generate instabilities quite absent in an old-fashioned manually controlled (brake) dynamometer. This gives a clue to investigating control instabilities: eliminate as much of the control system as possible by switching to open-loop control.

It will also be found helpful, if two controllers are 'fighting' each other, to ensure that they have widely differing time constants. In the common case of a speed-controlled engine coupled to a torque-controlled dynamometer, it is preferable that the latter should have a shorter response time than the former.

Some engines are inherently difficult to control because of the shape of their torque–speed characteristics. Turbocharged engines may have abrupt changes in the slope of the power–speed curve, plus sluggish response due to the time taken for speed changes in the turbocharger, which can make the optimization of the control system very difficult.

Inexperienced attempts to adjust a full three-term PID (proportional/integral/differential) controller can lead to problems and it is well to record the settings before starting to make adjustments so that one can at least return to the starting point. It is good practice to make sure that elementary sources of trouble, such as backlash and 'stick-slip' in control linkages, are eliminated before delving into control variables in software.

Choosing test control software

The world's engine test industry is served by a relatively small number of companies whose flagship product is a comprehensive suite of software designed specifically for controlling engine tests, acquiring test data and displaying results.

Several of these companies also make the major modules of instrumentation required for testing engines, such as dynamometers, engine fluid temperature controllers, fuel meters and exhaust emission equipment. However, many facilities are made up of modules of plant coming from several of these specialist suppliers in addition to locally made building services.

Such major software suites are the outcome of many man-years of work; it is not recommended that even a highly specialized user, however experienced, should undertake the task of producing their own test control software from scratch.

Prospective buyers of automated test systems should if possible inspect competitive systems at the sites of other users doing similar work. As with any complex system, the following general questions should be considered in no order of importance:

- How intuitive and supportive to the work does the software appear to be; or does the software provider understand the world in which the software has to function?
- Purchase cost is clearly important, but also consider the cost of ownership including technical support, upgrades and additional licences.
- Can the post-processing and reporting required be carried out directly by the new software or can data be transferred to other packages used on site?
- How good is the training and service support in the user's geographical location?
- In the case of international companies, does the software support multilanguage display of screen displays?
- Has the system been successfully used in the user's industrial sector?
- Can existing instrumentation or new third party instrumentation be used with the system without performance degradation?
- Is data security robust and can the post-processing, display and archiving required be accommodated?
- Can the data generated by the new system be stored in the existing data format?

Test cell computer roles and connected devices

This section is based generally on the assumption that each cell will have its individual PC in a hierarchical position above any of the devices described below. The subject of post-test data processing is covered in Chapter 19.

Editing and control of complex engine test sequences

The typical modern engine test programme consists of a long succession of stages of speed and torque settings together with the demand values sent to other devices within the test cell system. Together these form a complete test profile or 'sequence'. It is beyond the capacity of a human operator to run such test sequences accurately and repeatedly.

Although the terminology used by test equipment suppliers may vary, the principles involved in building up test profiles will be based on the same essential component instructions.

Test sequence 'editors' range from those that require the operator to have knowledge of software code to those that present the operator with an interactive 'form fill' screen. The nature and content of the questions shown on the screen will depend on the underlying logic which determines how the answers are interpreted. It is important that the user should understand the interaction logic of the test cell logic, otherwise there is a risk of calling up combinations of control that the particular system cannot run, though it may not be able to indicate where the error lies. For example, if the control mode is set at throttle: position and dynamometer: speed, the editor will 'expect' instructions in terms of these parameters; it will not in this case be able to accept a throttle control instruction in terms of torque when it 'needs' percentage open.

Each sequence will be made up of a series of stages, each either steady state, engine speed and torque constant, or transient, covering a move from one setting of speed and torque to another within a specified time or 'ramp rate'. Of course speed and torque may also remain constant and other individually controllable parameters may be varied.

Test sequence elements

In most cases, the test sequence editors embodied in a suite of software designed for the control of engine test beds are based on a simple 'form fill' layout. The following elements are involved:

- *Mode of control.* Some older analogue-based control systems require engine speed and load to be brought down to some minimum value between each stage, but modern digital controllers should be able to make a 'bumpless transfer' between modes of control.

- *Engine speed and torque or throttle position.* The way in which these parameters are set will depend on the control mode. A good sequence editor will present only viable options.
- *Ramp rate.* This is the acceleration required or the time specified for transition from one state to the next.
- *Duration or 'end condition' of stage.* The duration of the stage may be defined in several ways:
  - at a fixed time after the beginning of the stage;
  - on a chosen parameter reaching a specified value;
  - on reaching a specified logic condition, e.g. on completion of a fuel consumption measurement.
- *Choice of next stage.* At the completion of each stage (stage x), the editor will 'choose' the next stage. Typical instructions governing the choice are:
  - run stage $x + 1$;
  - rerun stage x a total of y times (possible combinations of 'looped' stages may be quite complex and include 'nested' loops);
  - choose next stage on basis of a particular analogue or digital state being registered (conditional stage).
- *Events to take place during a stage.* Examples are the triggering of ancillary events, such as fuel consumption measurement or smoke density measurement. These may be programmed in the same way as duration or end condition, above.
- *Nominated alarm table.* This is a set of alarm channels that are activated during the stage in which they are 'called'. The software and wired logic must prevent such programming from overriding 'global' safety alarms.

Data acquisition and the transducer chain

In this section, we consider data collection from single transducers sensing pressure or temperature; torque and speed are covered in Chapter 8 and data from complex instruments is covered later in this chapter.

The information defining the state of any measured parameter travels via cable from the transducer, through signal conditioning, to a storage device and one or more displays. During this journey it is subject to a number of possible sources of error and corruption covered in Chapter 10.

A well-integrated system will have every detail of the chain specified to suit the requirements and inherent characteristics of the parameter being measured.

Calibration of the signal chain

All professional engine test software suites will contain calibration routines for all common transducers and instruments. Such routines will always allow fixing of zero and full span points and compensation for non-linearity of output.

For the most demanding calibration the complete chain from transducer to final data store and indication needs to be included, but for normal work it is neither practical nor cost/time effective. Common practice is to use certified transducers and an electrical signal from a calibrated instrument which is injected into the measuring system at the point at which the transducer is plugged in. Several measuring points throughout the transducer's range are simulated and checked against the display.

Transducer boxes and distributed I/O

In order to measure the various temperatures and pressures of the UUT (engine or vehicle), transducer probes have to be attached and their signal cables connected into the data acquisition system. In many cases the transducer probes will be intrusive and fitted via test points sealed with compression fittings. In addition to the single probe cables, there will be cables from other instruments such as optical encoders requiring connection within the test cell in order for the signals, in raw form, to be transmitted to the signal conditioning device.

Modern practice is to reduce the vulnerability of signal corruption by conditioning the signal as near as possible to the transducer and to transmit them to the computer via a parallel SCSI or IEEE 1394 high speed serial bus; such implementations, generally known as distributed I/O, can greatly reduce complexity of installation.

Older installations will have discrete cables in a loom from the transducer box, through the cell wall to a connection strip in the control cabinet near the control desk. This network of short transducer cables needs to be marshalled and collected at a transducer box near the engine.

Transducer boxes will contain some or all of the following features:

- external, numbered or labelled, thermocouple, PRT or pressure line sockets;
- external labelled sockets for dedicated channels or instruments;
- power supply sockets for engine mounted devices, such as an ECU.

Transducer boxes may have upwards of 50 cables or pipes connected to them from all points of the engine under test; this needs good housekeeping by clipping in sublooms to prevent an unsightly, and difficult to use, tangle.

The transducer box can be positioned:

1. On a swinging boom positioned above the engine, which gives short cable lengths and an easy path for cables between it and the control room. However, unless positioned with care it will be vulnerable to local overheating from radiation or convection from the engine; in these cases, forced ventilation of the enclosed boom duct and transducer box should be provided.
2. On a pedestal alongside the engine but this may give problems in running cable back to the control room.
3. On the test cell wall with transducer cables taken from the engine via a lightweight boom.

The choice of instruments and transducers

It is not the intention to attempt a critical study of the vast range of instrumentation that may be of service in engine testing. The purpose is rather to draw the attention of the reader to the range of choices available and to set down some of the factors that should be taken into account in making a choice.

Table 11.2 lists the various measurements. Within each category the methods are listed in approximately ascending order of cost.

Note that:

- Many of the simpler instruments, e.g. spring balances, manometers, Bourdon gauges, liquid-in-glass thermometers, cannot be integrated with data logging systems. This may or may not be important.
- Accuracy costs money. Most transducer manufacturers supply instruments to several levels of accuracy with steeply increasing cost. In a well-integrated system a common level is required; over- or underspecification of individual parts compared with the target standard either wastes money or compromises the overall performance.
- It is cheaper to buy a stock item rather than to specify special features.
- Some transducers, particularly force and pressure, can be destroyed by overload. Ensure adequate capacity or overload protection.
- Always read the maker's catalogue with care, taking particular note of accuracies, overload capacity and fatigue life.

Time intervals and speed

Time can be measured to accuracy greater than is necessary in most engine testing, but the precise location of events in time is very much more difficult.[1] When linked events that occur very close together in time are sensed in different ways, with different instruments, it can be very difficult to establish the exact order or to detect simultaneity.

In general, a tachometer should only be used for speed measurement when the reading is not to be used as one component in calculating another quantity such as power. It is better practice to count the revolutions.

A single impulse trigger, such as a hole in the rim of the flywheel, is satisfactory for counting purposes but not ideal for locating top dead centre (see Chapter 14 for detailed coverage of t.d.c. measurement). Multiple impulses picked up from a starter ring gear or flywheel drillings used by the engine's ECU may be acceptable. Many dynamometers are fitted with a 60-tooth wheel and inductive pick-up as standard for use by their own control system.

For precise indications of instantaneous speed and speed crank angle down to tenths of a degree an optical encoder is necessary. Mounting arrangements of an encoder on the engine may be difficult and if, as is usually the case, drive is taken from the non-flywheel end of the crankshaft torsional effects in the crankshaft may

Table 11.2 *Common instrumentation and transducers for frequently required measurements*

Measurement	Principal applications	Method
Time interval	Rotational speed	Tachometer
		Single impulse trigger
		Starter ring gear
		Shaft encoder
Force, quasistatic	Dynamometer torque	Dead weights and spring balance
		Hydraulic load cell
		Strain gauge transducer
Force, cyclic	Stress and bearing load investigations	Strain gauge transducer
		Piezoelectric transducer
Pressure, quasistatic	Flow systems; lubricant, fuel, water, pressure charge, exhaust	Liquid manometer
		Bourdon gauge
		Strain gauge transducer
Pressure, cyclic	In-cylinder, inlet, exhaust events. Fuel injection	Strain gauge transducer
		Capacitative transducer
		Piezoelectric transducer
Position	Throttle and other controls	Mechanical linkage and pointer, counter
		LVDT transducer
		Shaft encoder
		Stepper motor
Displacement, cyclic	Valve lift, injection needle lift	Inductive transducer
		Hall effect transducer
		Capacitative transducer
Acceleration	Engine balancing, NVH	Strain gauge accelerometer
		Piezoelectric accelerometer
Temperature	Cooling water, lubricant, inlet air, exhaust, in-cylinder, mechanical components	Liquid-in-glass
		Vapour pressure
		Liquid-in-steel
		Thermocouple
		PRT
		Thermistor
		Electrical resistance
		Optical pyrometer
		Suction pyrometer

lead to errors which may differ from cylinder to cylinder (see Chapter 14 for a discussion of the problems involved).

Force, quasistatic

The strain gauge transducer has become the almost universal method of measuring forces and pressures. The technology is very familiar and there are many sources of supply. There are advantages in having the associated amplifier integral with the transducer as the transmitted signals are less liable to corruption, but the operating temperature may be limited.

The combination of dead weights and spring balance is still quite satisfactory for many purposes.

Hydraulic load cells are simple and robust devices still fairly widely available, particularly used in portable dynamometers.

Force, cyclic

The strain gauge transducer has a limited fatigue life and this renders it unsuitable for the measurement of forces that have a high degree of cyclic variation, though in general there will be no difficulty in coping with moderate variations such as result from torsional oscillations in drive systems.

Piezoelectric transducers are immune from fatigue effects but suffer from the limitation that the piezoelectric crystal that forms the sensing element produces an electrical charge that is proportional to pressure change and which thus requires integration by a charge amplifier. By definition, the transducer is unsuitable for steady-state measurements but is effectively the universal choice for in-cylinder and fuel injection pressure measurements. Piezoelectric transducers and signal conditioning instrumentation are in general more expensive than the corresponding strain gauge devices.

Measurement of pressure

To avoid ambiguity, it is important to specify the mode of measurement when referring to a pressure value. The three modes are (Fig. 11.2):

- absolute;
- relative;
- differential.

Pressures measured on a scale which uses a complete vacuum (zero pressure as in a theoretical vessel containing no molecules) as the reference point are said to be *absolute* pressures.

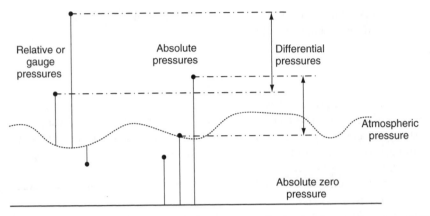

Figure 11.2 *Diagram showing relationship between absolute, relative and differential pressures*

Atmospheric pressure* at the surface of the earth varies, but is approximately 10^5 Pa (1000 mbar). This is an *absolute* pressure because it is expressed with respect to zero pressure. Most engine test instrumentation is designed to measure pressure values that are expressed with respect to atmospheric pressure, thus they indicate zero when their measurement is recording atmospheric pressure. In common parlance, relative pressures are referred to as *gauge* measurements. Because of the difference between an absolute pressure value and a gauge pressure, the variable value of atmospheric pressure engine test facilities needs to have a barometric measurement recorded as a reference in all tests.

Engine inlet manifold pressure will vary between pressures above and (generally) *below* the 'reference' barometric pressure. This is another form of relative pressure sometimes called a 'negative gauge pressure'.

In applications such as the measurement of flow through an orifice, where it is necessary to measure the difference in pressure between two places, the reference pressure may not necessarily be either zero or atmospheric pressure but one of the measured values; such measurements are of *differential* pressures.

Electronic pressure transducers

There are large ranges of strain gauge and piezoelectric pressure transducers made for use in engine testing, most having the following characteristics:

---

* For reference concerning the range of barometric pressures, the highest recorded atmospheric pressure, 108.6 kPa (1086 mbar) occurred in Mongolia, 19 December 2001. The lowest recorded atmospheric pressure, 87.0 kPa (870 mbar), occurred in the Western Pacific during Typhoon Tip on 12 October 1979.

- Pressure transducers normally consist of a metal cylinder made of a stainless steel, with a threaded connection at the sensing end and a signal cable attached at the other.
- The transducer body will contain some form of device able to convert the pressure sensed at the device inlet into an electrical output. The signal may be analogue millivolt, 4–20 mAmp or a digital output via a CAN bus serial communications interface.
- The output cable and connective circuitry will depend on the transducer type which may require 4, 3 or 2 wire connection one of which will be a stabilized electrical supply.
- Each transducer will have an operating range that has to be carefully chosen for the individual pressure channel, they will often be destroyed by overpressurization and be ineffective at measuring pressures below that range.
- Each transducer will be capable of dealing with specific pressurized media; therefore, use with fuels or special gases must be checked at the time of specification.

Capacitative transducers, in which the deflection of a diaphragm under pressure is sensed as change in capacity of an electrical condenser, may be the appropriate choice for low pressures, where a strain gauge or piezoelectric sensor may not be sufficiently sensitive.

Calibration of the complete pressure channel may be achieved by use of a certified, portable calibrator while some special pressure transducers such as barometric sensors and differential pressure transducers may need annual off-site certification.

Pressure sensors used in the engine testing environment, because of their expense and vulnerability to damage, are normally mounted within the transducer box with flexible tubes connecting them via self-sealing couplings to their measuring point.

Other methods of pressure measurement

The liquid manometer has a good deal to recommend it as a device for indicating low pressures. It is cheap, effectively self-calibrating and can give a good indication of the degree of unsteadiness arising from such factors as turbulence in a gas flow. It is recommended that manometers are mounted in the test cell with the indicating column fixed against the side of the control room window. Precautions regarding the use of mercury have reduced its common use in engine testing from common to rare.

The traditional Bourdon gauge with analogue indicator is the automatic choice for many fluid service sensing points, such as compressed air receivers.

Displacement

Inductive transducers are used for 'non-contact' situations, in which displacement is measured as a function of the variable impedance between a sensor coil and a moving conductive target.

The Hall effect transducer is another non-contact device in which the movement of a permanent magnet gives rise to an induced voltage in a sensor made from gold foil. It has the advantage of very small size and is used particularly as an injector needle lift indicator.

Capacitative transducers are useful for measuring liquid levels, very small clearances or changes due to wear.

Acceleration/vibration

There are two basic types of accelerometer, based respectively on strain gauge and piezoelectric sensors. Piezoelectric sensors should not be used for frequencies of less than about 3 Hz, but tend to be more robust than strain gauge units.

Temperature measurement – thermocouples

Most of the temperatures measured during engine testing do not vary significantly at a high rate of change, nor is the highest degree of accuracy cost effective. The types of thermocouple commonly used in engine testing are shown in Table 11.3.

For the majority of temperature channels, the most commonly used transducer is the type K thermocouple, having a stainless steel grounded probe and fitted with its own length of special thermocouple cable and standard plug. To produce a calibrated temperature reading, all thermocouples require to be fitted within a system having a 'cold junction' of known temperature with which there own output is compared. Thermocouples of any type may be purchased with a variety of probe and cable specifications depending on the nature of their installation; they should be considered as consumable items and spares kept in stock.

Temperature measurement – PRTs

When accuracy and long-term consistency in temperature measurement higher than that of thermocouples, or for measuring temperatures below 0°C, is required, then platinum resistance thermometers (PRT) are to be recommended.

The PRT works on the principle of resistance through a fine platinum wire as a function of temperature. PRTs are used over the temperature range −200°C to 750 °C.

Table 11.3 *Commonly used thermocouple types*

Type	Internal materials	Temperature range
Type T	Cu-Ni/Cu	0°C to +350°C
Type J	Fe-Ni/Cu	0°C to +800°C
Type K	Ni/Cr-Ni/Al/Mn	−40°C to +1200°C

Because of their greater durability or accuracy over time, it is common practice to use PRTs in preference to thermocouples within device control circuits. They require different signal conditioning from thermocouples and, for engine rigged channels, are fitted with dedicated plugs and sockets within the transducer box.

Thermistors

Thermistors (thermally sensitive resistors) have the characteristic that, dependent on the nature of the sensing 'bead', they exhibit a large change in resistance, either increasing or decreasing, over a narrow range of temperature increase. Because they are small and can be incorporated within motor windings, they are particularly useful as safety devices. Thermistors are generally available in bead or surface mounting form and in the range from −50°C to 200°C.

Pyrometers

Optical or radiation pyrometers, of which various types exist, are non-contact temperature measuring devices, used for such purposes as flame temperature measurement and for specialized research purposes, such as the measurement of piston ring surface temperatures by sighting through a hole in the cylinder wall. They are effectively the only means of measuring very high temperatures.

Finally, suction pyrometers, which usually incorporate a thermocouple as temperature-sensing device, are the most accurate available means of measuring exhaust gas temperatures,

Other temperature-sensing devices

Once a standard tool in engine testing, liquid-in-glass thermometers are now very rarely seen; however, they are cheap, simple and easily portable. On the other hand, they are fragile, not easy to read and have a relatively large heat capacity and a slow response rate. They retain some practical use during services commissioning and sensing of plant in remote locations. The interface between the body, solid, liquid or gas, of which the temperature is to be measured and the thermometer bulb needs careful consideration.

Vapour pressure and liquid-in-steel thermometers present the same interface problems are not as accurate as high-grade liquid-in-glass instruments, but are more suitable for the test cell environment and are more easily read.

Mobile, hand-held thermal imaging devices are more useful for fault finding and safety work than for collection of test data other than during installed on-site work with large stationary engines.

Like all other aspects of temperature measurement, there is much guidance available in the literature and on the internet.[2]

Smart instrumentation and devices

The test cell control and data acquisition computer will, in addition to taking in individual transducer signals, also have to switch on and off and acquire data from complex modules of instrumentation, a list of which includes

- fuel consumption devices: gravimetric or mass flow;
- oil consumption;
- engine 'blow-by';
- exhaust gas emission analysis equipment;
- combustion analysis (indicating) instrument.

The integration of these devices within the control system will require special software drivers to allow for communication covering basic control, data acquisition and calibration routines.

The interface between device and test cell computer may be by way of conventional analogue or digital I/O, by serial interface such as RS232 or IEEE or, increasingly, through a local area network (LAN) (see Chapter 19).

Computerized calibration procedures

Software routines will be provided to allow the operator to calibrate the various measuring devices and systems. Often these routines follow a 'form fill' format that requires the operator to type in confirmation of test signals in the correct order, permitting zero, span and intermediate and ambient values to be inserted in the calibration calculations. Such software should not only lead the operator through the calibration routines, but should also store and print out the results in such a way as to meet basic quality control and certification procedures. The appropriate linearization and conversion procedures required to turn transducer signals into the correct engineering units will be built into the software.

Control for endurance testing and 'unmanned' running

These tests are very expensive to run and call for a high level of performance and reliability in the control and data logging systems. Duplication of transducers and careful running of cables to prevent deterioration by wear or heat over long periods should be considered. The amount of data generated over 1000 hours of cyclic running can be enormous, so data filtering and compression are usually required.

Control for thermal shock and thermal cycling tests

These call for special programming arrangements, 'end condition' of stage being often defined by the attainment of a given temperature. There will often be the need to operate fluid control valves in strict sequence in order to direct engine coolant flows to or from chilled water buffer stores. Since this type of testing may be expected to induce a failure in some part of the engine system the failure monitoring system must be designed accordingly.

Control for transient or dynamic engine testing

For some purposes, such as the analysis of the performance of electronic engine management systems (see also Chapter 19), it may be necessary to record data at a speed at which accurate 'time stamping' of individual events becomes difficult. 'Time skew' can arise when the time taken to record all the values corresponding to an event taking place at a given instant is significant relative to the interval between successive events. High speed data recording can produce unwanted detail, for instance cyclic variations in crankshaft speed when rotational speed is sensed on the basis of very short measuring intervals.

Any vehicle user is aware that the performance of an engine is to a considerable extent judged by its characteristics when changing state. A cloud of black smoke from a truck pulling away from traffic lights or hesitant throttle response when accelerating are just two of many examples of inadequate performance that is a direct consequence of unsatisfactory transient behaviour on the part of the engine.

It is perhaps worth pointing out that while, in what may be described as 'classical' steady state testing, the transitions between successive test conditions are unimportant, in transient testing the transient states *are* the test sequences.

The terms *transient* or *dynamic* testing are somewhat subjective terms and cover testing that is, in practice, almost confined to automotive engines, since industrial and marine engines are rarely subject to the almost continuous variations in load and speed that are characteristic of the automotive engine environment. Test sequences that are neither transient nor dynamic are usually referred to as 'steady state' and although that is scientifically incorrect, the term covers sequences where the engine is taken to a set condition and held at that operating point for a finite time, normally in excess of 5 seconds, but in some cases any period above 100 ms, while measurements are taken.

Up to the 1990s, the majority of formalized transient engine testing was associated with emissions testing carried out in accordance with the kind of test procedure described in Chapter 16.

Since that time the work associated with engine and drive train calibration or mapping and track simulation in the motorsport industry has greatly increased, as has the sophistication of the models, which may be required to simulate both the road load and the vehicle power train; such test sequences can be considered as dynamic. Importantly dynamometer technology has provided the tools that allow for speed and torque changes in the test-bed drive line to match those experienced on the road and track to ever better resolution.

What may be considered a highly dynamic test sequence in one industrial sector may be considered positively pedestrian in another. Therefore, in this chapter transient and dynamic are classified in terms of four characteristic time scales where the times quoted represent the time in which the control instruction is sent and the change in force at the a.c. dynamometer air gap is sensed:

- <2 ms is 'high dynamic';
- <10 ms is dynamic;
- <50 ms is transient;
- >100 ms is steady state not so much defined by the period at with the engine state is held but also because the data are required at an operation point rather than during the transition between points.

Under these classifications falls the wide range of work required of engine test cells.

The test engineer involved in transient testing, which is not controlled by self-adjusting, boundary-seeking software tools of the type discussed in Chapter 19, is faced with two challenge: firstly to ensure that the test sequence accurately models the conditions experienced by the engine in service and secondly to ensure that the sequence is precisely repeatable.

The favoured modern tool is the use of a direct coupled four-quadrant, a.c. dynamometer with computer control; such a machine can be designed to react sufficiently rapidly to simulate all the transient conditions experienced by the engine on the road.

In the range up to 0.2 s, we are concerned essentially with torsional vibrations in the engine–transmission–road wheel complex ranging from two-mass engine–vehicle judder, with a frequency typically in the range 5–10 Hz, to much higher frequency oscillations involving the various components of the power train. Subtle effects can occur as a result of interaction with the engine mountings. Rotation of the engine on its mountings not only affects the dynamics of the whole system but may disturb the throttle linkage, with consequent variations in the throttle position.

This kind of power train investigation, however, involves the engine to a secondary degree; in fact, for such work the engine may now be replaced by a lower inertia electric motor. These tests are mainly concerned with the development of transmission systems and are thus rather outside the scope of this book.

In the transient test time scale work is characteristic of gear shifts; correct behaviour in this range is critical in a system called upon to simulate these events. This is a particularly demanding area, as the profile of gearshifts is extremely variable. At one extreme we have the fast automatic changes of a race car gearbox, at the other gear changes in the older commercial vehicle, which may take more than 2 s, and in between the whole range of individual driver characteristics, from aggressive to timid.

Whatever the required profile, the dynamometer must impose zero torque on the engine during the period of clutch disengagement and this requires precise following of the 'free' engine speed. The accelerating/decelerating torque required is proportional to dynamometer inertia and the rate of change of speed:

$$T = Id\omega/dt$$

where $T =$ torque required in Nm, $I =$ dynamometer inertia in kg m^2, and $d\omega/dt =$ rate of speed change in rad/s. This requirement is a major factor in the choice of dynamometers for transient testing (see Chapter 8).

A gear change involves throttle closing, an engine speed change, up or down, of perhaps 2000 rev/min, and the reapplication of power. Ideally, at every instant during this process the rate of supply of fuel and air to the engine cylinders and the injection and ignition timings should be identical with those corresponding to the optimized steady state values for the engine load and speed at that instant; it will be clear that this optimizing process makes immense demands on the mapping of the engine management system.

A special problem of engine control in this area concerns the response of a turbocharged engine to a sudden demand for more power. An analogous problem has been familiar for many years to process control engineers dealing with the management of boilers. If there is an increase in demand, the control system increases the air supply before increasing the fuel input; on a fall in demand, the fuel input is reduced before the air. The purpose of this 'air first up–fuel first down' system is to ensure that there is never a deficiency of air for proper combustion.

In the case of a turbocharged engine only specialized means, such as variable turbine geometry, can increase the air flow in advance of the fuel flow. A reduction in value from the (optimized) steady state value, with consequently increased emissions, is an inevitable concomitant to an increase in demand. The control of maximum fuel supply during acceleration is a difficult compromise between performance and driveability on the one hand and a clean exhaust on the other.

References

1. BS 3403 *Specification for Indicating Tachometer and Speedometer Systems for Industrial, Railway and Marine Use.*
2. BS 1041 Parts 1 to 7 *Temperature Measurement.*

Further reading

BS 6174 *Specification for Differential Pressure Transmitters with Electrical Outputs.*
National Physical Laboratory. *Good Practice Guides.* http://www.npl.co.uk/.

12 Measurement of fuel, combustion air and oil consumption

Introduction

In this chapter a review is made of the instruments required to measure the consumption of the liquid fuel, combustion air and lubricating oils during engine running in a test bed. Until the advent of instruments capable of measuring 'instant' air and fluid flow, the standard methods recorded cumulative consumption over a period of time, or measured the number of revolutions over which a known mass of liquid fuel was consumed. While these methods and instruments are still viable, more and more testers are required to record the actual transient consumption during test sequences that simulate 'real-life' driving conditions.

Modern engine designs incorporate a number of fuel spill-back or fuel return strategies that make the design of test cell consumption systems more complex than before. In simple terms, it is necessary now to measure the amount of fuel entering the metering and conditioning system, rather than leaving it, for the engine. The systems frequently have to control the pressure of the fuel return line and remove vapour bubbles and the heat energy picked up through the engine's fuel rail and pressure regulator circuit. Liquid fuel conditioning and consumption measurement is no longer a matter of choosing the appropriately sized instrument, but rather of designing a complex pipe circuit incorporating those instruments, heat exchangers and the engine.

Measurement of the air consumption of the engine is covered in principle and practice.

Measurement instruments for liquid fuel consumption

For all liquid fuel consumption measurement, it is critical to control fuel temperature within the metered system as far as is possible. Therefore, most modern cells have a closely integrated temperature control and measurement system. The condition of the fuel returned from the engine can cause significant problems as it may return at pulsing pressure, considerably warmer than the control temperature and containing vapour bubbles. This means that the volume and density of the fluid can be variable within the measured system, and that this variability has to be reduced as far as is possible by the overall metering system and the level of measurement uncertainty taken into account (see Chapter 7, section Engine fuel temperature control for a full discussion concerning the effect of fuel conditioning on fuel consumption measurement).

Cumulative flow meters

There are two generic types of fuel gauge intended for cumulative measurement of liquid fuels on the market:

- *Volumetric gauges*, which measure the number of engine revolutions taken to consume a known volume of fuel either from containment of known volume or as measurement of flow through a measuring device.
- *Gravimetric gauges*, which measure the number of engine revolutions taken to consume a known mass of fuel.

Figure 12.1 shows a volumetric gauge in which an optical system gives a precise time signal at the start and end of the emptying of one of a choice of calibrated

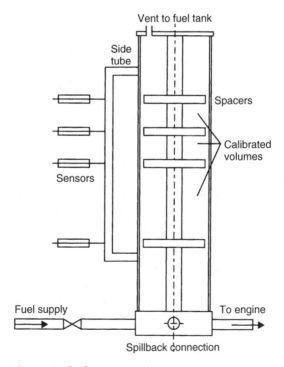

Figure 12.1 *Volumetric fuel consumption gauge*

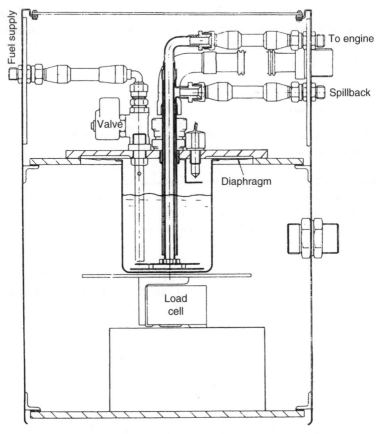

Figure 12.2 *Gravimetric, direct weighing, fuel gauge*

volumes. This signal actuates a counter, giving a precise value for the number of engine revolutions made during the consumption of the measured volume of fuel.

Figure 12.2 shows a gravimetric gauge designed to meter a mass rather than a volume of fuel, consisting essentially of a vessel mounted on a weighing cell from which fuel is drawn by the engine. The signals are processed in the same way as for the volumetric gauge. A typical specification being:

- measuring ranges: 0–150 kg/h, 0–200 kg/h, 0–360 kg/h;
- fuels: petrol and diesel fuels, special versions for 10 per cent methanol or ethanol;
- computer interface: serial: RS232C, 1–10 V analogue plus digital I/O;
- electrical supply required typically 0.5–2.5 kW for measuring and fuel conditioning.

A further type of gravimetric fuel gauge, in which a cylindrical float is suspended from a force transducer in a cylindrical vessel, exists. The change in flotation force is then directly proportional to the change of mass of fuel in the vessel. This design of gauge has the particular advantage that it is insensitive to vertical accelerations and is thus suitable for shipboard use.

These gravimetric gauges have the advantage over the volumetric type that the metered mass may be chosen at will, and a common measuring period may be chosen, independent of fuel consumption rate. The specific fuel consumption is derived directly from only three measured quantities: the mass of fuel consumed, the number of engine revolutions during the consumption of this mass and the mean torque. This is another reason for the inherently greater accuracy of cumulative fuel consumption measurement; rate measurements involve four or, in the case of volumetric meters, five measured quantities.

All gauges of this type have to deal with the problem of fuel spillback from fuel injection systems and Fig. 12.3 shows a circuit incorporating a gravimetric fuel gauge which deals with this matter. When a fuel consumption measurement is to be made, a solenoid valve diverts the spillback flow, normally returned to the header tank during an engine test, into the bottom of the fuel gauge. It is not satisfactory to return the spillback to a fuel filter downstream of the fuel gauge, since the air and vapour always present in the spilled fuel lead to variations in the fuel volume between fuel gauge and engine, and thus to incorrect values of fuel consumption.

Mass flow or consumption rate meters

There are many different designs of rate meter on the market and the choice of the most suitable unit for a given application is not easy. The following factors need to be taken into account:

- volumetric or gravimetric;
- absolute level of accuracy;
- turn-down ratio;
- sensitivity to temperature and fuel viscosity;
- pressure difference required to operate;
- wear resistance and tolerance of dirt and bubbles;
- analogue or impulse-counting readout;
- suitability for stationary/in-vehicle use.

Several designs of positive-displacement (volumetric) fuel gauges make use of a four-piston metering unit with the cylinders arranged radially around a single-throw crankshaft. Crankshaft rotation is transmitted magnetically to a pulse output flow transmitter. Cumulative flow quantity and instantaneous flow rate are indicated and these meters are suitable for in-vehicle use. A high turn-down ratio is claimed and one design includes a pressure-sensing system to eliminate leakage errors. A disadvantage is the appreciable pressure drop, which may approach 1 bar, required to drive the metering unit. For large flow rates metering units employing meshing helical rotors are usual.

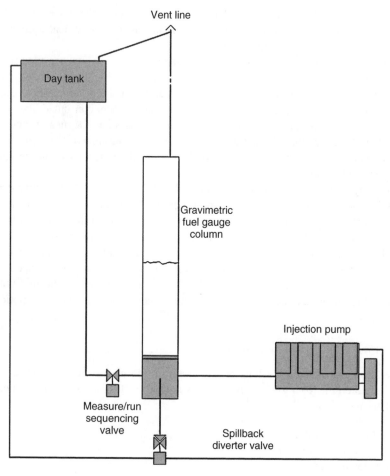

Figure 12.3   *Spillback and associated valves in a gravimetric fuel gauge circuit*

Some mass flow meters make use of the Coriolis effect, in which the fuel is passed through a vibrating U-tube, Fig. 12.4.

Some mass flow meters are able to measure spillback, but with others it may be necessary to use two sensing units, one in the flow line and one in the spill line; this increases cost and complexity.

The use of a rate meter in series with a cumulative meter may be considered, where the accuracy of a cumulative measurement device with the ability of a rate meter to deal with transient conditions is combined.

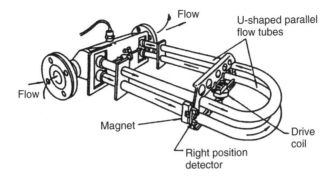

Figure 12.4 *Coriolis effect flow meter*

Fuel consumption measurements: gaseous fuels

For these fuels, consumption measurements are made by gas flow meters. Metering of gases is a more difficult matter than liquid metering. The density of a gas is sensitive to both pressure and temperature, both of which must be known when, as is usually the case, the flow is measured by volume. Also the pressure difference available to operate the meter is limited. The traditional domestic gas meter contains four chambers separated by bellows and controlled by slide valves. Successive increments in volume metered are quite large, so that instantaneous or short-term measurements are not possible.

Other types, capable of indicating smaller increments of flow, make use of rotors having sliding vanes or meshing rotors. In the case of natural gas supplied from the mains, flow measurement is a fairly simple matter. Pressure and temperature at the meter must be measured and accurate data on gas properties (density, calorific value, etc.) will be available from the supplier.

Measurement of liquefied petroleum gas consumption is less straightforward, since this is stored as a liquid under pressure and is vaporized, reduced in pressure and heated before reaching the engine cylinders. The gas meter must be installed in the line between the 'converter' and the carburettor. It is essential to measure the gas temperature at this point to achieve accurate results.

The effect of fuel condition on engine performance

The effect of the condition (pressure, temperature, humidity and purity) of the combustion air that enters the engine is discussed later in this chapter. In the case of a spark ignition engine the maximum power output is more or less directly proportional to the mass of oxygen contained in the combustion air, since the air/fuel ratio at full throttle is closely controlled.

In the case of the diesel engine the position is different and less clear cut. The maximum power output is generally determined by the maximum mass rate of fuel oil flow delivered by the fuel injection pump operating at the maximum fuel stop position. However, the setting of the fuel stop determines the maximum volumetric fuel flow rate and this is one reason why fuel temperature should always be closely controlled in engine testing.

Fuel has a high coefficient of cubical expansion, lying within the range 0.001 to 0.002 per °C (compared with water, for which the figure is 0.00021 per °C). This implies that the mass of fuel delivered by a pump is likely to diminish by between 0.1 and 0.2 per cent for each degree rise in temperature, a by no means negligible effect.

A further fuel property that is affected by temperature is its viscosity, significant because in general the higher the viscosity of the fuel the greater the volume that will be delivered by the pump at a given rack setting. This effect is likely to be specific to a particular pump design; as an example, one engine manufacturer regards a 'standard' fuel as having a viscosity of 3 cSt at 40°C and applies a correction of $2\frac{1}{2}$ per cent for each centistoke departure from this base viscosity, the delivered volume increasing with increasing viscosity.

As a further factor, the specific gravity (density) of the fuel will also affect the mass delivered by a constant volume pump, though here the effect is obscured by the fact that, in general, the calorific value of a hydrocarbon fuel falls with increasing density, thus tending to cancel the change out.

Measurement of lubricating oil consumption

Oil consumption measurement is of increasing importance since it is an important component of the total engine emissions but it is one of the most difficult measurements associated with engine testing within the test cell. Oil consumption rate is very slow relative to the quantity in circulation. In a typical vehicle engine of lubricant capacity of 6 litres, the oil consumption rate will be in the region of zero to 25 cm³/h or up to 0.5 per cent of the volume in circulation. This means that test durations tend to be of long duration before meaningful results are obtained.

With the exception of total loss systems, such as are used for cylinder lubrication of large marine diesel engines, no entirely satisfactory method exists. One method is to adapt the engine for dry sump operation. The sump is arranged to drain to a separate receiver, the contents of which are monitored, either by weighing or by depth measurement. An alternative method is to draw off oil down to a datum sump level and weigh it before and after the test period.

Difficulties with all systems include:

- The quantity of oil adhering to the internal surfaces of the engine is very sensitive to temperature, as is the volume in transit to the receiver. This can give rise to large apparent variations in consumption rate.

- Volumetric changes due to temperature soak of the engine and fluids.
- Apparent consumption is influenced by fuel dilution and by any loss of oil vapour.
- The rate of oil consumption tends to be very sensitive to conditions of load, speed and temperature. There is also a tendency to medium-term variations in apparent rate, due to such factors as 'ponding up' of oil in return drains and accumulation of air or vapour in the circuit.

For a critical analysis of the dry sump method, see SAE paper 880098,[1] which also describes a statistical test procedure aimed at minimizing these random errors.

An important but rather specialized technique for measuring oil consumption is by using specially blended oil containing either a sulphur compound or a radioactive isotope. Continuous measurement of sulphur dioxide or sensing of the isotope in the exhaust allows calculation of the oil consumption rate; special sampling procedures can be used to differentiate burned and unburned fractions.

Measurement of crankcase blow-by

In all reciprocating internal combustion engines, there is a flow of gas into and out of the clearances between piston top land, ring grooves and cylinder bore. In an automotive gasoline engine these can amount to 3 per cent of the combustion chamber volume. Since in a spark-ignition engine the gases consist of unburned mixture which emerges during the expansion stroke too late to be burned, this can be a major source of HC emissions and also represents a loss of power. Some of this gas will leak past the rings and piston skirt in the form of blow-by into the crankcase. It is then vented back into the induction manifold and to this extent reduces the HC emissions and fuel loss, but has an adverse effect on the lubricant.

Blow-by flow is highly variable over the full range of an engine's performance and life, therefore accurate blow-by meters will need to be able to deal with a wide range of flows and with pulsation. Instruments based on the orifice measurement principle, coupled with linearizing signal conditioning (the flow rate is proportional to the square root of the differential pressure), are good at measuring the blow-by gas in both directions of flow which can occur when there is heavy pulsation between pressure and partial vacuum in the crankcase.

Within the gas flow there may be carbon and other 'dirt' particles. The sensitivity of the measuring instrument to this dirt will depend on the application; sensitivity is shown in Table 12.1.

Crankcase blow-by is a significant indicator of engine condition and should preferably be monitored during any extended test sequence. An increase in blow-by can be a symptom of various problems such as incipient ring sticking, bore polishing or deficient cylinder bore lubrication.

Table 12.1   Types of blow-by meters

Type	Typical accuracy	Dirt sensitivity	Lowest flow (l/min)
Positive impeller	2% FS	Medium	Approximately 6
Hot wire	2%	High	Approximately 28
Commercial gas meter	1% FS	High	Claimed 0.5
Vortex	1%	High	Approximately 7
Flow through an orifice	1%	Low	Claimed 0.2

Measurement of air consumption and gas flows

The accurate measurement of the air consumption of an internal combustion engine is a matter of some complexity but of great importance. The theory also has relevance to gas flow measurement in emissions testing, therefore the subject is covered in some detail in the following text.

The influence of the condition of the air entering the engine on various aspects of engine performance is discussed and 'correction factors' defining these effects quantitatively are derived. The theory of various methods of measurement is given and the limitations of each method described.

Properties of air

Air is a mixture of gases with the following approximate composition:

Major constituents	By mass	By volume
Oxygen, O_2	23%	21%
Nitrogen, N_2	77%	79%

plus trace gases and water vapour of variable amount, usual range 0.2–2.0 per cent of volume of dry air.

The amount of water vapour present depends on temperature and prevailing atmospheric conditions. It can have an important influence on engine performance, notably on exhaust emissions, but for all but the most precise work its influence on air flow measurement may be neglected.

The relation between the pressure, specific volume and density of air is described by the gas equation

$$p_a \times 10^5 = \rho R T_a \tag{1}$$

where R, the gas constant, has the value for air

$$R = 287\,\mathrm{J/kgK}$$

A typical value for air density in temperate conditions at sea level would be

$$\rho = 1.2\,\mathrm{Kg/m^3}$$

Air consumption, condition and engine performance

The internal combustion engine is essentially an 'air engine' in that air is the working fluid; the function of the fuel is merely to supply heat. There is seldom any particular technical difficulty in the introduction of sufficient fuel into the working cylinder, but the attainable power output is strictly limited by the charge of air that can be aspirated.

It follows that the achievement of the highest possible volumetric efficiency is an important goal in the development of high-performance engines, and the design of inlet and exhaust systems, valves and cylinder passages represents a major part of the development programme for engines of this kind.

The standard methods of taking into account the effects of charge air condition as laid down in European and American Standards are complex and difficult to apply and are mostly used to correct the power output of engines undergoing acceptance or type tests. A simplified treatment, adequate for most routine test purposes, is given below.

The condition of the air entering the engine is a function of the following parameters:

- pressure;
- temperature;
- moisture content;
- impurities.

For the first three factors standard conditions, according to European and American practice, are:

- atmospheric pressure 1 bar (750 mmHg);
- temperature 25°C (298 K);
- relative humidity 30%.

Atmospheric pressure

Since the volumetric efficiency of an engine tends to be largely independent of the air supply pressure, the mass of air consumed tends to vary directly with the density, which is itself proportional to the absolute pressure, other conditions remaining unchanged. Since the standard atmosphere = 1 bar we may write:

$$\rho_t = \rho_n P \qquad (2)$$

where:

ρ_t = density under test conditions, kg/m^3
ρ_n = density under standard conditions, kg/m^3
P = atmospheric pressure under test conditions, bar

It follows that a change of 1 per cent or 7.5 mmHg corresponds to a change in the mass of air entering the engine of 1 per cent. For most days of the year the (sea level) atmospheric pressure will lie within the limits 750 mmHg ± 3 per cent, say between 775 and 730 mmHg, with a corresponding percentage variation in charge air mass of 6 per cent.

It is common practice to design the test cell ventilation system to maintain a small negative pressure in the cell, to prevent fumes entering the control room, but the level of depression is unlikely to exceed about 50 mm water gauge (50 Pa), equivalent to a change in barometer reading of less than 0.37 mmHg. Clearly the effect on combustion air flow, if the engine is drawing air from within the cell, is negligible.

The barometer also falls by about 86 mmHg (11.5 kPa) for an increase in altitude of 1000 m (the rate decreasing with altitude), see Table 12.2. This indicates that the mass of combustion air falls by about 1 per cent for each 90 m (300 ft) increase in altitude, a very significant effect.

Variations in charge air pressure have an important 'knock-on' effect: the pressure in the cylinder at the start of compression will in general vary with the air supply pressure and the pressure at the end of compression will change in the same proportion. This can have a significant effect on the combustion process.

Table 12.2   *Variation in atmospheric pressure with height above sea level*

Altitude (m)	Fall in pressure (bar)
0	0
500	0.059
1000	0.115
1500	0.168
2000	0.218
3000	0.312
4000	0.397

Air temperature

Variations in the temperature of the air supply have an effect of the same order of magnitude as variations in barometric pressure within the range to be expected in test cell operation. Air density varies inversely with its absolute temperature:

$$\rho_t = \rho_n \bullet \frac{298}{(t_t + 273)} \tag{3}$$

where t_t = test temperature.

However, the temperature at the start of compression determines that at the end. In the case of a naturally aspirated diesel engine with a compression ratio of 16:1 and an air supply at 25°C, the charge temperature at the start of compression would typically be about 50°C. At the end of compression the temperature would be in the region of 530°C, increasing to about 560°C for an air supply temperature 10°C higher. The level of this temperature can have a significant effect on such factors as the NO_x content of the exhaust, which is very sensitive to peak combustion temperature. The same effect applies, with generally higher temperatures, to turbocharged engines.

Compared with the effects of pressure and temperature, the influence of the relative humidity of the air supply on the air charge is relatively small, except at high air temperatures. The moisture content of the combustion air does, however, exert a number of influences on performance. Some of the thermodynamic properties of moist air have been discussed under the heading of psychometry (Chapter 5), but certain other aspects of the subject must now be considered.

The important point to note is that unit volume of moist air contains less oxygen in a form available for combustion than the same volume of dry air under the same conditions of temperature and pressure. Moist air is a mixture of air and steam: while the latter contains oxygen it is in chemical combination with hydrogen and is thus not available for combustion. The European and American Standards specify a relative humidity of 30 per cent at 25°C.

This corresponds to a vapour pressure 0.01 bar, thus implying a dry air pressure of 0.99 bar and a corresponding reduction in oxygen content of 1 per cent when compared with dry air at the same pressure. Figure 12.5 shows the variation in dry air volume expressed as a percentage adjustment to the standard condition with temperature for relative humidities ranging from 0 (dry) to 100 per cent (saturated).

The effect of humidity becomes much more pronounced at higher temperatures: thus at a temperature of 40°C the charge mass of saturated air is 7.4 per cent less than for dry air. The effect of temperature on moisture content (humidity) should be noted particularly, since the usual method of indicating specified moisture content for a given test method, by specifying a value of relative humidity, can be misleading. Figure 12.5 indicates that at a temperature of 0°C the difference between 0 per cent (dry) and 100 per cent (saturated) relative humidity represents a change of only 0.6 per cent in charge mass; at −10°C it is less than 0.3 per cent. It follows that it is

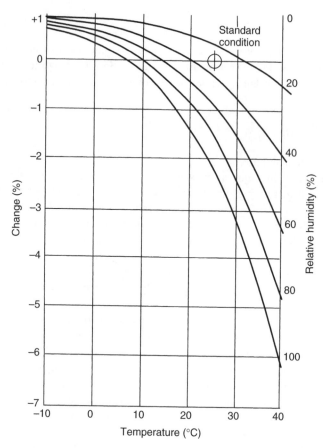

Figure 12.5 *Variation of air charge mass with relative humidity at different temperatures*

barely worthwhile adjusting the humidity of combustion air at temperatures below (perhaps) 10°C.

The moisture content of the combustion air has a significant effect on the formation of NO_x in the exhaust of diesel engines. The SAE procedure for measurement of diesel exhaust emissions gives a rather complicated expression for correcting for this. This indicates that, should the NO_x measurement be made with completely dry air, a correction of the order of +15 per cent should be made to give the corresponding value for a test with moisture content of 60 per cent relative humidity.

Finally, it should be mentioned that pollution of the combustion air can result in a reduction in the oxygen available for combustion. A likely source of such pollution may be the ingestion of exhaust fumes from other engines or from neighbouring industrial processes, such as paint plant, and care should also be taken in the siting of air intakes and exhaust discharges, to ensure that under unfavourable conditions there cannot be unintended exhaust recirculation. Exhaust gas has much the same density as air and a free oxygen content that can be low or even zero, so that approximately 1 per cent of exhaust gas in the combustion air will reduce the available oxygen in almost the same proportion.

These various adjustments or 'corrections' to the charge mass are of course cumulative in their effect and the following example illustrates the magnitudes involved. Consider:

A, a hot, humid summer day with the chance of thunder
B, a cold, dry winter day of settled weather

Condition	**A**	**B**
Pressure	0.987 bar, 740 mmHg	1.027 bar, 770 mmHg
Temperature	35°C	10°C
Relative humidity	80%	40%
Pressure	$\frac{1}{0.987} = 1.0135$	$\frac{1}{1.027} = 0.9740$
Temperature	$\frac{308}{298} = 1.0336$	$\frac{283}{1.027} = 0.9497$
Relative humidity from Fig. 12.4	+3.48% = 1.0348	0%
Total adjustment	1.0840, +8.4%	0.9233, −7.7%

This 'adjustment' indicates the factor by which the observed power should be multiplied to indicate the power to be expected under 'standard' conditions. We see that under hot, humid, stormy conditions, the power may be reduced by as much as 15 per cent compared with the power under cold anticyclonic conditions, a by no means negligible adjustment.

It should be pointed out that the power adjustment calculated above is based on the assumption that charge air mass directly determines the power output, but this would only be the case if the air/fuel ratio were rigidly controlled (as in spark ignition engines with precise stoichiometric control). In most cases a reduction, for example, in charge air mass due to an increase in altitude will result in a reduction in the air/fuel ratio, perhaps with an increase in exhaust smoke, but not necessarily in a reduction in power. The correction factors laid down in the various standards take into account the differing responses of the various types of engine to changes in charge oxygen content.

It should also perhaps be pointed out that the effects of water injection are quite different from the effects of humidity already present in the air. Humid air is a mixture of air and steam: the latent heat required to produce the steam has already been supplied. When, as is usually the case, water is injected into air leaving the turbocharger at a comparatively high temperature, the cooling effect associated with the evaporation of the water achieves an increase in charge density which much outweighs the decrease associated with the resulting steam.

The airbox method of measuring air consumption

The simpler methods of measuring air consumption involve drawing the air through some form of measuring orifice and measuring the pressure drop across the orifice.[2] It is good practice to limit this drop to not more than about 125 mm H_2O (1200 Pa). For pressures less than this, air may be treated as an incompressible fluid, with much simplification of the air flow calculation.

The velocity U developed by a gas expanding freely under the influence of a pressure difference Δp, if this difference is limited as above, is given by:

$$\frac{\rho U^2}{2} = \Delta p; \ U = \sqrt{\frac{2\Delta p}{\rho}} \tag{4}$$

Typically, air flow is measured by means of a sharp-edged orifice mounted in the side of an airbox, coupled to the engine inlet and of sufficient capacity to damp out the inevitable pulsations in the flow into the engine, which are at their most severe in the case of a single cylinder four-stroke engine, Fig. 12.6. In the case of turbocharged engines, the inlet air flow is comparatively smooth and a well-shaped nozzle without an airbox will give satisfactory results.

The air flow through a sharp-edged orifice takes the form sketched in Fig. 12.7. The coefficient of discharge of the orifice is the ratio of the transverse area of the flow at plane a (the vena contracta) to the plan area of the orifice. Tabulated values of C_d are available in BS 1042,[3] but for many purposes a value $C_d = 0.60$ may be assumed.

We may easily derive the volumetric flow rate of air through a sharp-edged orifice as follows:

Flow rate = coefficient of discharge × cross-sectional area of orifice × velocity of flow

$$Q = C_d \frac{\pi d^2}{4} \sqrt{\frac{2\Delta p}{\rho}} \tag{5}$$

Noting from eq. (3) that

$$\rho = \frac{p_a \times 10^5}{RT_a}$$

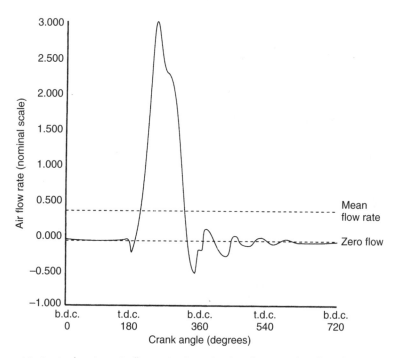

Figure 12.6   *Induction air flow, single cylinder, four-stroke diesel engine*

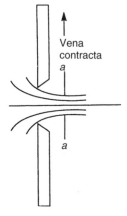

Figure 12.7   *Flow through a sharp-edged orifice*

and assuming that Δp is equivalent to h mm H$_2$O, we may write

$$Q = C_d \frac{\pi d^2}{4} \sqrt{\frac{2 \times 9.81 h \times 287 T}{p_a \times 10^5}} \tag{6a}$$

$$Q = 0.1864 \, C_d d^2 \sqrt{\frac{hT_a}{p_a}} \ \mathrm{m^3/s} \tag{6b}$$

To calculate the mass rate of flow, note that:

$$\dot{m} = \rho Q = \frac{p_a Q}{RT_a}$$

giving, from eq. (6a):

$$\dot{m} = C_d \frac{\pi d^2}{4} \sqrt{\frac{2 \times 9.81 h \times 10^5}{287 T_a}} \tag{7a}$$

$$\dot{m} = 64.94 \, C_d d^2 \sqrt{\frac{hp_a}{T_a}} \tag{7b}$$

Equations (6b) and (7b) give the fundamental relationship for measuring air flow by an orifice, nozzle or venturi.

Sample calculation

If

$C_d = 0.6$;
$d = 0.050$ m;
$h = 100$ mmH$_2$O;
$T_a = 293$ K (20°C);
$p_a = 1.00$ bar.

Then

$Q = 0.04786 \, \mathrm{m^3/s}$ (1.69 ft^3/s)
$\dot{m} = 0.05691 \, \mathrm{kg/s}$ (0.1255 lb/s)

To assist in the selection of orifice sizes, Table 12.3 gives approximate flow rates for orifices under the following standard conditions:

Table 12.3 *Approximate air flow rates for orifices sizes 10 to 150 mm*

Orifice dia. (mm)	$Q\,(m^3/s)$	$\dot{m}\,(kg/s)$
10	0.002	0.002
20	0.008	0.009
50	0.048	0.057
100	0.19	0.23
150	0.43	0.51

$h = 100\ \text{mmH}_2\text{O}$
$T_a = 293\ \text{K}\ (20°\text{C})$
$p_a = 1.00\ \text{bar}$

A disadvantage of flow measurement devices of this type is that the pressure difference across the device varies with the square of the flow rate. It follows that a turndown in flow rate of 10:1 corresponds to a reduction in pressure difference of 100:1, implying insufficient precision at low flow rates. It is good practice, when a wide range of flow rates is to be measured, to select a range of orifice sizes, each covering a turndown of not more than 2.5:1 in flow rate.

Air consumption of engines approximate calculation

The air consumption of an engine may be calculated from:

$$V = \eta_v \frac{V_s}{K} \frac{n}{60}\ \text{m}^3/\text{s} \qquad (8)$$

where $K = 1$ for a two-stroke and 2 for a four-stroke engine.

For initial sizing of the measuring orifice η_v, the ratio of the volume of air aspirated per stroke to the volume of the cylinder may be assumed to be about 0.8 for a naturally aspirated engine and up to about 2.5 for supercharged engines.

Sample calculation

Single cylinder four-stroke engine, swept volume 0.8 litre, running at a maximum of 3000 rev/min, naturally aspirated:

$$V = 0.8\ \frac{0.0008}{2}\ \frac{3000}{60} = 0.16\,\text{m}^3/\text{s}$$

Suitable orifice size, from Table 12.3, 30 mm.

Connection of airbox to engine inlet

It is essential that the configuration of the connection between the airbox and the engine inlet should model as closely as possible the configuration of the air intake arrangements in service. This is because pressure pulsations in the inlet can have a powerful influence on engine performance, in terms of both volumetric efficiency and pumping losses.

The resonant frequency of a pipe of length L, open at one end and closed at the other, $= a/4L$, where 'a' is the speed of sound, roughly 330 m/s. Thus an inlet connection 1 m long would have a resonant frequency of about 80 Hz. This corresponds to the frequency of intake valve opening in a four cylinder four-stroke engine running at 2400 rev/min. Clearly such an intake connection could disturb engine performance at this speed. In general, the intake connection should be as short as possible.

The viscous flow air meter

The viscous flow air meter, invented by Alcock and Ricardo in 1936, was for many years the most widely used alternative to the airbox and orifice method of measuring air flow. In this device the measuring orifice is replaced by an element consisting of a large number of small passages, generally of triangular form. The flow through these passages is substantially laminar, with the consequence that the pressure difference across the element is approximately proportional to the velocity of flow, rather than to its square, as is the case with a measuring orifice.

This has two advantages. Firstly average flow is proportional to average pressure difference, implying that a measurement of average pressure permits a direct calculation of flow rate, without the necessity for smoothing arrangements. Secondly, the acceptable turndown ratio is much greater.

The flow meter must be calibrated against a standard device, such as a measuring orifice.

Lucas–Dawe air mass flow meter

This device depends for its operation on the corona discharge from an electrode coincident with the axis of the duct through which the air is flowing. Air flow deflects the passage of the ion current to two annular electrodes and gives rise to an imbalance in the current flow that is proportional to air flow rate.

An advantage of the Lucas–Dawe flow meter is its rapid response to changes in flow rate, of the order of 1 ms, making it well suited to transient flow measurements, but it is sensitive to air temperature and humidity, and requires calibration against a standard.

Positive displacement flow meters

As rotation occurs, successive pockets of air are transferred from the suction to the delivery side of the flow meter, and flow rate is proportional to rotor speed. Some of these flow meters operate on the principle of the Roots blower. Advantages of the positive displacement flow meter are accuracy, simplicity and good turndown ratio. Disadvantages are cost, bulk, relatively large pressure drop and sensitivity to contamination in the flow.

Hot wire or hot film anemometer devices

The principle of operation of these popular devices is based on the cooling effect caused by gas flow on a heated wire or film surface. The heat loss is directly proportional to the air mass flow rate, providing the flow through the device is laminar. The advantages of these designs are their reliability and good tolerance to contaminated air flows. However, calibration at site is virtually impossible so they are supplied with maker's certification. Probably the best known device of this type in European test cells is the Sensyflow™ range made by ABB.

Notation

Atmospheric pressure	p_a bar
under test conditions	p bar
Atmospheric temperature	T_a K
Test temperature	t_t°C
Density of air	ρ kg/m^3
under standard conditions	ρ_n kg/m^3
under test conditions	ρ_t kg/m^3
Pressure difference across orifice	Δ_p Pa, h mmH$_2$O
Velocity of air at contraction	U m/s
Coefficient of discharge of orifice	C_d
Diameter of orifice	d m
Volumetric rate of flow of air	Q m^3/s
Mass rate of flow of air	$\dot{m}$ kg/s
Constant, 1 for two-stroke and 2 for four-stroke engines	K
Engine speed	n rev/min
Number of cylinders	N_c
Swept volume, total	V_s m^3
Air consumption rate	V m^3/s
Volumetric efficiency of engine	η_v
Volume of airbox	V_b m^3
Gas constant	R J/kg K

References

1. Johren, P.-W. and Newman, B.A. (1988) Evaluating the oil consumption behaviour of reciprocating engines in transient operation, SAE Paper 880098.
2. Kastner, L.J. (1947) The airbox method of measuring air consumption, *Proc. I. Mech. E.*, **157**.
3. BS 1042 *Measurement of Fluid Flow in Closed Conduits: Section 1.1*, (AS 2360.1.2-1993) *Specification for Square-edged Orifice Plates, Nozzles and Venturi Tubes Inserted in Circular Cross-section; Conduits Running Full; Section 1.4, Guide to the Use of Devices Specified in Sections 1.1 and 1.2.*

Further reading

BS 5514 Parts 1 to 6 *Reciprocating Internal Combustion Engines: Performance.*

BS 7405 *Guide to Selection and Application of Flowmeters for the Measurement of Fluid Flow in Closed Conduits.*

Plint, M.A. and Böswirth, L. (1978) *Fluid Mechanics: A Laboratory Course*, Griffin, London.

13 Thermal efficiency, measurement of heat and mechanical losses

Introduction

This chapter deals with the measurements and calculations necessary to determine the energy balance, thermal performance and mechanical losses of an internal combustion engine. A brief account is given of the basic theory in order to provide a framework for an interpretation of these observations and as background reading for Chapter 14, where modern combustion analysis is discussed. A notation table is included at the end of the chapter.

One ultimate measure of the performance of an internal combustion engine is the proportion of the heat of combustion of the fuel that is turned into useful work at the engine coupling. The thermal efficiency at full load of internal combustion engines ranges from about 20 per cent for small gasoline engines up to more than 50 per cent for large slow-running diesel engines, which are the most efficient means currently available of turning the heat of combustion of fuel into mechanical power.

It is useful to have some idea of the theoretical maximum thermal efficiency that is possible, as this sets a target for the engine developer. Theoretical thermodynamics allows us, within certain limitations, to predict this maximum value. The proportion of the heat of combustion that is not converted into useful work appears elsewhere: in the exhaust gases, in the cooling medium and as convection and radiation from the hot surfaces of the engine. In addition, there may be appreciable losses in the form of unburned or late burning fuel. It is important to be able to evaluate these various losses. Of particular interest are losses from the hot gas in the cylinder to the containing surfaces, since these directly affect the indicated power of the engine. The so-called 'adiabatic engine' seeks to minimize these particular losses.

Some of the heat carried away in the exhaust gas may be converted into useful work in a turbine or used for such purposes as steam generation or the production of hot water.

Fundamentals

Calorific value of fuels

The calorific value of a fuel is defined in terms of the amount of heat liberated when a fuel is burned completely in a calorimeter. Detailed methods and definitions are given in Ref. 1, but for the present purpose the following is sufficient.

Since all hydrocarbon fuels produce water as a product of combustion, part of these products (the exhaust gas in the case of an i.c. engine) consists of steam. If, as is the case in a calorimeter, the products of combustion are cooled to ambient temperature, this steam condenses, and in doing so gives up its latent heat. The corresponding measure of heat liberated is known as the higher or gross calorific value (also known as gross specific energy). If no account is taken of this latent heat we have the lower or net calorific value (also known as net specific energy). Since there is no possibility of an internal combustion engine making use of the latent heat, it is the invariable practice to define performance in terms of the lower calorific value C_1. Table 13.1 shows values of the lower calorific value and density for some typical fuels.

Gaseous fuels

These fuels, which have favourable emissions characteristics, are becoming of increasing importance:[2,3]

- Natural gas (NG), also sometimes described as compressed natural gas (CNG) and, when transported in bulk at very low temperature, as liquefied natural gas (LNG);

Table 13.1   *Properties of liquid fuels*

	Lower calorific value (MJ/kg)	*Stoichiometric air/fuel ratio*	*Density (kg/l)*
Gasoline	43.8	14.6	0.74
Gas oil	42.5	14.8	0.84
Methanol	19.9	6.46	0.729
Ethanol	27.2	8.94	0.79
Light fuel oil	40.6		0.925
Medium fuel oil	39.9	14.4	0.95
Heavy fuel oil	39.7		0.965

* BS 5514 Part 1 *Reciprocating Internal Combustion Engines*: Performance specifies a standard lower calorific value for distillate fuels as 42.7 MJ/kg.

Table 13.2 *Approximate properties of typical gaseous fuels[3]*

Natural gas (North Sea gas)	
Methane CH_4	93.3%
Higher hydrocarbons	4.6%
N_2, CO_2	2.1%
Lower calorific value*	48.0 MJ/kg
Stoichiometric air/fuel ratio	14.5 : 1
Approximate density (gas at 0°C)	0.79 kg/m^3
Liquefied petroleum gas (LPG)	
Propane C_3H_8	90%
Butane C_4H_{10}	5%
Unsaturates	5%
Lower calorific value	46.3 MJ/kg
Stoichiometric air/fuel ratio	15.7 : 1
Approximate density (gas at 0°C)	2.0 kg/m^3

* It should be noted that some gas suppliers commonly quote higher as opposed to lower calorific value. In the case of methane, with its high H/C ratio, the difference is nearly 10%.

- natural gas consists mainly of methane but, having evolved from organic deposits, invariably contains some higher hydrocarbons and traces of N_2 and CO_2. Composition varies considerably from field to field and the reader involved in work on natural gas should establish the particulars of the gas with which he is concerned.
- North Sea gas has become a standard gaseous fuel in the UK and its approximate properties are shown in Table 13.2.

Liquefied petroleum gas (LPG or LP-gas)

LPG is a product of the distillation process of crude oil or a condensate from wet natural gas. It consists largely of propane and, unlike natural gas, can be stored in liquid form at moderate pressures. In view of its good environmental properties (low unburned hydrocarbon emissions, low CO, virtually no particulate emissions and no sulphur), it is finding favour as a vehicle engine fuel. NO_x emissions tend to be higher than for gasoline, owing to its high combustion temperature. It has a high octane number, RON 110, which permits higher compression ratios. Approximate properties are shown in Table 13.2.

Ideal standard cycles: effect of compression ratio

Many theoretical cycles for the internal combustion engine have been proposed, some of them taking into account such factors as the exact course of the combustion process, the variation of the specific heat of air with temperature and the effects of dissociation of the products of combustion at high temperature. However, all these cycles merely modify the predictions of the cycle we shall be considering, generally in the direction of reduced attainable efficiency, without much changing the general picture; for a detailed discussion, see Heywood.[4]

The air standard cycle, also known as the Otto cycle, is shown in Fig. 13.1. It consists of four processes, forming a complete cycle, and is based on the following assumptions:

1. The working fluid throughout the cycle is air, and this is treated as a perfect gas.
2. The compression process 1–2 and the expansion process 3–4 are both treated as frictionless and adiabatic (without heat loss).
3. In place of heat addition by internal combustion phase 2–3 of the process is represented by the addition of heat from an external source, the volume remaining constant.
4. The exhaust process is replaced by cooling at constant volume to the initial temperature.

It will be evident that these conditions differ considerably from those encountered in an engine; nevertheless the thermodynamic analysis, which is very simple, gives useful indications regarding the performance to be expected from an internal combustion engine, in particular with regard to the influence of compression ratio.

The air standard cycle efficiency is given by:

$$\eta_{as} = 1 - \frac{1}{R^{\gamma-1}} \tag{1a}$$

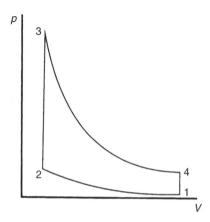

Figure 13.1   *Air standard cycle*

The course of events in an engine cylinder departs from this theoretical pattern in the following main respects:

1. Heat is lost to the cylinder walls, reducing the work necessary to compress the air, the rise in temperature and pressure during combustion, and the work performed during expansion.
2. Combustion, particularly in the diesel engine, does not take place at constant volume, resulting in a rounding of the top of the diagram, point 3, and a reduction in power. A better standard of reference for the diesel engine is the limited pressure cycle, Fig. 13.2, for which the efficiency is given by the expression:

$$\eta = 1 - \frac{1}{R^{\gamma-1}} \left[\frac{\alpha \beta^{\gamma} - 1}{\alpha \gamma (\beta - 1) + \alpha - 1} \right] \tag{1b}$$

where

$$\alpha = \frac{p3}{p2_a} \quad \beta = \frac{V3_b}{V3_a}$$

this reduces to eq. (1a), when $\beta = 1$.

3. The properties of air and of the products of combustion, do not correspond to those of an ideal gas, resulting in a smaller power output than predicted.
4. The gas exchange process is ignored in the standard cycle.

Figure 13.3 shows the variation of air standard cycle efficiency with compression ratio and shows the range of this ratio for spark ignition and diesel engines. It is clearly desirable to use as high a compression ratio as possible. Also shown is the

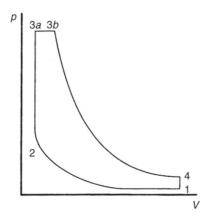

Figure 13.2 *Limited pressure cycle*

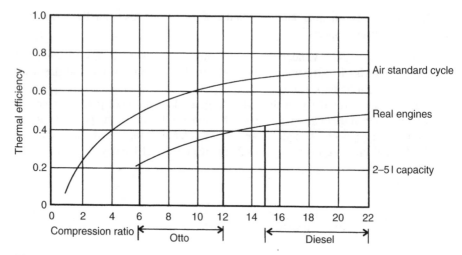

Figure 13.3   *Variation of air standard cycle efficiency with compression ratio*

approximate indicated thermal efficiency to be expected from gasoline and diesel engines of 2–5-litre swept volume. Larger engines and in particular large slow-speed diesel engines can achieve significantly higher efficiencies, mainly because heat losses from the cylinder contents become less in proportion as the size of the individual cylinders increases.

The energy balance of an internal combustion engine

The distribution of energy in an internal combustion engine is best considered in terms of the steady flow energy equation, combined with the concept of the control volume. In Fig. 13.4 an engine is shown, surrounded by the control surface. The various flows of energy into and out of the control volume are shown.

In

- fuel, with its associated heat of combustion;
- air, consumed by the engine.

Out

- power developed by the engine;
- exhaust gas;
- heat to cooling water or air;
- convection and radiation to the surroundings.

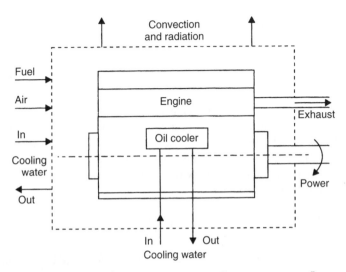

Figure 13.4 *Control volume of an i.c. engine showing energy flows*

The *steady flow energy equation* gives the relationship between these quantities, and is usually expressed in kilowatts:

$$H_1 = P_s + (H_2 - H_3) + Q_1 + Q_2 \tag{2}$$

in which the various terms have the following meanings:

H_1 = combustion energy of fuel = $\dot{m}_f C_L \times 10^3$
P_s = power output of engine
H_2 = enthalpy of exhaust gas* = $(\dot{m}_f + \dot{m}_a) C_p T_e$
H_3 = enthalpy of inlet air = $\dot{m}_a C_p T_a$
Q_1 = heat to cooling water[†] = $\dot{m}_w C_w (T_{2w} - T_{1w})$
Q_2 = convection and radiation.

This assumes that the specific heat of the exhaust gas, the mass of which is the sum of the masses of air and fuel supplied to the engine, is equal to that of air. This is not strictly true, but permits an approximate calculation to be made if the temperature of the exhaust gas is measured (exact measurement of exhaust temperature is no simple matter, see Chapter 20).

Note that it is not possible to show the indicated power directly in this energy balance since the difference between it and the power output P_s representing friction

---

* For detailed description of enthalpy see, for example, Refs 4 and 5.
[†] This may also include heat transferred to the lubricating oil and subsequently to the cooling water via an oil cooler.

and other losses appears elsewhere as part of the heat to the cooling water Q_1 and other losses Q_2.

Measurement of heat losses: heat to exhaust

If air and fuel flow rates, also exhaust temperature, are known, this may be calculated approximately, see H_2 above. For an accurate measurement of exhaust heat, use can be made of an exhaust calorimeter. This is a gas-to-water heat exchanger in which the exhaust gas is cooled to a moderate temperature and the heat content measured from observation of cooling water flow rate and temperature rise.

The expression for H_2 becomes:

$$H_2 = \dot{m}_c C_W (T_{2c} - T_{1c}) + (\dot{m}_f + \dot{m}_a) C_p T_{co} \tag{3}$$

The rate of flow of cooling water through the calorimeter should be regulated so that the temperature of the gas leaving the calorimeter, T_{co}, does not fall below about 60°C (333K). This is approximately the dew point temperature for exhaust gas: at lower temperatures the steam in the exhaust will start to condense, giving up its latent heat, see section on Calorific value of fuels.

Sample calculation: analysis of an engine test

Table 13.3 is an analysis, based on eq. (2), of one test point in a sequence of tests on a vehicle engine.

Table 13.3 *Energy balance of a gasoline engine at full throttle (four cylinder, four-stroke engine, swept volume 1.7 litre)*

Engine speed	3125 rev/min
Power output	$P_s = 36.8\,\text{kW}$
Fuel consumption rate	$\dot{m}_f = 0.00287\,\text{kg/s}$
Air consumption rate	$\dot{m}_a = 0.04176\,\text{kg/s}$
Lower calorific value of fuel	$C_L = 41.87 \times 10^6\,\text{J/kg}$
Exhaust temperature	$T_c = 1066\,\text{K}\ (793°\text{C})$
Cooling water flow	$\dot{m}_w = 0.123\,\text{kg/s}$
Cooling water inlet temperature*	$T_{1w} = 9.2°\text{C}$
Cooling water outlet temperature*	$T_{2w} = 72.8°\text{C}$
Inlet air temperature	$T_a = 292\,\text{K}\ (19°\text{C})$

* The engine was fitted with a heat exchanger. These are the temperatures of the primary cooling water flow to the exchanger.

Then noting that:

Specific heat of air at constant pressure	$C_p = 100 \text{ kJ/kg K}$
Specific heat of water	$C_w = 4.18 \text{ kJ/kg K}$
$H_1 = 0.00287 \times 41.87 \times 10^3$	$= 120.2 \text{ kW}$
P_s	$= 36.8 \text{ kW}$
$H_2 = (0.00287 + 0.04176) \times 1.00 \times 1066$	$= 47.6 \text{ kW}$
$H_3 = 0.04176 \times 1.00 \times 292$	$= 12.2 \text{ kW}$
$H_2 - H_3$	$= 35.4 \text{ kW}$
$Q_1 = 0.123 \times 4.18 (72.8 - 9.2)$	$= 32.7 \text{ kW}$
Q_2 (by difference)	$= 15.3 \text{ kW}$

We may now draw up an energy balance (quantities in kilowatts):

Heat of combustion H_1	120.2	Power output P_s	36.8 (30.6%)
		Exhaust $(H_2 - H_3)$	35.4 (29.5%)
		Other losses Q	15.3 (12.7%)
	120.2		120.2 (100%)

The thermal efficiency of the engine

$$\eta_{th} = \frac{P_s}{H_1} = 0.306$$

The compression ratio of the engine $R = 8.5$, giving:

$$\eta_{as} = 1 - \frac{1}{8.5^{1.4-1}} = 0.575$$

The mechanical efficiency of this engine at full throttle was approximately 0.80, giving an indicated thermal efficiency of

$$\frac{0.306}{0.8} = 0.3825$$

This is approximately two thirds of the air standard efficiency.

Sample calculation: exhaust calorimeter

In the test analysed in Table 13.3, the heat content of the exhaust was also measured by an exhaust calorimeter with the following result:

cooling water flow, $\dot{m}_c$	$= 0.139 \text{ kg/s}$
cooling water inlet temperature, T_{1c}	$= 9.2°C$
cooling water outlet temperature, T_{2c}	$= 63.4°C$
exhaust temperature leaving calorimeter, T_{co}	$= 355 \text{ K } (82°C)$

Then from eq. (3):

$$H_2 = 0.139 \times 4.18\,(63.4 - 9.2) + (0.00287 + 0.04176) \times 1.00 \times 355 = 47.3\,\text{kW}$$

$H_3 = 12.2\,\text{kW}$, as before, giving heat to exhaust $H_2 - H_3 = 35.1\,\text{kW}$. This shows a satisfactory agreement with the approximate value derived from the exhaust temperature and air and fuel flow rates.

Energy balances: typical values

Figure 13.5 shows full power energy balances for several typical engines. Results for the gasoline engine analysed above are shown at (a).

Table 13.4 shows energy balances in terms of heat losses per unit power output for various engine types. This will be found useful when designing such test cell services as cooling water and ventilation. They are expressed in terms of kW/kW power output.

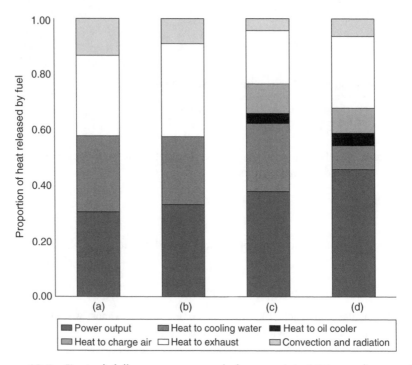

Figure 13.5 *Typical full power energy balances: (a) 1.71 gasoline engine (1998); (b) 2.51 naturally aspirated diesel engine; (c) 200 kW medium speed turbocharged marine diesel; (d) 7.6 MW combined heat and power unit*

Table 13.4 *Energy balance (kW per kW power output)*

	Automotive gasoline	Automotive diesel	Medium speed heavy diesel
Power output	1.0	1.0	1.0
Heat to cooling water	0.9	0.7	0.4
Heat to oil cooler			0.05
Heat to exhaust	0.9	0.7	0.65
Convection and radiation	0.2	0.2	0.15
Total	3.0	2.6	2.2

These proportions depend on the thermal efficiency of the engine and are only an approximate guide.

The role of indicated power in the energy balance

It should be observed that the indicated power output of the engine does not appear in any of our formulations of the energy balance. There is a good reason for this: the difference between indicated and brake power represents the friction losses in the engine and the power required to drive the auxiliaries, and it is impossible to allocate these between the various heat losses in the balance.

Most of the friction losses between piston and cylinder will appear in the cooling water; bearing losses and the power required to drive the oil pump will appear mostly in the oil cooler, while water pump losses will appear directly in the cooling water. An exact analysis is problematic and beyond the scope of this book.

Sample calculation: prediction of energy balance

Chapter 2 deals with the concept of the engine test cell as a thermodynamic system and the recommendation is made that at an early stage in designing a new test cell an estimate should be made of the various flows: fuel, air, water, electricity, heat and energy into and out of the cell.

In such cases it is usually necessary to make some assumptions as to the full power performance of the largest engine to be tested. This is possible on the basis of information given earlier and in this chapter and an example follows.

Prediction of energy balance

Taking the example of a 250 kW turbocharged diesel engine at full power. Assume specific fuel consumption = 0.21 kg/kWh (LCV 40.6 MJ/kg; thermal efficiency 0.42). Then following the general recommendations of Table 13.4, we may make an estimate, summarized in Table 13.5.

The corresponding thermodynamic system is shown in Fig. 13.6. This also shows rates of flow of fuel, air and cooling water, based on the following estimates:

Fuel flow
Assume fuel density 0.9 kg/litre

$$\text{Fuel flow} = \frac{250 \times 0.21}{0.9} = 581 \, \text{l/hour} \ (52.5 \, \text{kg/hour})$$

Table 13.5   *Energy balance, 250 kW turbocharged diesel engine*

In		Out	
Fuel	592 kW	Power	250 kW (42.2%)
		Heat to cooling water	110 kW (18.6%)
		Heat to oil cooler	15 kW (2.5%)
		Heat to exhaust	177 kW (29.9%)
		Convection and radiation	40 kW (6.8%)
	592 kW		592 kW

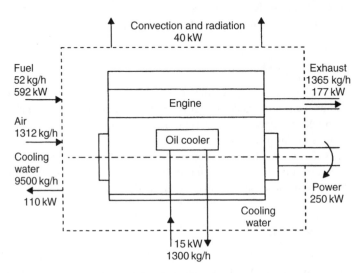

Figure 13.6   *Control volume, 250 kW diesel engine showing energy and fluid flows*

Induction air flow
Assume full load air/fuel ratio $= 25 : 1$
Air flow $= 250 \times 0.21 \times 25 = 1312.5 \, \text{kg/h}$
Taking air density as $1.2 \, \text{kg/m}^3$
Air flow $= 1094 \, \text{m}^3/\text{h}$ ($0.30 \, \text{m}^3/\text{s}$, $10.7 \, \text{ft}^3/\text{s}$)

Cooling water flow
Assume a temperature rise of 10°C through the jacket and oil cooler. Then since the specific heat of water $4.18 \, \text{kJ/kg} \, °\text{C}$ and $1 \, \text{kWh} = 3600 \, \text{kJ}$:
flow to jacket + oil cooler

$$\frac{125 \times 60}{4.18 \times 10} = 180 \, \text{kg/min} \; (180 \, \text{l/min})$$

Exhaust flow
Sum of fuel flow + induction air flow $= 1312.5 + 52.5 = 1365 \, \text{kg/h}$.

Energy balance for turbocharger[6]

In the case of turbocharged engines, it is useful to separate the energy flows to and from the turbocharger and associated air cooler from those associated with the complete engine. In this way an energy balance may be drawn up covering the following:

- exhaust gas entering the turbine from engine cylinders;
- induction air entering the compressor;
- cooling water entering the air cooler;
- exhaust leaving the turbine;
- induction air leaving the cooler;
- cooling water leaving the cooler.

A separate control volume contained within the control volume for the complete engine may be defined and an energy balance drawn up.

Summary

Stages in calculation of energy balance from an engine test
1. Obtain information on the lower calorific value of the fuel.
2. From a knowledge of compression ratio of engine calculate air standard cycle efficiency as a yardstick of performance.
3. For one or a number of test points measure:
   - fuel consumption rate;
   - air flow rate to engine;

- exhaust temperature;
- power output;
- cooling water inlet and outlet temperatures and flow rate.
4. Calculate the various terms of the energy balance and the thermal efficiency from this data.

Prediction of energy balance for a given engine

1. Record type of engine and rated power output.
2. Calculate fuel flow rate from known or assumed specific fuel consumption.
3. Draw up energy balance using the guidelines given in Table 13.4.

Measurement of mechanical losses in engines

This section is devoted to a critical study of the various methods of determining the mechanical efficiency of an internal combustion engine.

It is a curious fact that, in the long run, all the power developed by all the road vehicle engines in the world is dissipated as friction: either mechanical friction in the engine and transmission, rolling resistance between vehicle and road or wind resistance.

Mechanical efficiency, a measure of friction losses in the engine, is thus an important topic in engine development and therefore engine testing. It may exceed 80 per cent at high power outputs, but is generally lower and is of course zero when the engine is idling.

Under mixed driving conditions for a passenger vehicle between one third and one half of the power developed in the cylinders is dissipated either as mechanical friction in the engine, in driving the auxiliaries such as alternator and fan, or as pumping losses in the induction and exhaust tracts. Since the improvement of mechanical efficiency is such an important goal to engine and lubricant manufacturers, an accurate measure of mechanical losses is of prime importance. In fact the precise measurement of these losses is a particularly difficult problem, to which no completely satisfactory solution exists.

Fundamentals

The starting point in any investigation of mechanical losses should ideally be a precise knowledge of the power developed in the engine cylinder. This is represented by the indicator diagram, Fig. 13.7, which shows the relation between the pressure of the gas in the cylinder and the piston stroke or swept volume (also see Chapter 14). For a four-stroke engine, account must be taken of both the positive area A_1, representing the work done on the piston during the compression and expansion strokes, and

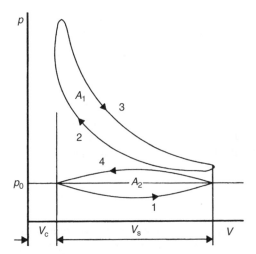

Figure 13.7 *Indicator diagram, four-stroke engine. 1, induction; 2, compression; 3, expansion; 4, exhaust*

the negative area A_2, representing the 'pumping losses', the work performed by the piston in expelling the exhaust gases and drawing the fresh charge into the cylinder.

In the case of an engine fitted with a mechanical supercharger or an exhaust gas turbocharger, there are additional exchanges of energy between the exhaust gas and the fresh charge, but this does not invalidate the indicator diagram as a measure of power developed in the cylinder.

The universally accepted measure of indicated power is the indicated mean effective pressure (i.m.e.p.). This represents the mean positive pressure exerted on the piston during the working strokes after allowing for the negative pressure represented by the pumping losses.

The relation between i.m.e.p. and indicated power of the engine is given by the expression:

$$P_i = \frac{\overline{p_i} V_s n}{60K} \times 10^{-1} \text{ kW} \tag{4}$$

here $n/60K$ represents the number of power strokes per second in each cylinder.

The useful power output of the engine may be represented by the brake mean effective pressure

$$P = \frac{\overline{p_b} V_s n}{60K} \times 10^{-1} \text{ kW} \tag{5}$$

The mechanical losses in the engine plus power to drive auxiliaries are represented by $(P_i - P)$ and may be represented by the friction mean effective pressure

$$\bar{p}_f = \bar{p}_i - \bar{p}_b$$

It may not be realistic, for several experimental reasons, to place much faith in measurements of indicated power based on the indicator diagram. Several other techniques, each having their limitations, will be discussed and the critical determination of top dead centre (t.d.c.) is covered in detail in Chapter 14.

Motoring tests

One method of estimating mechanical losses involves running the engine under stable temperature conditions and connected to a four-quadrant dynamometer. Ignition or fuel injection are then cut and the quickest possible measurement made of the power necessary to motor the engine at the same speed.

Sources of error include:

- Under non-firing conditions, the cylinder pressure is greatly reduced, with a consequent reduction in friction losses between piston rings, cylinder skirt and cylinder liner and in the running gear.
- The cylinder wall temperature falls very rapidly as soon as combustion ceases, with a consequent increase in viscous drag that may to some extent compensate for the above effect.
- Pumping losses are generally much changed in the absence of combustion.

Many detailed studies of engine friction under motored conditions have been reported in the literature, for a summary see Ref. 7. These usually involved the progressive removal of various components: camshaft and valve train, oil and fuel pumps, water pump, generator, seals, etc., in order to determine the contribution made by each element.

The Morse test

In this test, the engine is run under steady conditions and ignition or injection is cut off in each cylinder in turn: it is of course only applicable to multicylinder engines.

On cutting out a cylinder, the dynamometer is rapidly adjusted to restore the engine speed and the reduction in power measured. This is assumed to be equal to the indicated power contributed by the non-firing cylinder. The process is repeated for all cylinders and the sum of the reductions in power is taken to be a measure of the indicated power of the engine.

A modification of the Morse test[8] makes use of electronically controlled unit injectors, allowing the cylinders to be disabled in different ways and at different frequencies, thus keeping temperatures and operating conditions as near normal as possible.

The Morse test is subject, though to a less extent, to the sources of error described for the motoring test.

The Willan's line method

This is applicable only to unthrottled compression ignition engines. It is a matter of observation that a curve of fuel consumption rate against torque or b.m.e.p. at constant speed plots quite accurately as a straight line up to about 75 per cent of full power, Fig. 13.8. This suggests that for the straight line part of the characteristic, equal increments of fuel produce equal increments of power; combustion efficiency is constant.

At zero power output from the engine, all the fuel burned is expended in overcoming the mechanical losses in the engine, and it is a reasonable inference that an extrapolation of the Willan's line to zero fuel consumption gives a measure of the friction losses in the engine.

Strictly speaking, the method only allows an estimate to be made of mechanical losses under no-load conditions. When developing power the losses in the engine will undoubtedly be greater.

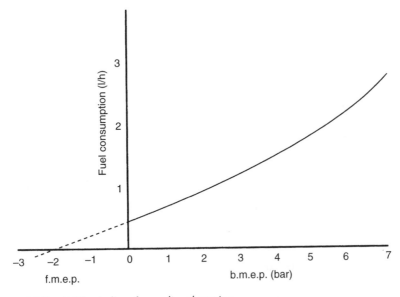

Figure 13.8 *Willan's line for a diesel engine*

Summary

The four standard methods of estimating mechanical losses in an engine and its auxiliaries have been briefly described. No great accuracy can be claimed for any of these methods and it is instructive to apply as many of them as possible and compare the results (for a critical assessment, see Ref. 9). Measurement of mechanical losses in an engine is still something of an 'art'.

While no method can be claimed to give a precise absolute value for mechanical losses they are, of course, quite effective in monitoring the influence of specific changes made to a particular engine.

The intention is to give non-expert readers the essential background to this area of engine performance testing so that the use of modern tools used in engine indicating and combustion analysis described briefly in Chapter 14 can be better appreciated.

Notation

Indicated mean effected pressure (i.m.e.p.)	$\bar{p}_i$ bar
Brake mean effected pressure (b.m.e.p.)	$\bar{p}_b$ bar
Friction mean effected pressure (f.m.e.p.)	$\bar{p}_f$ bar
Swept volume of engine	V_s litre
Engine speed	n rev/min
Constant: 1 for two-stroke, 2 for four-stroke engines	K
Indicated power	P_i kW
Engine power output	P kW

Thermal efficiency	
Lower calorific value	C_L MJ/kg
Compression ratio	R
Ratio of specific heat of air	$\gamma\,(=1.4)$
Air standard efficiency	η_{as}
Mass flow rate of fuel	$\dot{m}_f$ kg/s
Mass flow rate of inlet air	$\dot{m}_a$ kg/s
Power output of engine	P_s kW
Specific heat of air at constant pressure	C_p kJ/kg K
Specific heat of water	C_w kJ/kg K
Ambient temperature	T_a K
Exhaust temperature	T_e K
Cooling water inlet temperature	T_{1w} K
Cooling water outlet temperature	T_{2w} K
Mass flow rate of cooling water	$\dot{m}_w$ kg/s

For exhaust calorimeter

Mass flow rate of cooling water	$\dot{m}_C$ kg/s
Cooling water inlet temperature	T_{1C} K
Cooling water outlet temperature	T_{2C} K
Temperature of exhaust leaving calorimeter	T_{CO} K

References

1. BS 7420 *Guide for Determination of Calorific Values of Solid, Liquid and Gaseous Fuels* (including Definitions).
2. *A.S.T.M. Special Technical Publication 525* (1972) *LP-Gas Engine Fuels*, American Society for Testing and Materials.
3. Goodyear, E.M. (1980) *Alternative Fuels*, Cranfield, Macmillan, London.
4. Heywood, J.B. (1988) *Internal Combustion Engine Fundamentals*, McGraw-Hill, Maidenhead.
5. Eastop, T.D. and McConkey, A. (1993) *Applied Thermodynamics for Engineering Technologists*, Longman, London.
6. Watson, N. and Janota, M.S. (1982) *Turbocharging the Internal Combustion Engine*, Wiley-Interscience, New York.
7. Martin, F.A. (1985) *Friction in Internal Combustion Engine Bearings*, I. Mech. E. Paper C 67/85.
8. Haines, S.N.M. and Shields, S.A. (1989) The determination of diesel engine friction characteristics by electronic cylinder disablement, *Proc. I. Mech. E. Part A*, **203** (A2), 129–138.
9. Ciulli, E. (1993) A review of internal combustion engine losses. Part 2: Studies of global evaluations, *Proc. I. Mech. E. Part D*, **207** (D3), 229–240.

14 The combustion process and combustion analysis

Introduction

This chapter deals with the process of combustion in the internal combustion engine and the equipment available to the test engineer for its analysis. The processes taking place in the cylinder and combustion chamber are central to the performance of the internal combustion engine; in fact, the rest of the engine may be regarded merely as a device for managing these processes and for extracting useful work from them.

The term 'combustion analysis' (CA) has become somewhat interchangeable with that of 'engine indicating' (EI), the latter term covers both combustion analysis and a number of other engine phenomena that occur at the same timescale as combustion, such as injector needle movement. In this book the term 'engine indicating' is used to cover the general task of high speed engine data analysis and the term 'combustion analysis' is used in its specific sense.

In many test cells in the world, throughout the 1990s and up to the present date, combustion analysis equipment may be a temporary visitor; brought in when required and operated in parallel with the test bed automation system. Temporary installations may give problems and inconsistencies due to signal interference and it is now common to find dedicated engine calibration cells with engine indicating equipment as a permanent part of an optimized installation.

Currently there is no ASAM standard for the storage of raw CA data, so they tend to be stored within structured binary files designed by the makers of the analysis equipment.

There may be fewer than 10 specialist suppliers of complete IE hardware and software suites worldwide and probably only half these are also specialist CA transducer manufacturers. Any search of the internet will result in discovery of many universities and private test-houses that have assembled their own analysis systems from commercially available components.

If the instrumentation and software suites of two or more subsystems are from different suppliers, it is vital that the communications protocols are mutually understood and that one competent party is responsible for integration. Any engine control unit communication link would normally be designed to an industrial standard, such as the current ASAP-3, but third party testers may find that such outputs as alarm codes require manufacturers' documentation that is not in the public domain.

To fully exploit the capabilities of the leading engine calibration systems, the cells are often set up to run unmanned for prolonged periods. The system alarms and interlocks, including remote alarm indication, must be carefully considered for such cells.

As with many other aspects of engine testing, the correlation of results produced on different systems, some of which are not well integrated, ranges from difficult to self-deluding.

Fundamental influences on combustion

At a fundamental level, the performance of an internal combustion engine is largely determined by the events taking place in the combustion chamber and engine cylinder. These are influenced by a large number of factors:

- configuration of the combustion chamber and cylinder head;
- flow pattern (swirl and turbulence) of charge entering the cylinder, in turn determined by design of induction tract and inlet valve size, shape, location, lift and timing;
- ignition and injection timing, spark plug position and characteristics, injector and injection pump design, location of injector;
- compression ratio;
- air/fuel ratio;
- fuel properties;
- mixture preparation;
- exhaust gas recirculation (EGR);
- degree of cooling of chamber walls, piston and bore.

For those readers not familiar with the details of the combustion process of the spark ignition (SI) or compression ignition (CI) engine, they are summarized below.

Combustion in the gasoline engine

1. A mixture of gas and vapour is formed, either in the induction tract, in which the fuel is introduced either by a carburettor or injector, or in the cylinder in the case of port injection.
2. Combustion is initiated at one or more locations by an electric spark.
3. After a delay, the control of which still presents problems, a flame is propagated through the combustible mixture at a rate determined, among other factors, by the air motion in the cylinder.
4. Heat is released progressively with a consequent increase in temperature and pressure. During this process the bulk properties of the fluid change as it is transformed from a mixture of air and fuel to a volume of combustion products.

Undesirable effects, such as preignition, excessive rates of pressure rise, late burning and detonation or 'knock' may be present.

5.   Heat is transferred by radiation and convection to the surroundings.
6.   Mechanical work is performed by expansion of the products of combustion.

The development of the combustion chamber, inlet and exhaust passages and fuel supply system of a new engine involve a vast amount of experimental work, some on flow rigs that model the geometry of the engine, most of it on the test bed.

Combustion in the diesel engine

1.   Air is drawn into the cylinder, without throttling, but frequently with pressure charging. Compression ratios range from about 14:1 to 22:1, depending on the degree of supercharge, resulting in compression pressures in the range 40–60 bar and temperatures from 700 to 900°C.
2.   Fuel is injected at pressures that have increased in recent years from 600 to 1500 bar or higher. Charge air temperature is well above the autoignition point of the fuel. Fuel droplets vaporize, forming a combustible mixture which ignites after a delay which is a function of charge air pressure and temperature, droplet size and fuel ignition quality (cetane number).
3.   Fuel subsequently injected is ignited immediately and the progress of combustion and pressure rise is to some extent controlled by the rate of injection. Air motion in the combustion chamber is organized to bring unburned air continually into the path of the fuel jet.
4.   Heat is transferred by radiation and convection to the surroundings.
5.   Mechanical work is performed by expansion of the products of combustion.

Choice of combustion 'profile' involves a number of compromises. A high maximum pressure has a favourable effect on fuel consumption, but increases NO_x emissions, while a reduction in maximum pressure, brought about by retarding the combustion process, results in increased particulate emissions. Maximum combustion pressures can exceed 200 bar.

Unlike the spark-ignition engine, the diesel must run with substantial excess air to limit the production of smoke and soot. A precombustion chamber engine can run with a lambda ratio of about 1.2 at maximum power, while a direct injection (DI) engine requires a minimum excess air factor of about 50 per cent, roughly the same as the maximum at which a spark-ignition engine will run. It is this characteristic that has led to the widespread use of pressure charging, which achieves a reasonable specific output by increasing the mass of the air charge.

Large industrial or marine engines invariably use direct injection; in the case of vehicle engines, the indirect injection or prechamber engine is being abandoned in favour of direct injection because of the better fuel consumption and cold starting performance of the DI engine. Compact very high pressure injectors and electronic control of the injection process have made this development possible.

Effects on combustion process of air/fuel ratio

Anyone who attempts to 'tune' a gasoline engine must have a clear understanding of this subject to make any progress with the task.

A gasoline engine is capable of operating on a range of air-to-fuel ratios by weight from about 81 to 201, and weaker than this in the case of stratified charge and lean-burn engines. Several definitions are important:

- *Mixture strength*. A loose term usually identified with air/fuel ratio and described as 'weak' (excess air) or 'rich' (excess fuel).
- *Air/fuel ratio*. Mass of air in charge to mass of fuel.
- *Stoichiometric air/fuel ratio* (sometimes known as 'correct' air/fuel ratio). The ratio at which there is exactly enough oxygen present for complete combustion of the fuel. Most gasolines lie within the range 14:1 to 15:1 and a ratio of 14.51:1 may be used as a rule of thumb. Alcohols, which contain oxygen in their make-up, have a much lower value in the range 7:1 to 9:1.
- *Excess air factor or 'lambda'(λ) ratio*. The ratio of actual to stoichiometric air/fuel ratio. The range is from about 0.6 (rich) to 1.5 (weak). Lambda ratio has a great influence on power, fuel consumption and emissions.
- *The equivalence ratio*, ϕ, is the reciprocal of the lambda ratio and is preferred by some authors.

A basic test is to vary the air/fuel ratio over the whole range at which the engine is able to run, keeping throttle opening and speed constant. The results are often presented in the form of a 'hook curve', Fig. 14.1, which shows the relation between specific fuel consumption and b.m.e.p. over the full range of mixture strengths.

If such a test is carried out on an engine fitted with a quartz observation window the following changes are observed:

- At mixture strength corresponding to maximum power and over a range of weaker mixtures combustion takes place smoothly and rapidly with a blue flame which is extinguished fairly early in the expansion stroke.
- With further weakening combustion becomes uneven and persists throughout the expansion stroke. 'Popping back' into the induction manifold may occur.
- As we proceed towards richer mixtures, the combustion takes on a yellow colour, arising from incandescent carbon particles, and may persist until exhaust valve opening. This may lead to explosions in the exhaust system.

The following features of Fig. 14.1 call for comment:

- Point a corresponds to the weakest mixture at which the engine will run. Power is much reduced and specific fuel consumption can be as much as twice that corresponding to best efficiency.

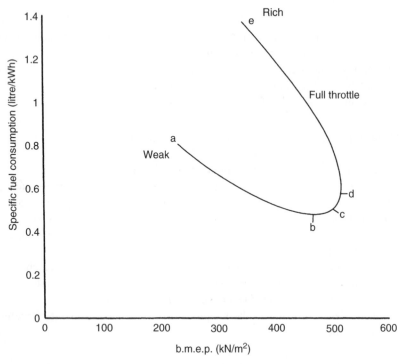

Figure 14.1   *Hook curve for gasoline engine*

- Point b corresponds to the best performance of the engine (maximum thermal efficiency). The power output is about 95 per cent of that corresponding to maximum power.
- Point c corresponds to the stoichiometric ratio.
- Point d gives maximum power, but the specific consumption is about 10 per cent greater than at the point of best efficiency. It will be evident that a prime requirement for the engine management system must be to operate at point b, except when maximum power is demanded.
- Point e is the maximum mixture strength at which the engine will run.

It is possible to produce similar curves for the whole range of throttle positions and speeds and hence to derive a complete map of optimum air and fuel flow rates as one of the bases for development of the engine management system, whether it be a traditional carburettor or a computer-controlled injection system.

If at the same time that the hook curve is produced the air flow rate is measured, the same information may be presented in the form of curves of power output and specific consumption against air/fuel ratio or λ, see Fig. 14.2, which corresponds to Fig. 14.1.

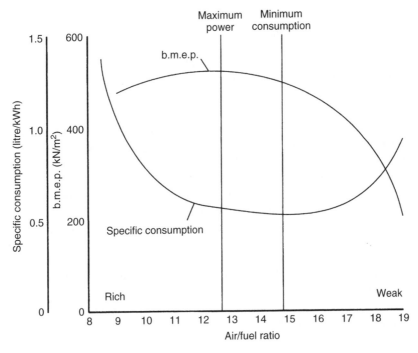

Figure 14.2 *Variation of power output and specific fuel consumption with air/fuel ratio*

Cylinder pressure diagrams

Many special techniques have been developed for the study of the combustion process. The oldest is direct observation of the process of flame propagation, using high-speed photography through a quartz window. More recent developments include the use of flame ionization detectors (FID) to monitor the passage of the flame and the proportion of the fuel burned, also hot-wire and laser Doppler anemometry.

The standard tool for the study of the combustion process is the cylinder pressure indicator; the different attributes of which are described later in this chapter. In addition to cylinder pressure, a variety of quantities must be measured, using appropriate transducers, either on a time base or synchronized with crankshaft position. They may include

- fuel line pressure and needle lift;
- cylinder pressure;
- ionization signals;
- crank angle;
- time;
- ignition system events;

- inlet and exhaust pressures;
- various (quasi-static) temperatures.

Each signal calls for individual treatment and appropriate recording methods. Data acquisition rates have increased in recent years and systems are available with a sampling rate of up to 1 MHz on 16 channels. This is equivalent to cylinder pressures taken at 0.1° intervals at 16 000 rev/mm.

Combustion analysis aims at an understanding of all features of the process, in particular of the profile of heat release. The test engineer concerned with engine development should be familiar with both the principles involved and the considerable problems of interpretation of results that arise. A comprehensive account of the theory of combustion analysis would much exceed the scope of this book, but a description of the essential features follows.

The aim of combustion analysis is to produce a curve relating mass fraction burned (in the case of a spark ignition engine) or cumulative heat release (in the case of a diesel engine) to time or crank angle. Derived quantities of interest include rate of burning or heat release per degree crank angle and analysis of heat flows based on the first law of thermodynamics.

Stone and Green-Armytage[1] describe a simplified technique for deriving the burn rate curve from the indicator diagram using the classical method of Rassweiler and Withrow.[2] This takes as the starting point a consideration of the process of combustion in a constant volume bomb calorimeter. Figure 14.3 shows a curve of pressure and rate of change of pressure against time. It is then assumed that mass fraction burned and cumulative heat release, at any stage of the process, are directly proportional to the pressure rise.

Combustion in the engine is assumed to follow a similar course, the difference being that the process does not take place at constant volume. There are three different effects to be taken into account:

- pressure changes due to combustion;
- pressure changes due to changes in volume;
- pressure changes arising from heat transfer to or from the containing surfaces.

Figure 14.4 shows a pressure–crank angle diagram and indicates the procedure. The starting point is the 'motored' or 'no burn' curve, shown dotted. This curve used to be obtained by interrupting the ignition or injection for a single cycle and recording the corresponding pressure diagram; certain engine indicators are able to do this. With the more common availability of motoring dynamometers it is possible to obtain the curve by motoring the unfired engine. An alternative, less accurate, method is to fit a polytropic compression line to the compression curve prior to the start of combustion and to extrapolate this on the assumption that the polytropic index n_c in the expression $pv^{n_c} = $ constant remains unchanged throughout the remainder of the compression stroke.

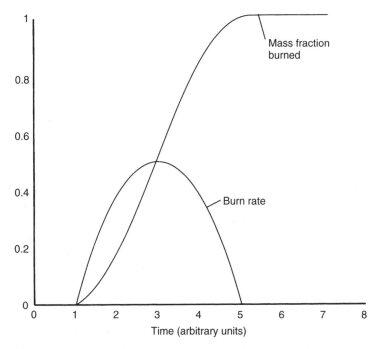

Figure 14.3 *Combustion in a constant volume bomb calorimeter*

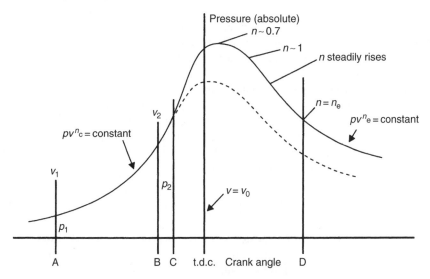

Figure 14.4 *Pressure–crank angle diagram showing derivation of mass fraction burned*

If A and B are two points on the compression line at which volumes and (absolute) pressures are respectively v_1, v_2 and p_1, p_2, then the index of compression is given by:

$$n_c = \frac{\log(p_1/p_2)}{\log(v_2/v_1)} \tag{1}$$

The value of the index n_c during compression is likely to be in the region of 1.3, to be compared with the value of 1.4 for the adiabatic compression of air. The lower value is the result of heat losses to the cylinder walls and, in the case of spark ignition engines, to the heat absorbed by the vaporization of the fuel. The start of combustion is reasonably well defined by point C, at which the two curves start to diverge.

It is now necessary to determine the point D, Fig. 14.4, at which combustion may be deemed to be complete. Various methods may be used, but perhaps the most practical one is a variation on the method described above for determining the start of combustion. Choose two points on the expansion line sufficiently late in the stroke for it to be reasonable to assume that combustion is complete but before exhaust valve opening: 90° and 135° after t.d.c. may be a reasonable choice. The polytropic index of expansion (in the absence of combustion) n_e is calculated from eq. (1). This is also likely to be in the region of 1.3, except in the special case when burning continues right up to exhaust valve opening, as in a spark-ignition engine burning a weak mixture.

The next step is to calculate the polytropic index of compression/expansion for successive intervals, typically one degree of crank angle, from the start of combustion. Once again eq. (1) is used, inserting the pressures and volumes at the beginning and end of each interval. The value of n varies widely as combustion proceeds, but towards the end of the process it converges on the value n_e determined above.

The point D at which the two indices become equal is generally ill-defined, as there is likely to be considerable scatter in the values derived for successive intervals, but fortunately the shape of the burn rate curve is not very sensitive to the position chosen for point D.

The curve of heat release or cumulative mass fraction burned is derived as follows. Figure 14.5a shows diagrammatically an element of the indicator diagram before t.d.c. but after the start of combustion. During this interval the pressure rises from p_1 to p_2, while the pressure rise corresponding to the (no combustion) index n_c is from p_1 to p_0. It is then assumed that the difference between p_2 and p_0 represents the pressure rise due to combustion. It is given by:

$$\Delta p = p_2 - p_0 = p_2 - p_1 \left(\frac{v1}{v2}\right)^{n_c} \tag{2}$$

It is now necessary to refer this pressure rise to some constant volume, usually taken as v_0, the volume of the combustion chamber at t.d.c.. It is assumed that the pressure rise is inversely proportional to the volume:

$$\Delta p_c = \Delta p \frac{(v_1 + v_2)}{2v_0} \tag{3}$$

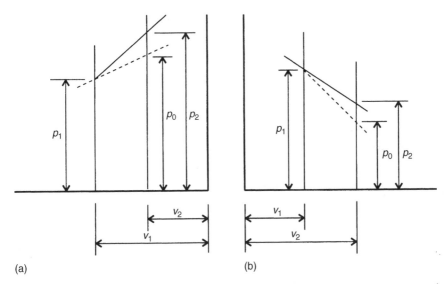

Figure 14.5 *Derivation of mass fraction burned: (a) compression; (b) expansion*

Essentially the same procedure is followed during the expansion process, Fig. 14.5b, except that here there is a pressure fall corresponding to the (no combustion) index of expansion n_e. Δp and Δp_c are calculated from eqs (2) and (3) as before.

Finally Δp_c is summed for the whole combustion period, and the resulting curve, Fig. 14.6, is taken to represent the relation between mass fraction burned or cumulative heat release and crank angle. The 'tail' of the curve, following the end of combustion, point D, will be horizontal if the correct value of n_e has been assumed. If n_e has been chosen too low, it will slope downwards and if too high upwards.

As with much combustion analysis work, there are a number of assumptions implicit in this method, but these do not seriously affect its value as a development tool. The technique is particularly valuable in the case of the DI diesel engine, in which it is found that the formation of exhaust emissions is very sensitive to the course of the combustion process, which may be controlled to some extent by changing injection characteristics and air motion in the cylinder. Figure 14.7 from Ref. 3 shows the time scale of a typical diesel combustion process.

Figure 14.8 shows an indicator diagram taken from a small single cylinder diesel engine and the corresponding heat release curve derived by means of a computer program modelled on the method of calculation described above.

The start of combustion, at 9° before t.d.c., is clearly defined and lags the start of injection by 6°. Combustion is complete by about 60° after t.d.c. and the derived value of the (no combustion) index of expansion in this case was 1.36 (smaller engines generally display a higher index, owing to the higher heat losses in proportion to

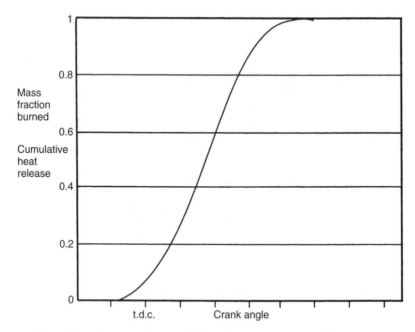

Figure 14.6   *Mass fraction burned plotted against crank angle*

volume). Figure 14.9 shows a similar trace taken from a larger automotive diesel by a combustion analysis instrument.

Total and instantaneous energy release

Many of the required results used by the engine developer are derived from the fundamental calculations relating to energy release in the cylinder. This can be expressed as total energy per cycle (expressed as a pressure value × m.e.p.) or in the calculation of instantaneous energy release with respect to crank angle (either as a rate or integral curve).

These calculations are well known and supported by established theory. Experimenters may adjust or adapt the calculations according to engine type or application. Any assumptions made in these calculations, in order to simplify them and reduce calculation time, can be adjusted according to preference and laboratory procedure. However, this can lead to the analysis becoming somewhat subjective, it is therefore important to know which theory and calculation method has been adopted when comparing results.

In this section, the basic principle behind these calculations will be discussed for the purpose of an overview; established publications shown in the reference section can be used as a source if more detailed information is needed.

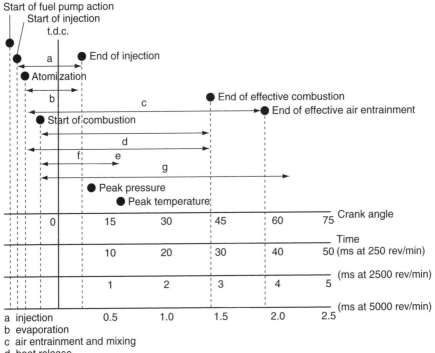

a injection
b evaporation
c air entrainment and mixing
d heat release
e heat transfer
f NO_x formation
g soot and CO formation

Figure 14.7 *Time scale of diesel combustion process*

Cyclic energy release – mean effective pressure (indicated, gross and pumping)

Cylinder pressure data can be used to calculate the work transfer from the gas to the piston. This is generally expressed as the indicated mean effective pressure (i.m.e.p.) and is a measure of the work output for the swept volume of the engine. The result is a fundamental parameter for determining engine efficiency as it is independent of speed, number of cylinders and displacement of the engine.

The i.m.e.p. calculation is basically the enclosed area of the high pressure part of the *pV* diagram and can be derived via integration:

$$\text{i.m.e.p.} = \int P.dV$$

This effectively gives energy released or gross m.e.p. (gross work over the compression and expansion cycle, g.m.e.p.). Integration of the low pressure (or gas exchange)

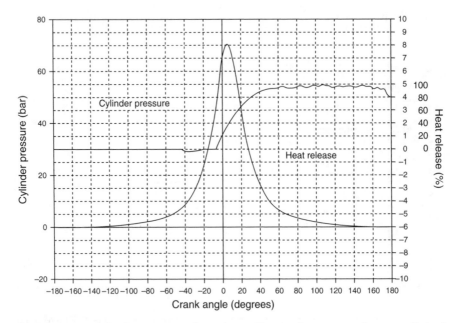

Figure 14.8   *Cylinder pressure diagram and heat release curve for a small single cylinder diesel engine*

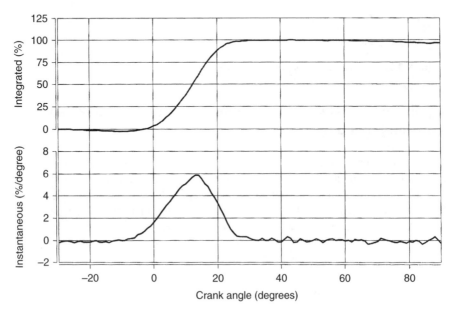

Figure 14.9   *Diagram showing instantaneous and integrated energy release vs crank angle. (Source: AVL)*

part of the cycle gives the work lost during this part of the process (pumping losses also known as p.m.e.p.). Subtraction of this value from the gross i.m.e.p. gives the net i.m.e.p. (n.m.e.p.) or actual work per cycle.

Therefore, it can be stated that:

n.m.e.p. = g.m.e.p. − p.m.e.p.

The most important factor to consider when measuring i.m.e.p. is that the t.d.c. position (see section 'Exact determination of true top dead centre position' below).

In order to optimize the calculation note that it is not necessary to acquire data at high resolution. Measurement resolution of a maximum of one degree crank angle is sufficient, higher than this does not improve accuracy and is a waste of system resources.

Also important for accurate i.m.e.p. calculation are the transducer properties; mounting location and stability of the transducer sensitivity during the engine cycle.

Instantaneous energy release

The cylinder pressure data, in conjunction with cylinder volume can be used to extrapolate the instantaneous energy release from the cylinder with respect to crank angle. There are many well-documented theories and proposals for this, but all are fundamentally similar and stem from the original work carried out by Rassweiler and Withrow.[2]

The calculations can range from the quite simple with many assumptions, to very sophisticated simulation models with many variables to be defined and boundary conditions to be set.

The fundamental principle to the calculation of energy release relies on the definition of the compression and expansion process via a polytropic exponent (ratio of specific heats under constant pressure and volume conditions for a given working fluid).

If this can be stated with accuracy then a motored pressure curve can be extrapolated for the fired cycle. The fired curve can then be compared with the motored curve and the difference in pressure between them, at each angular position, and with respect to cylinder volume at that position allows calculation of energy release.

A typical simplified algorithm to calculate energy release:

$$Q_i = \frac{K}{\kappa - 1} \left[\kappa \cdot p_i \cdot (V_{i+n} - V_{i-n}) + V_i \cdot (p_{i+n} - p_{i-n}) \right] \tag{4}$$

This calculation can be executed quickly within a computerized data acquisition system, but it does not account for wall or blow-by losses in the cylinder. Also, it assumes that the polytropic index is constant throughout the process. These compromises can affect accuracy in absolute terms but in relative measurement applications this simplified calculation is well proven and established in daily use. Under conditions where engine development times are short, algorithms have to be simplified such that they can be executed quickly.

Most importantly for the engine developer are the results which can be extracted from the curves which give important information about the progress and quality of combustion.

Computerized engine indicating technology and methodology

The modern engine indicating or combustion analyser creates data that are larger by at least an order of magnitude when compared with the log point data measured and stored by a typical test cell automation system. Therefore, appropriate methodology for handling, reducing and analysing these data is an essential function of EI equipment.

The combustion analysis device is the reciprocal of the test bed with respect to raw and calculated data. The ratio of formula to channels is far higher for a combustion analyser and an efficient way of creating and standardizing thermodynamic calculations is another important part of the overall efficiency of the process.

Key to the whole process is the measurement of pressure within the combustion chamber during the complete four-stroke or two-stoke process. This process in exact detail can be highly variable between cylinders and even within cycles inside the same cylinder hence the need for high levels of data analysis to enable mean performance levels to be determined.

The tools used to measure cylinder pressure are pressure transducers that work on the piezoelectric principle, wherein the change of pressure creates a change in circuit charge. The circuit itself is quite unlike any other sensor or transducer commonly seen in an engine test environment and is therefore important that users fully understand the special handling requirements of this technique.

Basic circuit and operation of pressure measurement chain

Piezoelectric combustion pressure sensors are always used with appropriate signal conditioning in combustion measurement applications. The signal conditioning required for a charge signal is a charge amplifier and this basically consists of a high-gain inverting voltage amplifier with a MOSFET (metal–oxide–semiconductor field-effect transistor) or JFET (junction field-effect transistor) at its input to achieve high insulation resistance.

The purpose of the charge amplifier is to convert the high impedance charge input into a usable output voltage. The basic circuit diagram is shown in Fig. 14.10. A charge amplifier consists of a high gain amplifier and a negative feedback capacitor (C_G). When a charge is delivered from a piezoelectric pressure transducer (PT), there is a slight voltage increase at the input of the amplifier (A). This increase appears

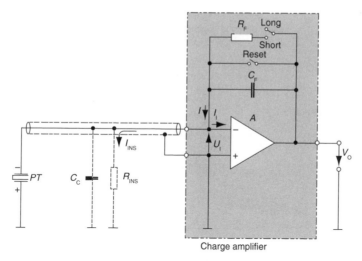

Charge amplifier

Figure 14.10   *Basic charge amplifier circuit (AVL List)*

at the output substantially amplified and inverted. The negatively biased negative feedback capacitor (C_F) correspondingly taps charge from the input and keeps the voltage rise small at the amplifier input.

At the output of the amplifier (A), the voltage (V_O) sets itself so that it picks up enough charge through the capacitor to allow the remaining input voltage to result in exactly V_O when amplified by A. Because the gain factor of A is very large (up to about 100 000), the input voltage remains virtually zero. The charge output from the pressure transducer is not used to increase the voltage at the input capacitances, it is drawn off by the feedback capacitor.

With sufficiently high open-loop gain, the cable and sensor capacitance can be neglected; therefore, changes in the input capacitance due to different cables with different cable capacitance (CC) have virtually no effect on the measurement result. This leaves the output voltage dependent only on the input charge and the range capacitance. That is,

$$V_O = Q/C_F$$

An output is produced only when a change in state is experienced, thus the piezo-electric transducer and charge amplifier cannot perform true static measurements. This is not an issue in normal combustion measurements, as they require a highly dynamic measurement chain to adequately capture all aspects of the phenomenon. Special considerations are required for calibration of the measurement chain and in data analysis.

While the modern charge amplifier measurement chain, Fig. 14.11, is a robust and well-proven measurement and the technique is almost universally adopted for

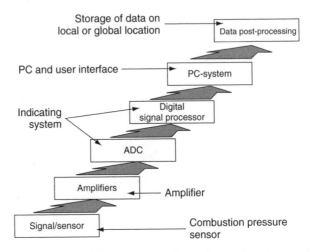

Figure 14.11   *Typical workflow process through combustion analysis*

combustion pressure measurement, there are important issues and measurement characteristics to consider:

- *Drift*. This is defined as an undesirable change in output signal over time that is not a function of the measured variable. In the piezoelectric measurement system, there is always an inherent drift in the output signal due to the working principle of the system. Electrical drift in a charge amplifier can be caused by low insulation resistance at the input.
- *Time constant*. The time constant of a charge amplifier is determined by the product of the range capacitor and the time constant resistor. It is an important factor for assessing the capability of a piezoelectric measurement system for measurement of very slow engine phenomena (cranking speed) without any significant errors due to the discharge of the capacitor. Many charge amplifiers have selectable time constants that are altered by changing the time constant resistor.

Drift and time constant simultaneously affect a charge amplifier's output. One or the other will be dominant. There are a number of methods available in modern amplifier technology to counteract electrical drift and modern charge amplifier technology contains an electronic drift compensation circuit.

If it is possible to maintain extremely high insulation values at the pressure transducer, amplifier input, cabling and associated connections, leakage currents can be minimized and high quality measurements can be made. In order to do this though, the equipment must be kept clean and free of dirt and grease to laboratory standards. In practice, a normal engine test cell environment does not provide this level of cleanliness and this method of preventing drift is not a practical proposition. Operation of the amplifier in short time constant mode means that the drift due to input offset voltage can be limited to a certain value and thus drifting into saturation can be prevented.

- *Filtering*. Electrical filtering is generally provided in the charge amplifier to eliminate certain frequencies from the raw measured data. Typically high pass filters are used to remove unwanted lower frequency components; for combustion measurement, this may be necessary in a knock measurement situation where only the high frequency components are of interest.

The high pass filter will allow optimization of the input range on the measurement system to ensure the best possible analogue to digital conversion of the signal. Low pass filters can be implemented to remove unwanted high frequency, interference signal content from the measurement signal such as structure-borne noise signals from the engines that are transmitted to the transducer.

It is an important point to bear in mind when using electrical filters that a certain phase shift will always occur and that this can cause errors which must be considered. For example, any phase shift has a negative effect on the accuracy of the i.m.e.p. determination when a low pass filter is used. The higher the engine speed, the higher the lowest permitted filter frequency. As a general rule, in order to avoid unacceptable phase shift, the main frequency of the cylinder pressure signal should not be more than 1 per cent of the filter frequency.

Ground loops in indication equipment

These are discussed in Chapter 10 and can be a significant problem for piezoelectric measurement chains (due to the low level signal) and can be extremely difficult to detect. It is therefore prudent to take appropriate precautions to prevent them and hence avoid their effects. The simplest way is to ensure all connected components are at a common ground via interconnection with an appropriate cable; this can be difficult to implement if equipment is not permanently installed at the test cell.

Ground isolated transducers and charge amplifiers can be purchased that will eliminate the problem, but it is important to note that only one or the other of these solutions should be used. If used together the charge circuit will not be complete and hence will not function!

Modern amplifier technology utilizes developments in digital electronics and software to correct or allow correction of signal changes occurring during the test period. The amplifier settings are software driven via the instrumentation PC, which means that settings can be adjusted during measurements without entering the test cell. Local intelligence at the amplifier provides automatic gain factor calculation from sensor sensitivity, pressure range and output voltage thus providing optimum adaptation of the input signal with nearly infinitely variable gain factors. This optimizing information can be stored on board the amplifier complete with sensor parameters.

Modern amplifiers are physically small which allows mounting very close to the actual sensors and this means short cables between sensors and amplifier and therefore the best protection from electromagnetic interference. The direct connection of piezo sensors without interconnection reduces signal drift and possible problems with grease and humidity on intermediate connectors.

Exact determination of true top dead centre position

The test engineer in possession of modern equipment continues to be faced with what appears to be a rather prosaic problem: the determination of the engine's top dead centre point.

This is a more serious difficulty than may be at first apparent. An electronic engine indicator records cylinder pressure in terms of crank angle, and measurements at each 1/10 of one degree of rotation are commonly available. However, in order to compute indicated power it is necessary to transform the cylinder pressure-crank angle data to a basis of cylinder pressure-piston stroke. This demands a very accurate determination of crank angle at the top dead centre position of the piston.

If the indicator records top dead centre 1° of crankshaft rotation ahead of the true position the computed i.m.e.p. will be up to 5 per cent greater than its true value. If the indicator records t.d.c. 1° late computed i.m.e.p. will be up to 5 per cent less than the true value.

Top dead centre can now be accurately determined, for a particular running condition, by using a special capacitance sensor such as the AVL 428 unit, which is mounted in a spark plug hole of one cylinder while running the engine on the remaining cylinders. The t.d.c. position thus determined is then electronically recorded by using the angle of rotation between t.d.c. and a reference pulse using a rotational encoder attached to the crankshaft. Another technique makes use of recording the cylinder pressure peak from a cylinder pressure transducer but without such special transducers and the required software to match the shaft angle, precise determination of geometrical t.d.c. is not easy. The usual and time-honoured method is to rotate the engine to positions equally spaced on either side of t.d.c., using a dial gauge to set the piston height, and to bisect the distance between these points. The rotation position of this even then had to be fixed on the flywheel in such a way as to be recognized by a pick-up.

This traditional method gives rise to several sources of error:

- the difficulty of carrying out this operation with sufficient accuracy;
- the difficulty of ensuring that the signal from the pick-up, when running at speed, coincides with the geometrical coincidence of pin and pick-up;
- torsional deflections of the crankshaft, which are always appreciable and are likely to result in discrepancies in the position of t.d.c. when the engine is running, particularly at cylinders remote from the flywheel.

In high performance engines, the t.d.c. position does not remain entirely constant during the test run due to physical changes in shape of the components; in these cases a t.d.c. position specific to a particular steady state may have to be calculated.

Integration of engine indication (EI) equipment within the test cell

Whatever the level of sophistication the physical equipment involved in engine calibration and EI work can be considered as a high speed data acquisition, storage, manipulation and display system. It will consist of three subsystems:

- Engine rigged transducers: some of which may require special engine preparation, such as machining of heads for pressure transducers and mounting of an optical speed encoder.
- Interconnection loom which, in the case of mobile systems, will have to pass through the firewall between cell and control room.
- Data acquisition and analysis equipment mounted in, and fully integrated with, the control room instrumentation.

High end engine calibration cells

Let us consider the operational problem faced by the cell operators of an OEM test cell carrying out the optimization of an engine control map. There are three distinct subsystems producing and displaying data while the engine is running.

The test cell automation system that is controlling the engine performance and monitoring and protecting, via alarm logic, both engine and test facility operation. The engine ECU that is controlling the fuelling and ignition using engine-mounted parameters to calculate essential variables that it is incapable of measuring directly such as torque from inlet manifold pressure.

The combustion analysis system that is using special (non-vehicular) transducers to directly measure cylinder pressure against crank shaft rotation and is able to derive and display such values as

- mean effective pressures, i.m.e.p., p.m.e.p., etc.;
- location of peak pressure and pressure rise rates;
- mass faction burn parameters;
- polytropic index;
- injector needle movement and injected fuel mass;
- heat release, etc.

To enable such instrumentation and data to efficiently fulfil its purpose in the minimum time, the operator of the cell needs to be highly skilled and experienced and very careful consideration needs to be given to the design of the experiments (DoE) being carried out on the test bed.

The task described is now so common that often there are insufficient highly skilled and experienced operators available. In response, the test equipment industry has produced integrated suites of instrumentation and software that, to a degree, can reduce the skill levels required to design the test sequences and to efficiently locate the boundary operating conditions of the test engine (see Chapter 19).

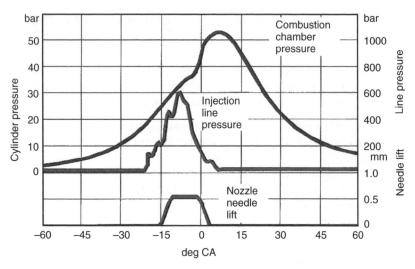

Figure 14.12   *Typical engine indicating display parameters*

Cells that carry out engine calibration work such as that described above have to be designed from the beginning as an integrated system if they are to work efficiently. All of the data having different and various native abscissa, such as time base, crank angle base and engine cycle base, communicated via different software interfaces (COM, ASAP, CAN bus, etc.), have to be correctly aligned, processed, displayed and stored which is no mean feat. In fully automated beds, it is common to have upwards of six 17-inch monitors on the operator's desk. In addition to the pressure transducers required for cylinder pressure, additional transducers are required to measure events and consequences of the combustion that occurs on the same timescale, such as injector operation and fuel rail pressure (see Fig. 14.12).

Middle market, occasional combustion analysis cells

The operator of and engine test required to carry out occasional map modification work performance testing will find that much of the combustion analysis equipment is often, although not always, roving trolley-mounted plant rather than a permanent fixture in the cell and control room. This means that unlike exhaust gas emission plant, the only real space requirement imposed is that of the EI equipment operator; this is because such equipment is invariably operated by, and in the care of, specialists. Test sequences designed to support IE work are, at this level of work, usually created with off-line software or problem areas of engine performance are located through manual control.

Control cables fitted to the ECU via large multiway plugs give practical rigging problems when having to be passed through the cell/control fire wall.

Typical engine indicating plant will have a 'pseudo-oscilloscope' display which will be separate from the default cell control system display, therefore desk space for two operators and an absolute minimum of two flat display screens should be catered for.

'Knock' sensing

One of the most important boundary conditions that an engine control system must be programmed to avoid is 'knock'. This destructive, spontaneous ignition of unburnt gas in the cylinder can be detected in the engine test cell by

- characteristic pressure pulse signature; the normal method used by EI equipment using specially rigged engines;
- an accelerometer tuned to look for the vibration caused by knock; the common method used by ECU since the transducer is appropriate for use in production vehicles;
- microphone and operator detection is still a valid method but probably only appropriate to specialist motor sport test facilities where low engine numbers and rapidly changing ECU maps do not allow for standard models to be formed.

Engine indicating pressure transducers (EIPT)

Cylinder pressure measurements play the key role in any engine indicating and combustion analysis work. The choice of pressure transducer from the wide range now commercially available, its correct mounting in the engine and its integration within the calibrated system are vital ingredients in obtaining optimum results.

Up to the early 1990s commercially available piezoelectric pressure transducers were more temperature sensitive both in terms of cyclic effects and absolute maximum working temperature than units available today. The temperature effects are due to exposure to the combustion process and most would fail if required to operate above 200°C, measured at the diaphragm. Consequently, most engine indicating pressure transducers (EIPTs) were water cooled. Water-cooled EIPTs will operate in most i.c. engine combustion chambers, including turbocharged units, but it is vital that the cooling system meets the following conditions:

- The transducer cooling system, while in use, must be integrated with the control system shutdown circuit so as to ensure that the cooling is running before engine start and during the complete termination of testing and cooldown of the engine. Transient dips in supply pressure will risk permanent loss of the transducers.
- The water should be distilled (deionized) and filtered. The transducers have very small passages that will become ineffective or blocked by scale.

- The deionized water should be supplied at as low and constant pressure as possible to ensure flow, so as to avoid changes in internal transducer pressure that could corrupt output signal.

Uncooled EIPTs are being developed continually and currently will operate to over 400°C which covers the majority of naturally aspirated automotive work outside specialist and motor-sport development work.

Mounting of the EIPT and special designs

EIPTs are made in various sizes being threaded at the diaphragm end in sizes ranging from 6 up to 18 mm. They are normally inserted into the combustion chamber so that their tips are flush with the parent combustion chamber material. This gives rise to problems of finding space and material in the engine head structure to support the transducer, so special transducers which take the form of an instrumented spark plug for gasoline engines, or glow plugs for diesel engines, have been developed.

Speed/crank angle sensors

For all tests involving engine indicating work, the engine will have to be rigged at its free crank-shaft end with an incremental optical shaft encoder. The alignment of the drive of these devices has to be undertaken with care and using the correct components to suit the application. The devices consist of a static pick-up that is able to read positional data from an engine shaft-mounted disk that is etched with typically 3600 equally spaced lines producing that number of square wave pulses per revolution in the device output. There may be a number of missing lines which enable a discrete angular position to be established as an event trigger. The disks are normally made of a glass material, except in cases where very high shock forces are present where steel disks with lower pulse counts may be required. Some crank angle encoders have a second output that determines direction of rotation.

Result calculations for combustion analysis

Calculation of critical results from the large store of raw combustion data is an essential part of the analysis and forms the basis of intelligent data reduction. Result calculation and a detailed analysis of the combustion process can be performed off-line, but this is not always convenient as the process takes time. Advances in technology allow some systems to carry out result calculation to a high degree of accuracy on-line.

These systems can display the calculated data during measurement so that the operator or engineer can view the data and react accordingly. Alternatively, these results can be transferred immediately to the test bed for data integration (into the automation system data model) or for control purposes as in tests where the control mode used is speed vs i.m.e.p. for cold start calibration work

Results derived from raw measured data

The actual results which are derived from the raw measured data and calculated curves can be grouped logically into direct and indirect results.

Direct results

These are derived directly from the raw cylinder pressure curve. They are simple results to calculate which can be returned quickly. Most of them can also be calculated before the end of the complete engine cycle. Generally, they are less sensitive and any signal error will create a result calculation error of similar magnitude. Typical direct result calculations are

- maximum pressure and position;
- maximum pressure rise and position;
- knock detection;
- misfiring;
- combustion noise analysis;
- injection/ignition timing;
- cyclic variation of above values.

These results will be acquired from data with acquisition resolution appropriate to the task. Resolution of pressure and pressure rise will typically be at one degree crank angle, certain values will need higher resolutions due to higher frequency components of interest (combustion knock or noise) or the need for accurate determination of angular position (injection timing).

Indirect results

These results are more complex as they are derived indirectly from the raw pressure curve data. They are reliant on additional information like cylinder volume, plus other parameters, such as the polytropic exponent. These results are much more sensitive to correct set-up of the equipment and the error in a signal acquired is multiplied considerably when passed into a calculation. Hence the result error can be greater by an order of magnitude. Typical indirect results are

- heat release calculation (dQ, integral);
- indicated mean effective pressure;
- pumping losses, p.m.e.p.;
- combustion temperature;

- burn rate calculation;
- friction losses;
- energy conversion;
- gas exchange analysis;
- residual gas calculation;
- cyclic variation of above values;

and commonly are interpreted as:

Result	Interpretation
Mass burn fraction 0–5%	Start of combustion, early flame growth
Mass burn fraction 50%	Correlation with MBT for gasoline engine
Mass burn fractions 0–5% and 10–90%	End of combustion, turbulent flame growth
Max cylinder pressure and angle	Cylinder head loading, mechanical limits
g.m.e.p., n.m.e.p., p.m.e.p.	Mechanical efficiency and friction losses
Coefficient of variance of i.m.e.p.	Stability indicator

Most of these results, although derived through complex calculations, do not place a heavy demand on the acquisition hardware, nor do they require high resolutions with respect to crank angle (typically one degree is sufficient) or analogue-to-digital conversion (12–14 bit). The main demand is for sufficient calculating power to execute the algorithms and return results and statistics as quickly as possible.

Summary

As discussed in this section, extraction of result values from the measured and calculated curves is an important function for efficient data analysis and handling. For the combustion analysis system important features are:

- real-time or on-line calculation and display of all required results for screen display or transfer to the test bed via software interface or as analogue voltage values;
- features in the user interface for easy parameterization of result calculations, or for definition of new, customer-specific result calculation methods without the requirement for detailed programming knowledge;
- flexibility in the software interface for integration of the combustion analyser to the test bed allowing remote control of measurement tasks, fast result data transfer for control mode operation, tagging of data files such that test bed and combustion data can be correlated and aligned in post processing;

- intelligent data model which allows efficient packaging and alignment of all data types relevant in a combustion measurement (example, crank angle data, cyclic data, time-based data, engine parameters, etc.).

Most importantly, engine indicating and combustion analysers are always very sensitive to correct parameterization and the fundamentals, that is:

- correct determination of t.d.c. (particularly for i.m.e.p.);
- correct definition of the polytropic exponent (particularly for heat release);
- correct and appropriate method of zero level correction.

Some readers with particular interests in engine indicating and mapping may find relevant information concerning data processing in Chapter 19.

Notation

Pressure, beginning of interval	p_1 bar
Pressure, end of interval	p_2 bar
Pressure, end of interval, no burn	p_0 bar
Volume, beginning of interval	v_1 m^3
Volume, end of interval	v_2 m^3
Volume at top dead centre	v_0 m^3
Pressure change due to combustion	Δp bar
Pressure change, normalized to t.d.c.	Δp_c bar
Polytropic index, general	n
Polytropic index, compression	n_c
Polytropic index, expansion (no burn)	n_e

References

1. Stone, C.R. and Green-Armytage, D.I. (1987) Comparison of methods for the calculation of mass fraction burnt from engine pressure-time diagrams, *Proc. I. Mech. E.*, **201** (D1).
2. Rassweiler, G.M. and Withrow, L. (1938) Motion pictures of engine flame propogation. Model for S.I. engines, *S.A.E. Jl.*, **42**, 185–204 (translation).
3. Bazari, Z. (1992) *A D.I. Diesel Combustion and Emission Predictive Capability for Use in Cycle Simulation*, S.A.E. Paper No. 920462.

15 The test department organization, health and safety management, risk assessment correlation of results and design of experiments

Introduction

This chapter seeks to describe the administrative and technical management tasks and organization of a medium-sized test department, suggesting how the constituent parts should fit together and interact. Responsibilities imposed by health and safety legislation and best practice are mentioned and a suggested format for risk assessments is included. The prime task of technical management is to ensure that the test equipment is used to its optimum efficiency which in many cases is organizationally interpreted as ensuring plant achieves maximum 'up-time'. The efficiency of many test facilities is judged by periodic cell up-time or 'shaft rotation' figures. But, however well motivated an organization is, it can nonetheless work as 'busy fools' if the tasks undertaken are not organized in such a way as to be time- and cost-efficient, therefore the subject of design of experiments has been included in the chapter.

Management and group roles

Although the organizational arrangements may differ a test facility will employ staff having three distinctly different roles:

1. facility staff and management charged with building, maintaining, developing and the installed plant, its support services and the building fabric;
2. the internal user group charged with designing and conducting tests, collecting data and disseminating information;
3. a quality group charged with audit to QA standards and instrument calibration.

Each group will have some responsibility for two funding streams, operational and project specific. In both cases the first essential is to understand the brief, which must always include an indication of time and cost. The growing dichotomy between the first two of these roles, caused by ever narrower specialization in the user group and increasing scarcity of multidisciplinary facility or project staff, was the original justification for the first edition of this book. It is interesting to note that one of the tasks that may tend to 'float' between groups is that of specification of new or modified facilities; referring back to Chapter 1, it is not unusual for test industry suppliers to negotiate with a purchase or facility 'customer' and deliver to a 'user' with a different requirement, the responsibility for avoiding such wasteful practices and having a common specified requirement is that of the first level of common management.

The listed tasks are not mutually exclusive and in a small test shop may be merged, although in all but the smallest department management of the QA task, which includes the all-important responsibility for calibration and accuracy of instrumentation, should be kept distinct from line management of the users of the facility.

Figure 15.1 shows the allocation of tasks in a large test department and indicates the various areas of overlapping responsibility, also the impact of 'quality management'. Periodical calibration of dynamometers, instruments, tools and the maintenance of calibration records may be directly in the hands of the quality manager or carried out by the facilities department under his supervision; either way the procedure or computerized process must be clearly laid down. Any test facility requiring registration to national or international quality assurance standards will have to provide and record a rigorous control of the calibration process for all of the instrumentation used.

Computerized data logging has reduced the role of the test cell technician, who used to spend much of his time reading instruments and filling in log sheets. Pressure on costs has reduced the number of technicians, who now spend more of their time rigging and setting up engines, while the graduate test engineer tends to spend more of his time in the control room in the immediate consideration of the data being generated. The profile of the staff using the control room* is of fundamental importance to the design and operation of the test department; while there is no 'correct' solution it has to be optimized for the particular needs of each site.

Health and safety and risk management

Health and safety (H&S) matters have already been mentioned several times in connection with test cell operation. Everyone employed or visiting a test facility has responsibilities (under the law in the UK) in this connection. Formal responsibility for H&S within a large organization will be that of a manager trained to ensure that policies of the company and legal requirements are adhered to by the supervisory organization.

---

* The word 'using' should be taken in the widest sense. Users of the control area, during normal operation, ranges from a single production worker to a technical conference of customers and sponsors. During calibration or service periods it may house internal and subcontract specialists; the managerial control of the space needs to be clear to all of the users.

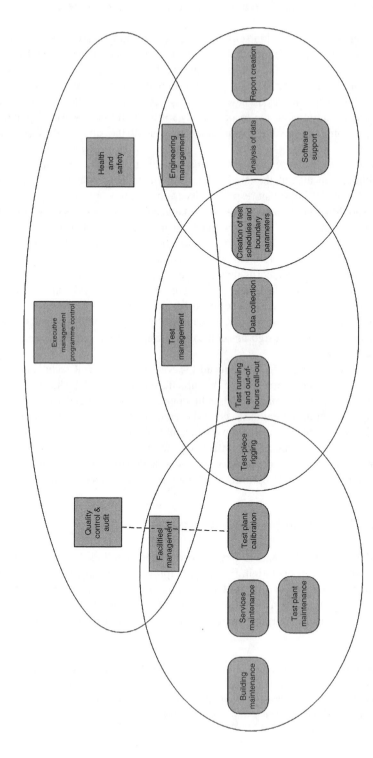

Figure 15.1 *Allocation of tasks within a management structure of a medium/large test facility*

Risk can be defined as the danger or potential of injury, financial loss or technical failure. While H&S managers will concentrate on the first of these, senior managers have to consider all three at the commencement of every new testing enterprise or task.

The legislatively approved manner of dealing with risk management is to impose a process by which, before commencement, a responsible person has to carry out and record a risk assessment. A form containing the minimum information required is shown in Fig. 15.2. Risk assessment is not just a single paper exercise, in a list of tasks, required by a change in work circumstances, it is a continuous task, particularly during complex projects where some risks may change by the minute before disappearing at task completion.

Staff involved in carrying out risk assessments need to understand that the object of the exercise is less about describing the risk and more about putting in place

RISK ASSESSMENT

Activity: *Description*

Hazard	Probability (P) Severity (S) Risk (Max 25)	Minimize risk by	Minimize residual risk by
Brief description of specific hazard	$P = 1{-}5$ $S = 1{-}5$ $1 =$ rare $5 =$ probable Risk $= S \times P$	*Description of primary actions required plus any specific responsibilities of job holders. Reference to attached procedures if required.*	*Secondary actions required, e.g. training or maintenance schedules*

Notes

Assessment carried out by:
Version and date:
Distribution:

Figure 15.2 *A basic risk assessment form and risk 'scoring' method*

realistic actions and procedures in order to eliminate or reduce the potential effects of the hazard.

Both risks of injury (acute) and risks to health (chronic), such as exposure to carcinogenic materials, should be considered in the risk assessments.

There are important events within the life cycle of test facilities when H&S processes and risk assessment should be applied:

- planning and prestart** stages of a new or modified test facility: both project specific and operational;
- at the change of any legislation covering explicitly or implicitly the facility;
- service, repair and calibration periods by internal or subcontract staff;***
- a significantly different test object or test routine such as those requiring unmanned running or new reference fuels;
- addition of new instrumentation.

The formal induction of new staff joining a test facility workforce and the regular review of the levels of training required with its development are important parts of a comprehensive H&S and environmental policy. In any company that carries out an annual appraisal of staff the subject of training will be under review by both management and staff members; where no such policy exists, training should be the formal responsibility of line management.

Planning, executing and reporting the test programme

It must never be forgotten that the sole purpose of research and development testing is to prove, or improve, the performance of the engine or component that is the subject of the test. The sole purpose of most other tests is to check that the performance of the engine or complete system conforms to some specified standard.

It follows that the data assembled during the test is only of value in so far as it contributes to the achievement of these purposes. It is only too easy, now that facilities are so readily available for the collection of colossal quantities of data, to regard the accumulation of data as an activity of value in its own right. It is essential that the manager or engineer responsible for the operation of a test facility should keep this very much in mind. In the experience of the authors nothing is more helpful in this respect than the discipline of keeping a proper log book and of writing properly structured reports.

** In the UK under CDM regulations, the customer has responsibilities for advising the contractors of any pre-existing conditions of the site that may affect any risk assessment they carry out.

*** The provision of a risk assessment by a subcontractor does not abrogate the responsibility of the site management for H&S matters related directly or indirectly to the work being done by the subcontractor. The quality of the assessment and the adherence to the processes required need to be checked.

The whole idea of a written log book may be considered by many to be old fashioned in an age in which the computer has taken over. Computers, however, are not able to report those subjective recordings of a trained technician, nor are the data they hold always available to those who require it in the unplanned absence of the writer.

The log book is also a vital record of all sorts of peripheral information on such matters as safety, maintenance, suspected faults in equipment or data recording and, last but most important, as an immediate record of 'hunches' and intuitions arising from a consideration of perhaps trivial anomalies and unexpected features of performance. It must be obvious that to be of real benefit the log book must be read and valued by the supervisory chain.

Test execution, analysis and reporting

The execution, analysis and reporting of a programme of tests and experiments are difficult arts and involve a number of stages:

- First of all, the experimental engineer must understand the questions that his experiment is intended to answer and the requirements of the 'customer' who has asked them.
- There must be an adequate understanding of the relevant theory.
- The necessary apparatus and instrumentation must be assembled and, if necessary, designed and constructed.
- The experimental programme must itself be designed, with due regard to the levels of accuracy required and with an awareness of possible pitfalls, misleading results and undetected sources of error.
- The test programme is executed, the engineer keeping a close watch on progress.
- The test data are reduced and presented in a suitable form to the 'customer' and to the level of accuracy required.
- The findings are summarized and related to the questions the programme was intended to answer.

Finally, the records of the test programme must be put together in a coherent form so that in a year's time, when everyone concerned has forgotten the details, it will still be possible for a reader to understand exactly what was done. Test programmes are very expensive and often throw up information the significance of which is not immediately apparent, but which can prove to be of great value at a later date.

Formal reports may follow a similar logical sequence:

- objective of experimental programme;
- essential theoretical background;
- description of equipment, instrumentation and experimental method;
- calculations and results;
- discussion, conclusions and recommendations.

In writing the report, the profile of the 'customer' must be kept in mind. A customer who is a client from another company will require rather different treatment from one who is within the same organization. There will be common characteristics; the customer

- will be a busy person who requires a clear answer to specific questions;
- will probably not require a detailed account of the equipment, but will need a clear and accurate account of the instrumentation used and the experimental methods adopted;
- will be concerned with the accuracy and reliability of the results;
- must be convinced by an intelligent presentation that the problem has been understood and the correct answers given.

Although English is commonly used in the engine test industry worldwide, native writers of English should be aware of readers for whom English is a second or third language and avoid idioms and, most importantly, give the full meaning of acronyms.

Cell to cell correlation

It is usual for the management of a test department to wish to be reassured that all the test stands in the test department give the same answer, and it is not unusual for an attempt to be made to answer this question by the apparently logical procedure of testing the same engine on all the beds. The outcome of such test based on detailed comparison of results is invariably a disappointment, and can lead to expensive and unnecessary disputes between the test facility management and the supplier of the test equipment.

It cannot be too strongly emphasized that an engine, however sophisticated its management system, is not suitable for duty as a standard source of torque.

In Chapter 10, we discuss the very substantial changes in engine performance that can arise from changes in atmospheric (and hence in combustion air) conditions, in addition engine power output is highly sensitive to variations in fuel, lubricating oil and cooling water temperature and it would be necessary to equalize these very carefully if meaningful results were to be hoped for.

Finally, it is unlikely that a set of test cells will be totally identical: apparently small differences in such factors as the layout of the ventilation air louvres and in the exhaust system can have a significant effect on performance.

A fairly good indication of the impossibility of using an engine as a standard in this way is contained in the standard, BS 5514, discussed below. This standard lists the 'permissible deviation' in engine torque as measured repeatedly during a single test run on a single test bed as 2 per cent. This apparently wide tolerance is no doubt based on experience and, by implication, invalidates the use of an engine to correlate dynamometer performance.

There is no substitute for the careful and regular calibration of all the machines and while cell to cell correlation testing is widely practised, the results have to be judged on the basis of the degree of control over all critical variables external to the engine combustion chambers and by staff with practical experience in this type of exercise.

Acceptance and type tests, power test codes and correction factors

Most of the work carried out in engine test installations is concerned with purely technical and engineering matters. Acceptance and type tests involve much more: they are concerned with the interface between the engineering and commercial worlds and between manufacturer and customer. The reputation of the manufacturer is directly involved and any ambiguity in the test procedures or in the interpretation of the results can lead to the loss of goodwill and even to litigation. This is particularly the case when the customer is a government department or one of the defence services.

It is therefore not surprising that tests of these kinds are the subject of extensive regulations, in many cases having statutory force, and not always easy to interpret. It is essential that the test facility manager who finds himself concerned with this kind of work should make himself thoroughly familiar with these regulations; the more important English language documents are described briefly below.

Acceptance tests

These can range from the briefest acceptable running-in period plus a check on the general operation of the engine, carried out in accordance with the engine manufacturer's internal procedures and without the involvement of the customer, to elaborate and detailed tests carried out in the presence of the customer or his representative in order to ensure that the engine and its performance meet an agreed specification.

Type tests

Type tests are particularly associated with supplies to the Armed Forces, who invariably lay down the procedure in detail. A type test is designed to prove the whole performance of the engine under the conditions it may actually meet in service. It should cover everything necessary to prove beyond doubt or dispute that the required performance is met or exceeded. This means that a type test will involve prolonged running under conditions representative of service with the specified maintenance procedures and measurements of wear and of such factors as piston cleanliness.

It will be quite distinct from the relatively brief production tests which, while also checking that the required performance is achieved, are aimed primarily at ensuring that the quality of materials and workmanship and the production routine are up to the standard established by the type test.

Power test codes and correction factors

Several complex sets of rules are in general use for specifying the procedure for measuring the performance of an engine and for correcting this to standard conditions. The most significant for the English-speaking world are:

BS 5514[1]	Reciprocating internal combustion engines
ISO 3046[2]	Reciprocating internal combustion engines (identical to BS 5514)
BS AU 141[3]	Road vehicle diesel engines
Lloyd's *Rulebook*[4]	(primarily concerned with marine engines and installations)

(American) SAE standards

SAE J1995[5]	Engine power test code – spark ignition and compression ignition – gross power rating
SAE J1349[6]	Engine power test code – spark ignition and compression ignition – net power rating.

BS 5514/ISO 3046

Perhaps the best starting point for the test engineer who finds himself involved in type testing or elaborate acceptance tests is a study of this standard (in either version). It is in six parts, briefly summarized below.

Part 1. Standard reference conditions, declarations of power, fuel and lubricating oil consumptions and test methods

In this part, standard atmospheric conditions are specified as follows:

- atmospheric pressure 1 bar (=750 mmHg);
- temperature 25°C (298 K);
- relative humidity 30 per cent.

The American SAE standards specify the same conditions (for marine engines standard conditions are specified as 45°C (113°F) and 60 per cent relative humidity).

Specific fuel consumptions should be related to a lower calorific value of 42 700 kJ/kg: alternatively the actual LCV should be quoted.

The various procedures for correcting power and fuel consumption to these conditions are laid down and a number of examples are given. These procedures are extremely complicated and not easy to use: they involve 45 different symbols, 17 equations and 10 look-up tables. The procedures in the SAE standards listed above are only marginally less complex. Rigorous application of these rules involves a good deal of work and for everyday test work the 'correction factors' applied to measurements of combustion air consumption (the prime determinant of maximum engine power output) described in Chapter 12 will be found quite adequate.

Definitions are given for a number of different kinds of rated power: continuous, overload, service, ISO, etc., and a long list of auxiliaries which may or may not be driven by the engine is provided. It is specified that in any declaration of brake power the auxiliaries operating during the test should be listed and, in some cases, the power absorbed by the auxiliary should be given.

The section on test methods gives much detailed advice and instruction, including tables listing measurements to be made, functional checks and tests for various special purposes. Finally, part 1 gives a useful check list of information to be supplied by the customer and by the manufacturer, including the contents of the test report.

Part 3. Test measurements

This part discusses 'accuracy' in general terms, but consists largely in a tabulation of all kinds of measurement associated with engine testing with, for each, a statement of the 'permissible deviation'. The definition of this term is extremely limited: it defines the range of values over which successive measurements made during a particular test are allowed to vary for the test to be valid. These limits are by no means tight and are now outside the limits required of most R&D test facilities: e.g. ±3 per cent for power, ±3 per cent for specific fuel consumption; they thus have little relevance to the general subject of accuracy.

Part 4. Speed governing

This part, which has been considerably elaborated in the 1997 issue, specifies four levels of governing accuracy, M1 to M4, and gives detailed instructions for carrying out the various tests.

Part 5. Torsional vibrations

This part is mainly concerned with defining the division of responsibility between the engine manufacturer, the customer and the supplier of the 'set', or machinery to be driven by the engine, e.g. a generator, compressor or ship propulsion system. In general, the supplier of the set is regarded as responsible for calculations and tests.

Part 6. Specification of overspeed protection

This part defines the various parameters associated with an overspeed protection system. The requirements are to be agreed between engine manufacturer and customer and it is recommended that reset after overspeed should be manual.

Part 7. Codes for engine power

This part defines various letter codes, e.g. ICN for ISO standard power in English, French, Russian and German.

It will be clear that this standard gives much valuable guidance regarding many aspects of engine testing.

Statistical design of experiments

Throughout this book it has been the aim to give sufficient information concerning often complex subjects to enable the reader to deal with the more straightforward applications without reference to other sources. In the case of the present topic this is scarcely possible; all that can be done is to explain its significance and indicate where detailed guidance may be found.

The traditional rule of thumb in engineering development work has been to change one thing at a time. It was thought that the best way to assess the effect of a design change was to keep everything else fixed. There are disadvantages:

- It is very slow.
- The information gained refers only to the one factor.
- Even if we have hit on the optimum value it may change when other factors are changed.

While many modern design of experiment (DoE) techniques and algorithms capable of optimizing multivariate problems are embedded in modern software suites that are designed to assist in engine mapping and calibration (see Chapter 19), it is considered worthwhile describing one established method of approach.

Consider a typical multivariate development problem: optimizing the combustion system of a direct injection diesel engine, Fig. 15.3. The volume of the piston cavity is predetermined by the compression ratio but there is room for manoeuvre in:

Overall diameter D
Radius r
Depth d

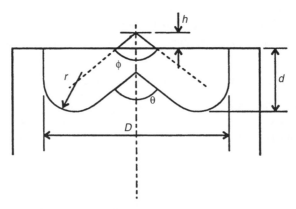

Figure 15.3   *Diesel engine combustion system*

Angle of central cone θ
Angle of fuel jets ϕ
Height of injector h
Injection pressure p

not to mention other variables such as nozzle hole diameter and length and air swirl rate.

Clearly any attempt to optimize this design by changing one factor at a time is likely to over-run both the time and cost allocations and the question must be asked whether there is some better way.

In fact, techniques for dealing with multivariate problems have been in existence for a good many years. They were developed mainly in the field of agriculture, where the close control of experimental conditions is not possible and where a single experiment can take a year. Published material from statisticians and biologists naturally tended to deal with examples from plant and animal breeding and did not appear to be immediately relevant to other disciplines.

The situation began to change in the 1980s, partly as a result of the work of Taguchi, who applied these methods in the field of quality control. These ideas were promoted by the American Supplier Institute and the term 'Taguchi methods' is often applied loosely to any industrial experiment having a statistical basis. A clear exposition of the technique as applied to the sort of work with which the engine developer is concerned is given by Grove and Davis[7] who applied these methods in the Ford Motor Company.

Returning to the case of the diesel combustion chamber, suppose that we are taking a first look at the influence of the injector characteristics:

Included angle of fuel jets ϕ
Height of injector h
Injection pressure p

For our initial tests we choose two values, denoted by + and −, for each factor, the values spanning our best guess at the optimum value. We then run a series of tests in accordance with Table 15.1. This is known as an orthogonal array, the feature of which is that if we write +1 and −1 for each entry the sum of the products of any two columns, taken row by row, is zero.

The final column shows the result of the tests in terms of the dependent variable, the specific fuel consumption. The chosen values were:

$\Phi- = 110°$
$\Phi+ = 130°$
$h- = 2\,\mathrm{mm}$
$h+ = 8\,\mathrm{mm}$
$p- = 800\,\mathrm{bar}$
$p+ = 1200\,\mathrm{bar}$

Table 15.1   *Test program for three parameters*

Run	ϕ	h	p	Specific consumption (g/kWh)
1	−	−	+	207
2	+	−	+	208
3	−	+	+	210
4	+	+	+	205
5	−	−	−	218
6	+	−	−	216
7	−	+	−	220
8	+	+	−	212

One way of presenting these results is shown in Fig. 15.4 in which each of the four pairs of results is plotted against each factor in turn. We can draw certain conclusions from these plots. It is clear that injection pressure is the major influence on fuel consumption and that there is some interaction between the three factors. There are various statistical procedures that can extract more information and guidance as to how the experimental programme should continue.

The first step is to calculate the main effect of each factor. Line B of Fig. 15.4a shows that in this case the effect of changing the injection pressure from $p-$ to $p+$ is to reduce the specific consumption from 212 to 205 g/kWh. The main effect is defined as one half this change or −3.5. The average main effect for the four pairs of tests is shown in Table 15.2.

A further characteristic of importance concerns the degree of interaction between the various factors: the interaction between ϕ and p is known as the $\phi \times p$ interaction.

Referring Fig. 15.4a, it is defined as one half the difference between the effect of p on the specific consumption for $\phi+$ and the effect for $\phi-$, or

$$1/2[(205 - 202) - (210 - 220)] = +1.5$$

It would be zero if the lines in Fig. 15.4a were parallel. It may be shown that the inverse interaction $p \times \phi$ has the same value.

It may be shown that the sign of an interaction + or − is obtained by multiplying together the signs of the two factors concerned. Table 15.2 shows the calculated values of the three interactions in our example. It will be observed that the strongest interaction, a negative one, is between h and ϕ. This makes sense, since an increase in h coupled with a decrease in ϕ will tend to direct the fuel jet to the same point in the toroidal cavity of the combustion chamber.

This same technique for planning a series of tests may be applied to any number of factors. Thus an orthogonal table may be constructed for all seven factors in Fig. 15.3 and main effects and interactions calculated. The results will identify the

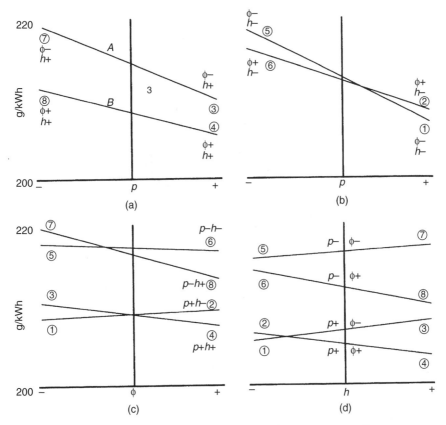

Figure 15.4 *Relationship between parameters showing main effects*

Table 15.2 *Main effects and interactions*

Run	φ	h	p	φ × h	φ × p	h × p
1	−	−	+	+	−	−
2	+	−	+	−	+	−
3	−	+	+	−	−	+
4	+	+	+	+	+	+
5	−	−	−	+	+	+
6	+	−	−	−	−	+
7	−	+	−	−	+	−
8	+	+	−	+	−	−
Main effect				Interaction		
	−1.75	−0.25	−4.5	−1.5	+0.75	+0.25

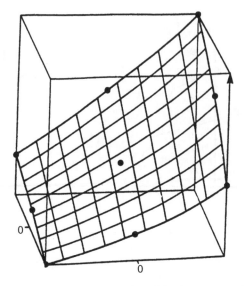

Figure 15.5 *Engine map, torque, speed and fuel consumption*

significant factors and indicate the direction in which to move for an optimum result. Tests run on these lines tend to yield far more information than simple 'one variable at a time' experiments. More advanced statistical analysis, beyond the scope of the present work, identifies which effects are genuine and which are the result of random variation.

A more elaborate version of the same method uses factors at three levels +, 0 and −. This is a particularly valuable technique for such tasks as the mapping of engine characteristics, since it permits the derivation of the coefficients of quadratic equations that describe the surface profile of the characteristic. Figure 15.5, reproduced from Ref. 1, shows a part of an engine map relating speed, torque and fuel consumption. The technique involves the choice of three values, equally spaced, of each factor. These are indicated in the figure.

It is hoped that the reader will gain some impression of the value of these methods and gain some insight into the three-dimensional graphs that frequently appear in papers relating to engine testing, from the above brief treatment.

Summary

The management structure of a test department is described and the planning, execution and reporting of test programmes are discussed.

The correction of engine test results to international standards has been discussed and the statistical design of experiments is briefly introduced.

References

1. BS 5514 Parts 1 to 6 *Reciprocating Internal Combustion Engines: Performance.*
2. ISO 3046 *Reciprocating Internal Combustion Engines: Performance.*
3. BS AU 141a *Specification for the Performance of Diesel Engines for Road Vehicles.*
4. *Rulebook,* Chapter 8, Lloyd's Register of Shipping, London.
5. SAE Standard: *Engine power test code – spark ignition and compression ignition – gross power rating,* SAE J1995 Jun 90.
6. SAE Standard: *Engine power test code – spark ignition and compression ignition – net power rating,* SAE J1349 Jun 90 (www.sae.org/certifiedpower).
7. Grove, D.M. and Davis, T.P. (1992) *Engineering Quality and Experimental Design,* Longmans, London.

16 Exhaust emissions

Introduction

It is probably true to say that, in the twenty-first century, a majority both of engine and vehicle development and of routine testing is concerned with environmental legislation directed primarily towards the limitation and control of engine emissions. Testing may be divided between that directed towards development of engine and vehicle systems having the lowest possible environmental impact and that required by environmental legislation (homologation and certification) in order that a vehicle may be sold in a particular world market.

Environmental legislation is particularly prolific in the field of vehicle emissions, but other areas, which include small two-stoke engines, medium-speed stationary engines, marine engines and the engines of railway locomotives, are also receiving increasing attention. The subject of engine and vehicle emissions is extremely wide reaching and continues to generate a vast number of technical papers each year, a summary of which is well outside the scope of this book, but some are listed at the end of this chapter.

A gasoline-powered vehicle in 2005 was capable of emitting between 1 and 3 per cent of the carbon monoxide (CO) of a comparable sized unit in 1970, similar dramatic reductions are to be seen in vehicle emissions of hydrocarbons (HC), oxides of nitrogen (NO_x) and diesel particulates (all measured in mg/km).

Emission legislation in 2005 had divided into that developed in the United States and a generally more stringent form developed by the European Community (EU). Just over 50 per cent of the vehicles being sold in 2005 were being calibrated to EU emission standards. Asia-Pacific markets, with the exception of Taiwan, implement EU standards and some countries, such as Australia, currently allow both standards.

The recent history of such legislation has, naturally, been characterized by legislation setting future limits of exhaust emission components in advance of the engine and test technologies required to meet them, therefore it is impossible to summarize the list of components and systems that the test engineer will be using in the future. However, legislative approval of test technologies and techniques normally runs somewhat slower than the development of instrumentation since the cost and organizational complexity of developing and establishing a standard test procedure, even in a single country, are very significant; so once a procedure is established it is essential that it should remain unchanged for as long a period as possible. This probably means current technology will continue to be used for some years but will mean that laboratories in future will be using the new technologies, such as mass spectroscopy, for engine development while reserving other cell configurations for certification.

The various harmful results of atmospheric pollution on the environment in general, and human health in particular, are now widely recorded and all readers of this edition will be aware of them. The human health problems caused by polyaromatic hydrocarbons (PAH) are a more recent discovery and have given rise to limits being set in Europe (Directive 2003/17/EC) and various national regulatory bodies. The US Environmental Protection Agency (EPA) list of the PAH group includes 16 named compounds, the best known are benzene, formaldehyde and 1,3-butadiene, many of which are known or suspected carcinogens. There has also been increased concern about the very small particulates, in the order of $0.1 \mu m$ and smaller, so-called 'nanoparticles' below 50 nm in size, that are able to penetrate to the surface of the lung (see Principles of particulate emissions measurement, below).

A reminder of the basic chemical transformation may be of interest to the reader.

Basic chemistry of internal combustion engine emissions

The two processes under review and shown in Fig. 16.1 are the complete and incomplete combustion of the hydrocarbon fuel in air (treated here as an oxygen/nitrogen mixture). The production of carbon dioxide in this process is of concern since it is classed as a 'greenhouse' gas, but can only be reduced by an increase in the overall efficiency of the engine and vehicle thereby producing lower fuel consumption per work unit.

The emission gases and particulates covered by legislation are produced by incomplete combustion and are

- carbon monoxide (CO), a highly toxic odourless gas;
- carbon (C), experienced in the form of smoke;

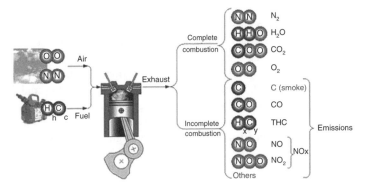

Figure 16.1 *Gaseous components of combustion processes*

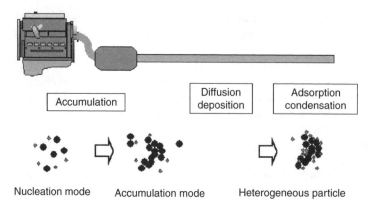

Figure 16.2 *Phases of particle formation*

* hydrocarbons (HC or <u>Total</u><u>HC</u>) formed by unburnt fractions of the original liquid fuel;
* nitric oxide (NO) and nitrogen dioxide (NO_2), together considered as NO_x.

The physics and chemistry covering the development of particles between being formed in the combustion chamber and circulated in the atmosphere is complex and has to be emulated within the systems designed to measure exhaust emissions. The main phases of particle formation are shown in Fig. 16.2; it is the products of these processes that produce the gases and particles that may be absorbed by life forms 'on the street'.

Emissions from spark ignition engines

Perhaps the most characteristic feature of exhaust emissions, particularly in the case of the spark ignition engine, is that almost every step that can be devised in order to reduce the amount of any given pollutant has undesirable side effects, most frequently an increase in some other pollutant.

Figure 16.3 shows the effect of changing air/fuel ratio on the emission of the main pollutants: CO, NO_x and unburned hydrocarbons. It will be apparent that one line of attack is to confine operation to a narrow window, around the stoichiometric ratio, and a major thrust of vehicle engine development since the mid-1990s has concentrated on the so-called 'stoichiometric' engine, used in conjunction with the three-way exhaust catalyser, which converts these pollutants to CO_2 and nitrogen.

We have already seen in Chapter 14 that a spark ignition engine develops maximum power with a rich mixture, $\lambda \sim 0.9$, and best economy with a weak mixture, $\lambda \sim 1.1$, and before exhaust pollution became a matter of concern it was the aim of the engine designer to run as close as possible to the latter condition except when maximum power was demanded. In both conditions the emissions of CO, NO_x and

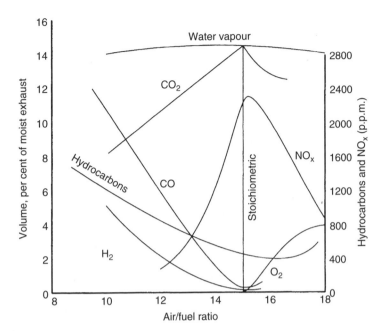

Figure 16.3   *Relation between exhaust emissions and air/fuel ratio for gasoline engines*

unburned hydrocarbons are high: however, the catalyser requires precise control of the mixture strength to within about ±5 per cent of stoichiometric.

This requirement has been met by the introduction of fuel injection systems with elaborate arrangements to control mixture strength (lambda closed-loop control). In addition, detailed development of the inlet passages in the immediate neighbourhood of the inlet valve and the adoption of four-valve heads has been aimed at improving the homogeneity of the mixture entering the cylinder and developing small-scale turbulence: this improves the regularity of combustion, itself an important factor in reducing emissions. Other measures to improve the uniformity of the mixture include preheating of the intake air and steps to reduce the extent of the liquid fuel film on the intake passages.

Another area of development that significantly affects emissions concerns the ignition system. Spark plug design and location, duration and energy of spark, as well as the use of multiple spark plugs, all affect combustion and hence emissions, while spark timing has a powerful influence on fuel consumption, as well as emissions. Here again conflicting influences come into play. Delaying ignition after that corresponding to best efficiency reduces the production of NO_x and unburned hydrocarbons, due to the continuation of combustion into the exhaust period, but increases fuel consumption.

Exhaust gas recirculation (EGR), involving the deflection of a proportion of the exhaust into the inlet manifold, reduces the NO_x level, mainly as a result of reduction in the maximum combustion temperature. The level of NO_x production is very sensitive to this temperature (this incidentally has discouraged development of the so-called adiabatic engine, in which fuel economy is improved by reducing in-cylinder heat losses). It also acts against the pursuit of higher efficiency by increasing compression ratios.

EGR control calibration requires that the volume and content of the recirculated portion of the exhaust flow is measured as a distinct channel for the gas analyser system.

An alternative line of development is concerned with the lean burn engine, which operates at lambda values of 1.4 or more, where NO_x values have fallen to an acceptable value, and fuel consumption is acceptable. However, HC emissions are then high, the power output per unit swept volume is reduced and running under light load and idling conditions tends to be irregular. The engine depends for its performance on the development of a stratified charge, usually with in-cylinder fuel injection.

Development of exhaust after-treatment systems is a continuing process, at present particularly concerned with reducing the time taken for the catalyser to reach operating temperature, important because emissions are particularly severe during cold starts.

Emissions from diesel engines

The composition of the emissions from a diesel engine, like that of the spark-ignition engine, depends on the engine design, its operating condition and the composition of the fuel used, the latter property being particularly important since it is very variable worldwide. The sulphur content of diesel fuel is responsible for sulphur dioxide, SO_2, in the exhaust. Permitted levels of sulphur in fuels for road vehicles have been drastically reduced and have given rise to incidental problems with fuel injection equipment arising from the reduced lubricity of the fuel that has been, and continues to be, the subject of engine testing in cells and in real-life operation. In the developing world, sulphur contents tend to be much higher and an obstacle to reduction is the substantial cost of the necessary refinery modifications.

The diesel engine presents rather different problems from the spark-ignition engine because it always operates with considerable excess air, so that CO emissions are not a significant problem, and the close control of air/fuel ratio, so significant in the control of gasoline engine emissions, is not required. On the other hand, particulates are much more of a problem and NO_x production is substantial.

The indirect injection (prechamber) engine performs well in terms of NO_x emissions; however, the fuel consumption penalty associated with indirect injection has resulted in a general move to direct injection, associated with four-valve cylinder heads with a central fuel injector. This development is associated with a sharp increase in fuel injection pressures, now commonly 1500 bar or more, and the 'shaping' of the fuel injection pulse. The 'common rail' fuel injection system, where fuel is supplied at a uniform high pressure and injector needle movement controlled electrically, is now the most common fuelling system in small automotive applications.

NO_x emissions are very sensitive to maximum cylinder temperature and to the excess air factor. This has prompted the use of increased levels of turbocharging with improved after-cooling, as well as the use of retarded injection which results in reduced peak pressures and temperatures but, beyond a certain point, in increased fuel consumption.

Diesel particulate emissions

Individual diesel exhaust particles may be highly complex coagulations of compounds, as shown in Fig. 16.4 below. Around 90 per cent of particles emitted by a modern automotive diesel may be below 1 μm in size which challenges the tools used to measure their presence.

Emission legislation, certification and test processes

This is an immensely complicated and rapidly evolving field: all but the briefest summary would be impossible within the limits of this book, and would in any case be out of date almost as soon as it was published. The reader requiring detailed knowledge is advised to consult Ref. 1. This publication summarizes international standards originating from the USA, the UN Economic Commission for Europe (ECE) and the European Union (EU), together with national standards for some 18 individual countries.

Since so much of the operational strategy of automotive test facilities is determined by current and planned environmental protection legislation, it is necessary for test

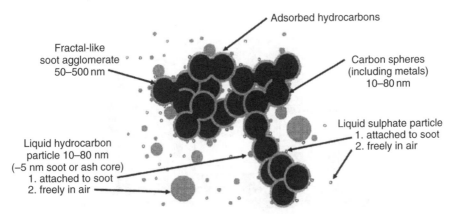

Figure 16.4 *Schematic representation of diesel particles (Kulijk–Foster)*

engineers to be aware of the legislative framework under which their home and export markets work. A single source of up-to-date information on worldwide emission legislation may not exist in the public domain, therefore the industry has to support specialists in particular areas of commercial interest in order to keep up to date within geographical zones.

Vehicle emission legislation dates from 1966 when the California Air Resources Board (CARB) produced tailpipe emission standards for hydrocarbons (HC) and carbon monoxide (CO) within the State of California. The Environmental Protection Agency (EPA) and the Clean Air Act were introduced by the US Government in the 1960s. European and Japanese legislation followed from 1970. The increasing public awareness of the environmental and human health problems resulting from vehicular emissions plus legislative barriers to important automotive markets began a development impetus that produced, and continues to produce, new emission-reducing technologies and requiring improved fuels. Catalytic converters required lead-free gasoline, thus the whole support infrastructure and pattern of vehicle registration has increasingly become dominated by emission legislation; the process whereby automotive technology, test technology and legislative requirements 'leapfrog' each other continues to this day.

The huge improvements in individual engine emissions between 1965 and 2005 have meant that now the refining and formulation of lubricating oils and fuels have to be improved worldwide in order to benefit from the modern combustion control and after-treatment technologies.

In spite of improved instrumentation and much reduced levels of pollutants produced by vehicles and engines, the basic form of much legislative testing has remained very similar to the original EPA methodology of the 1970s.

In general, all engine emission legislation consists of three component parts (Fig. 16.5):

1. Test cycle describes the operation of the tested vehicle or engine. For light duty vehicles, it simulates the actual driving on the road in that it defines a vehicle

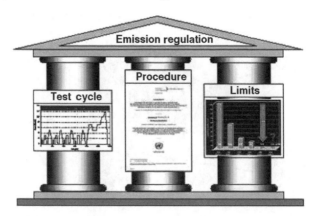

Figure 16.5   *The three pillars of emission legislation*

velocity profile over the test time. For heavy duty and off-road engines where only the engine is tested on an engine dynamometer, the test cycle defines a speed and torque profile over the test time.

2. Test procedure defines in detail how the test is executed, which measurement method and which test systems have to be used. It defines the test conditions and result calculations to apply. Probably the best known example of a legislative procedure is the use of sample bags in which an accumulation of diluted sample of exhaust gas resulting from a drive cycle is stored for analysis of content and concentration.

3. Test limits, which define the maximum allowed emission of the regulated components in the engine exhaust. For light duty vehicles, the limit is expressed in mass per driving distance (g/km); for heavy duty vehicles, the limits are expressed in mass per unit of work (g/kWh).

Light duty certification requires the whole vehicle to be tested and certified, while heavy duty engines are tested in a test cell operated in accordance with the stipulated test procedures.

The basic legislation methodology for cars and light truck engines was, and still is, based on driving the test vehicle through a prescribed 'drive cycle' on a chassis dynamometer (see Chapter 18), while storing a proportion of the exhaust gases, produced during the whole test, in bags so that the amount and proportions of the key pollutants can be determined.

The drive cycles were designed to simulate typical vehicle usage, but since they had to be driven as consistently as possible over a wide variety of vehicles, the rate of change in simulated road speeds is quite slow when compared with real life. It has only been since the beginning of the twenty-first century that instrumentation has been available that is truly capable of measuring some of the exhaust pollutants as they are produced during the transient phases of engine loads.

Human drivers using a drivers-aid display rather like an early generation computer game display are still employed, but increasingly test vehicles are driven using electromechanical robots.

Drive cycles have been subject to much international examination by national laboratories around the world, it being obvious that a 'typical' drive for a suburban commuter living in Los Angeles will be quite different to that of an inhabitant of Mumbai or Cairns; however, the majority of testing worldwide is done to drive cycles produced by the three dominating geographical producer areas.

It was understood from the early days of emission studies that some of the chemical reactions resulting from in-cylinder combustion continue, as the gases and particularly 'soot' particles, pass through the exhaust system and into the atmosphere. Therefore, the pollution products had to be subjected to realistic post-combustion mixing with air before being analysed. The complex chemistry and the need for correlation across many test sites worldwide has meant that much of the legislation concerning vehicle homologation is very prescriptive concerning equipment used. There is, therefore, a lag between the instrumentation developed for emission research and that 'allowed' for certification by legislation.

Legislation classifications

Much of the test subject classification in automotive emission legislation is by vehicle size rather than engine size and type. The main classifications being

- light duty – gasoline;
- light duty – diesel;
- heavy duty.

Each of these categories has legislative pollutant limits, test methodologies and instrumentation designed specifically for them.

In European emission legislation, there are both categories and weight classifications of vehicles which are shown below:

- *Category M:* Motor vehicles with at least four wheels and used for the carriage of passengers.
- *Category M1:* Vehicles used for the carriage of passengers and comprising no more than eight seats in addition to the driver's seat.
- *Category M2:* Vehicles used for the carriage of passengers and comprising no more than eight seats in addition to the driver's seat and a maximum mass not exceeding 5 tonnes.
- *Category M3:* Vehicles used for the carriage of passengers and comprising no more than eight seats in addition to the driver's seat and a maximum mass exceeding 5 tonnes.
- *Category N:* Motor vehicles with at least four wheels and used for the carriage of goods.
- *Category N1:* Vehicles used for the carriage of goods having a maximum mass not exceeding 3.5 tonnes.
- *Category N2:* Vehicles used for the carriage of goods having a maximum mass exceeding 3.5 tonnes and not exceeding 12 tonnes.
- *Category N3:* Vehicles used for the carriage of goods having a maximum mass exceeding 12 tonnes.

Weight classifications for light commercial vehicles:

- Class I $\quad$ RW $\leq 1305\,kg$
- Class II $\quad$ $1305 < RW \leq 1760\,kg$
- Class III $\quad$ $1760\,kg < RW$

Tables 16.1a, b and c list some examples of emission limits for conventional vehicles classified above. Note that the dates refer to new vehicle types; new vehicle registrations are 1 year later. Euro III and IV standards also apply to Class I light commercial vehicles $<1305\,kg$.

The classification of vehicles meeting various emission limits has given rise to a number of acronyms that are commonly used in technical press and papers:

Low emission vehicle	LEV
Ultra low emission vehicle	ULEV
Super ultra low emission vehicle	SULEV
Transitional low emission vehicle	TLEV
Zero emission vehicle	ZEV
Partial zero emission vehicle	PZEV

Table 16.1a *European emission limits for petrol cars in grams per kilometre*

Gasoline	As from	CO	THC	NO_x
Euro I	1/7/1992	4.05	0.66	0.49
Euro II	1/1/1996	3.28	0.34	0.25
Euro III	1/1/2000	2.30	0.20	0.15
Euro IV	1/1/2005	1.00	0.10	0.08

Table 16.1b *European emission limits for diesel cars in grams per kilometre*

Diesel	As from	CO	THC	NO_x	PM
Euro I	1/7/1992	2.88	0.20	0.78	0.14
Euro II	1/1/1996	1.06	0.19	0.73	0.10
Euro III	1/1/2000	0.64	0.06	0.50	0.05
Euro IV	1/1/2005	0.50	0.05	0.25	0.025

Table 16.1c *European emission limits for light commercial vehicles N2 Class in grams per kilometre*

N2	As from	Fuel type	CO	HC	NO_x	$HC + NO_x$	PM
Euro I	1/10/1994	All	5.17	–	–	1.4	0.19
Euro II	1/1/1998	Petrol	4	–	–	0.65	–
		Diesel	1.2	–	–	1.1	0.15
Euro III	1/1/2002	Petrol	4.17	0.25	0.18	–	–
		Diesel	0.8	–	0.65	0.72	0.07
Euro IV	1/1/2006	Petrol	1.81	0.13	0.1	–	–
		Diesel	0.63	–	0.33	0.39	0.04

In the 1970s the net was cast further and legislation covering small utility engines, being spark-ignition engines, both two- and four-stroke, of the type used in mowers, chainsaws and small generators was produced and continues to be developed. In Europe, small engines under 19 kW were covered in 2002 (Directive 2002/88/EC).

Legislation covering 'non-road' diesel engines was produced in Europe in 1997 (Directive 97/68/EC) and intended to cover off-road trucks, drilling rigs, mobile cranes, etc. The coverage is complex in detail and much is phased in over time to allow for replacement engines and specialist usages, but much of the test details are based on ISO 8178-1 which is an international standard for non-road engine applications, specifically:

- EPA CFR 40 Part 89 Non-road compression ignition engines;
- EPA CFR 40 Part 90 Non-road spark ignition engines;
- ECE 97/68 amended by ECE 202/88.

The engine test sequences based on ISO 8178 are sequences of steady state modes with different emission limits. A typical set-up for running these tests in a cell is shown in Fig. 16.6.

Engineers requiring details need access to the relevant government websites covering local legislation.

International agreement is clearly required for the control of marine plant since it can travel the world; such agreement is contained in the 1997 MARPOL Protocol Annex V1. Marine engine legislation is divided into three categories defined by size (displacement per cylinder) but covering the three distinctly different engine technologies (Table 16.2).

Category 3 has caused difficulty between environmental groups and legislators since it covers very large diesels running on residual fuels (see Chapter 7), the formulation of which is largely unregulated. Many countries apply rules based on

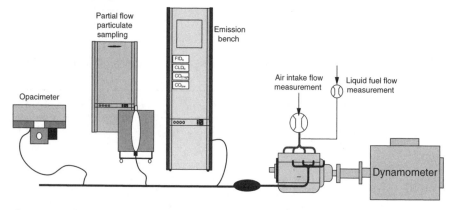

Figure 16.6 *Typical emission plant used in ISO 8178 engine emission cycles (diesel)*

Table 16.2 *Marine engine categories*

Category	Displacement per cylinder (D)	Basic engine technology
1	$D <5\,dm^3$, power $\geq 37\,kW$	Land-based non-road diesel
2	$D \geq 5\,dm^3$ and $<30\,dm^3$	Locomotive diesel engine
3	$D \geq 30\,dm^3$	Unique marine engine design

ISO 8178 for ships within coastal waters and habour which requires ships running on residual fuels to change over to light distillate oils before entering waters covered by such legislation.

Aftermarket emission legislation

In addition to the prime legislation covering the certification of new engines and vehicle systems in most countries of the developed world, there is emission legislation covering the condition of cars in the population as they age. Such tests are required annually and range from a single visual smoke check, through a check at fast idle for levels of CO and HC, to a test under light load on a rolling road that checks CO, CO_2 and HC. In the UK, the basic emission test (BET) for gasoline engines is as follows:

Engine warmed up
Fast idle 2500–3000 r.p.m., CO no more than 0.3 per cent, HC no more than 200
   p.p.m.
Lambda between 0.97 and 1.03
Normal idle 450–1500 r.p.m., CO no more than 0.5 per cent.

For diesel engines, the test checks for visible smoke during idle and transient states by using a calibrated smoke meter placed in the tailpipe, while the engine is subjected to a number of accelerations.

Homologation

Homologation is a term widely thought to be exclusively involved with certification to emission legislation; in fact it is another case of a word, having originally no particular engineering associations, being taken over and given a specialized meaning.

Knowledge of and compliance with the legislative requirements of different markets, covering exhaust emissions but also safety, fuel consumption rates, noise vibration and harshness, competitive benchmarking, etc., is an important aspect of the problem of gaining acceptance of a given product. For the engine or vehicle manufacturer, homologation is a complex and expensive area of activity since all major versions, and all derivatives, of the vehicle must meet the formal requirements that are in force in each country in which the vehicle is to be sold.

Homologation is the process of establishing and certifying this conformity, both for whole vehicles and for components.

Principles of particulate emissions measurement

Particulates, when they appear to the human observer, are called 'smoke'. Smoke colours are indicative of the dominant source of particulate:

- black = 'soot' or more accurately carbon, which typically makes up some 95 per cent of diesel smoke either in elemental, the majority, or organic form;
- blue = hydrocarbons, typically due to lubricating oil burning due to an engine fault;
- white = water vapour, typically from condensation in a cold engine or coolant leaking into the combustion chambers – white smoke is not detected by conventional smoke meters;
- brown = NO_2 may be detected in exhaust of heavy fuel engines.

There are essentially four methods in use for measuring particulate emissions and numerical results produced by these methods cannot readily be related one to the other:

- Opacimeters that measure the opacity of the undiluted exhaust by the degree of obscuration of a light beam. These devices are able to detect particulate levels in gas flow at lower levels than the human eye can detect. The value output is normally in percentage of light blocked by the test flow. Zero being clean purge air and 100 per cent being very thick black smoke. A secondary output giving an absorption factor 'k' m^{-1} may be given, which allows some degree of comparison between devices since it removes the effect of different distances between light source and sensor.
- Smoke meters that perform the measurement of the particulate content of an undiluted sample of exhaust gas by drawing it through a filter paper of specified properties and estimating the consequent blackening of the paper against a pristine paper. The value output is in some form of 'smoke number' specific to the instrument maker.
- Particulate samplers that measure the actual mass of particulates trapped by a filter paper during the passage of a specified volume of diluted exhaust gas. This requires climatically controlled laboratory weighing facilities.
- The ever lower particulate emissions from engines has required a new generation of 'microsoot' devices capable of detecting particulate levels down to typically 5 μg per m^3 of exhaust gas. These devices incorporate a laser and work on a photo-acoustic principle.

Principles of measurement and analysis of gaseous emissions

The description of instrumentation that follows, which is not exhaustive, lists the main types of instrument used for measuring the various gaseous components of exhaust emissions. Continuing developments are taking place in this field, notably in the evolution of instruments having the shortest possible response time.

Non-dispersive infrared analyser (NDIR)

Called 'non-dispersive' because all the polychromatic light from the source passes through the gas sample before going through a filter in front of the sensor, whereas 'dispersive' instruments, found in analytical laboratories, filter the source light to a narrow frequency band before the sample.

The CO_2 molecule has a very marked and unique absorptance band of infrared (IR) light that shows a dominant peak at the $4.26\,\mu m$ wavelength which the instrument sensor is tuned to detect and measure. By selecting filters sensitive to other wavelengths of IR, it is possible to detect other compounds such as CO and other hydrocarbons at around $3.4\,\mu m$ (see Fourier transform infrared analyser, below). Note that the measurement of CO_2 using an NDIR analyser is cross-sensitive to the presence of water vapour in the sample gas.

Fourier transform infrared analyser (FTIR)

This operates on the same principle as the NDIR, but performs a Fourier analysis of the complete infrared absorption spectrum of the gas sample. This permits the measurement of the content of a large number of different components. The method is particularly useful for dealing with emissions from engines burning alcohol-based fuels, since methanol and formaldehyde may be detected.

Chemiluminescence detector (CLD)

Chemiluminescence is the phenomenon by which some chemical reactions produce light. The reaction of interest to exhaust emissions is

$$NO + O_3 \rightarrow NO_2 + O_2 \rightarrow NO_2 + O_2 + photon$$

The nitrogen compounds in exhaust gas are a mixture of NO and NO_2, described as NO_x.

In the detector, the NO_2 is first catalytically converted to NO and the sample is reacted with ozone which is generated by an electrical discharge through oxygen, at low pressure in a heated vacuum chamber. The light is measured by a photomultiplier and indicates the NO_x concentration in the sample.

A great deal of development work continues to be carried out to improve chemical reaction times which are highly temperature-dependent, and so shorten instrument response times.

Flame ionization detector (FID)

The FID has a very wide dynamic range and high sensitivity to all substances that contain carbon. The operation of this instrument shown schematically in Fig. 16.7 depends on the production of free electrons and positive ions that takes place during the combustion of hydrocarbons. If the combustion is arranged to take place in an electric field, the current flow between anode and cathode is closely proportional to the number of carbon atoms taking part in the reaction. In the detector the sample is mixed with hydrogen and helium and burned in a chamber which is heated to prevent condensation of the water vapour formed. A typical, sample to measurement, response time is 1–2 s.

Fast FID, cutter FID and GC-FID

This is a miniaturized development of the FID instrument which is capable of a response time measured in milliseconds and may thus be used for in-cylinder and

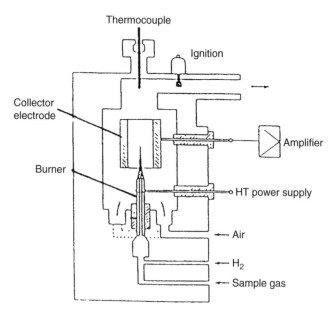

Figure 16.7 *Flame ionization detector*

exhaust port measurements. Cutter FID analysers and GC-FID analysers are used to measure methane CH_4. In the fast FID, non-methane hydrocarbons are catalysed into CH_4 or CO_2 and measured.

Paramagnetic detection (PMD) analyser

PMD analysers are used to measure oxygen in the testing of gasoline engines. They work due to the fact that oxygen has a strong paramagnetic susceptibility. Inside the measuring cell, the oxygen molecules are drawn into a strong inhomogeneous magnetic field where they tend to collect in the area of strongest flux and physically displace a balanced detector whose deflection is proportional to oxygen concentration. Since NO_x and CO_2 show some paramagnetic characteristic, the analyser has to be capable of calculating compensation for this interference.

Mass spectrometer

These devices, not yet specified in any emission legislation, are developing rapidly and can distinguish most of the components of automotive engine exhaust gases; currently they remain an R&D tool, but may represent the future technology of general emission measurement.

Response times

All these instruments are required to operate under one of two quite different conditions, depending on whether the demand is for an accurate measurement of a sample collected over a fairly long time interval, as in the case of the various statutory test procedures, or for an instantaneous measurement made under rapidly changing conditions, such as arise in true transient testing, the purpose of which is to study the performance of the engine and its control system in detail.

These two sets of requirements are conflicting: analysers for steady state work must be accurate, sensitive and stable and thus tend to have slow response times and be well damped. Analysers for transient work do not require such a high standard of accuracy but must respond very quickly, preferably within a few milliseconds. Reduction of response time is a prime object of instrument development in this field.

A further problem in transient testing is concerned with the time and distance lags associated with the positioning of the exhaust gas sampling points. As any search will reveal, this matter is discussed in many SAE papers concerning the reconstruction of the true signal from the instrument signals, taking into account sampling delays and instrument response characteristics.

Integration of exhaust emission instrumentation

The great majority of engine test work is concerned with the taking of measurements that are in principle quite simple, even though great skill may be needed to interpret them: forces, pressures, masses, flow rates, speeds, displacements, temperatures, oscillations. Where emissions measurements are concerned, we are forced to move into a totally different and very sophisticated field of instrumentation engineering. The apparatus makes use of subtle and difficult techniques borrowed from the field of physics.

It is very desirable, for all but the simplest garage-type emissions apparatus, that a technician should be specially trained to take responsibility for the maintenance and calibration of the instruments involved.

There are significant subsystems that require to be housed in engine or vehicle test facilities, each of which require some special health and safety considerations due to the handling of hot and toxic gases within a building space. These will be considered in the following text.

The full flow of exhaust gases has to be safely ducted and samples for analysis have to be removed in the correct amounts at critical points in any vehicular after-treatment system.

It is important that water, which is a product of the combustion process, particularly that of the gasoline and some biofuels, is not allowed to condense within the sampling system. Some of the compounds produced by the partial combustion of fuels are soluble in water, so it is possible to lose a significant amount of these compounds in the condensation on the equipment surfaces.

To prevent condensation in critical parts of the exhaust, the sampling system will be heated. For example, the sampling system for THC may consist of heated probe, sampling line filter and sampling pump, all of which are required to prevent the temperature of the equipment being lower than the dew point of the exhaust gas.

Calibration, span gases, storing gas distribution system

In order to function correctly exhaust gas analysers have to be calibrated regularly, using gases of known composition. The principle of calibration is similar to that of any transducer in that the zero point is set, in this case by purging with pure nitrogen (sometimes referred to as 'zero-air'), then the 100 per cent value is set with a gas of that composition, known as a 'span gas'. Good practice then dictates that a gas of an intermediate composition is then used as a check of system linearity. The whole routine may be highly automated by a calibration routine built into the analyser control unit. After calibration, the gas supply system is purged of at least NO_x to prevent degrading in the lines.

Many R&D analysers are capable of part-spanning to change the measuring sensitivity and can be calibrated by using either known span gas precisely diluted with a zero-air using a device known as a 'gas divider' or, more commonly, with

Table 16.3 *List of typical gases that may be used in the operation of basic engine emission certification cell*

Gases	Used as	Concentration	For analyser type
Synthetic air (2)	Burner air	21% O_2/79% N_2	FID
H_2/He mixture	Burner gas	40% H_2, 60% He	FID
C_3H_8 in synth. air	Test gas	99.95%	External bottle with CFO system
O_2	Operating gas	99.98% purity	CLD
CO in N_2	Span gas	50 ppm	CO_{low}
C_3H_8 in synth air	Span gas	100 ppm	FID
C_3H_8 in synth air	Span gas	10000 ppm	FID
NO in N_2	Span gas	100 ppm	CLD
NO in N_2	Span gas	5000 ppm	CLD
CO in N_2	Span gas	500 ppm	CO_{low}
CO in N_2	Span gas	10%	CO_{high}
CO_2 in N_2	Span gas	5%	CO_2
CO_2 in N_2	Span gas	20%	CO_2
O_2 in N_2	Span gas	20%	CO_2
Synthetic air (1)	Zero gas	oil free	FID
N_2	Zero gas	CO free	NO, CO_{low}, CO_{high}, CO_2, O_2

a bottled span gas of the correct composition. If using the gas divider method, it is important that the same gas (N_2 or artificial air) is used in the dilution as that which was used as the zero gas.

In addition to calibration or span gases, the facility will require operational gases such as hydrogen plus an oxygen source as fuel for any flame ionization (FID) and in some cases test gases for such routines as a critical flow check (see Constant Volume Sampling (CVS) systems).

Table 16.3 shows a possible minimum list of gases required in a modern emissions test facility.

All these gases have to be stored at high pressure, usually between 150 and 200 bar, and distributed, at a regulated pressure of about 4 bar, to the analysis devices within the facility. The pipe work system used in the distribution has to be chemically inert and medically clean to prevent compromising the composition of the calibration gases. The minimum number of 50 litre or 40 litre* gas storage bottles containing individual gases is probably 12, and a typical diesel/gasoline certification cell may have upwards of 20 gas lines feeding its analyser system. It is not uncommon in

---

* A 40 litre bottle operating at 150 bar has a capacity of around 6000 litres of gas at the analyser pressure and a 50 litre bottle (normal size for synthetic air and nitrogen) around 7500 litres.

mixed fuel facilities with two cells to have 20 calibration bottles and four operating gases connected at any one time.

Each gas bottle has to be delivered, connected, disconnected and removed in the cycle of use; therefore the positioning of the gas store has some important logistical considerations.

Clearly, the distance between the gas bottles, the pressure regulation station and analysers directly affects the cost of gas distribution. The practical requirements of routing upwards of twelve $\frac{1}{4}$-inch or 6-mm OD stainless steel pipes on a support frame system is an important design consideration in a new facility. The length of gas lines may be sufficient to create an unacceptable pressure drop which may require transmission at an intermediate pressure and regulation local to the analysers. Clearly, the location of the gas bottle store or stores can have important cost and operational implications on the facility.

The storage of the gas bottles should be in a well-ventilated secure store where the bottles are not subjected to direct solar heating and thus remain under 50°C. For operational reasons, in very cold climates the bottle store needs to be kept at temperatures that allow handling and operation without injury to the operator or regulation malfunction. The store commonly takes the form of one or more roofed 'lean-to' structures on the outside wall of the facility, in temperate climates having steel mesh walls and always positioned for easy access of delivery trucks. Storage of gas bottles above ground within buildings creates operational and safety problems that are best avoided.

Pressure regulators may be either wall mounted with a short length of high pressure (up to 200 bar) flexible hose connecting the cylinder or, more rarely, directly cylinder mounted. The regulators handling NO_x gases need to be constructed of stainless steel, while other gases can be handled with brass and steel regulators. Local regulations and planning permission requirements may apply.

The gas lines are usually made of stainless steel specifically made for the purpose; those transporting NO_x, sometimes all, are electropolished internally to minimize degradation and reaction between gas and pipe. Pipes have to be joined by orbital welding[**] or fittings made expressly for the purpose. The pipes should terminate near the emission bench position in a manifold of isolating valves and self-sealing connectors for the short flexible connector lines to the analyser made of inert Teflon or PTFE not more than 2 m long.

Gas analysers and emission benches

Although often used interchangeably, the term 'analyser' should refer to an individual piece of instrumentation dedicated to a single analytical task installed within an 'emission bench' that contains several such units. Most analysers fit within a standard

---

[**] Orbital welding uses a gas tungsten arc welding (GTAW) process that prevents oxidation and minimizes surface distortion of pipe interior.

Table 16.4 *Typical analyser measuring minimum and maximum ranges*

Analyser	Low range	High range
NDIR CO_{low}	0–100 p.p.m.	0–3000 p.p.m.
CO_{high}	0–1%	0–12%
CO_2	0–2%	0–20%
NO	0–100 p.p.m.	0–5000 p.p.m.
PMD O_2	0–10%	0–25%
FID heated THC	0–100 p.p.m.	0–10 000 p.p.m.
CLD NO_x	0–100 p.p.m.	0–5000 p.p.m.

19-inch rack and are typically 4 HU high; most are capable of being set for measuring different ranges of their specific chemical. The Table 16.4 shows what analysers might be installed within a high quality bench built to measure raw gas from engines tested under ISO 8178 cycles.

Constant volume sampling (CVS) systems

Clearly, the above methods give no indication, except by inference based on experience, of the actual mass of particulates present in a given volume of exhaust. In both diesel and gasoline testing we need to dilute the exhaust with ambient air to prevent condensation of water in the collection system. It is necessary to measure or control the total volume of exhaust plus dilution air and collect a continuously proportioned volume of the sample for analysis. The use of a full flow CVS system is mandatory in some legislation, particularly that produced by the EPA. It may be assumed that this will change over time with developing technology and 'mini-dilution systems' will become widely allowed.

CVS systems consist of the following major component parts:

- A tunnel inlet air filter in the case of diesel testing or a filter/mixing tee in the case of gasoline testing, which mixes the exhaust gas and dilution air in a ratio usually about 4:1. There is also a sample point to draw off some of the (ambient) dilution air for later analysis from a sample bag.
- The dilution tunnel, made of polished stainless steel and of sufficient size to encourage thorough mixing and to reduce the sample temperature to about 125°F (51.7°C). It is important to prevent condensation of water in the tunnel so in the case of systems designed for use in climatic cells, the air will be heated before mixing with gas.
- A proportion of the diluted exhaust is extracted by the bag sampling unit for storage in sample bags.
- The critical flow venturi controls and measures the flow of gas that the turbo-blower draws through the system.

- The gas sample storage bag array.
- The analyser and control system by which the mass of HC, CO, NO_x, CO_2 and CH_4 is calculated from gas concentration in the bags, the gas density and total volume, taking into account the composition of the dilution air component.
- In the case of diesel testing, part of the flow is taken off to the dilution sampler containing filter papers for determining the mass of particulates over the test cycle.

A typical layout of a CVS system based on a 2×2 chassis dynamometer is shown in Fig. 16.8. The arrangement shown is for a cell certifying vehicles, either diesel or gasoline, to Euro 2 or Euro 3 regulation. The system shown includes critical flow venturi (marked CFV) through which the gas/air is sucked by a fixed speed 'blower' fitted with an outlet silencer. Some systems control flow with a variable speed blower and a different type of 'subsonic' venturi.

US Federal regulations require that the flow and dilution rates can be tested by incorporation of a critical flow orifice (CFO) check. This requires a subsidiary stainless steel circuit that precisely injects 99.95 per cent propane as a test gas upstream of the mixing point, the CVS system dilutes this gas according to the flow rate setting allowing the performance to be checked at a stabilized temperature.

The full flow particulate tunnel for heavy vehicle engines is a very bulky device. The modules listed above may be dispersed within a test facility because of the constraints of the building. The tunnel has to be in the cell near the engine and there are some legislative requirements concerning the distance from the tail pipe and the dilution point, ranging from 6.1 m for light duty to about 10 m for heavy duty systems.

Similarly, there are limits in some legislation concerning the transit time for HC samples from the sample probe to the analyser of 4 s, which will determine the position within the building spaces of the FID unit.

The analysers may be in the control room, the venturi and sample bags within another room adjoining the control room. The turboblower is commonly placed on the roof since they often produce high noise levels. Alternatively, for light vehicle engines the analysers, CVS control and bag storage may be packaged in one, typically four-bay, instrument cabinet.

The so-called 'mini-dilution tunnel' has been developed to reduce the problems inherent in CVS systems within existing test facilities. However, users must check if its use is authorized by the legislation to which they are working.

It is clearly necessary in this case to measure precisely the proportion of the total exhaust flow entering the tunnel. This is achieved by the use of a specially designed sampling probe system that ensures that the ratio between the sampling flow rate and the exhaust gas flow rate is kept constant

In all CVS systems, exhaust gases may be sent to the analysers for measurement either from the accumulated volume in the sample bags or directly from the vehicle, so-called 'modal' sampling.

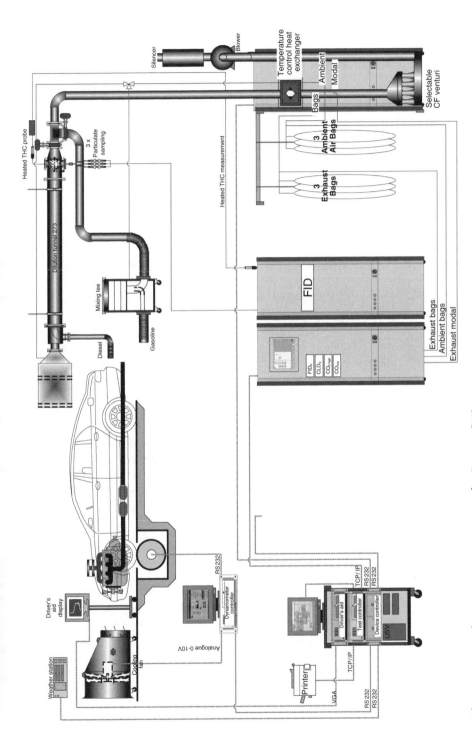

Figure 16.8 The major components of a Euro 3 CVS system

In an advanced test facility such as those involved with SULEV development, there will be three to five tapping points created in the vehicle system, from which gas may be drawn for analysis such as

1.   exhaust gas recirculation sample (EGR);
2.   before the vehicle catalytic converter (pre-CAT);
3.   after the vehicle catalytic converter (post-CAT);
4.   tail pipe sample (modal);
5.   diluted sample (sample bags).

From 2 and 3 the efficiency of the catalyst may be calculated.

In addition to the tapping points listed above, there may be several more taking samples from the CVS tunnel for particulate measurement.

Health and safety implications and interlocks for CVS systems

The following three fault conditions are drawn from experience and should be included in any operational risk analysis of a new facility.

Where the exhaust dilution air is drawn from the test cell and in the fault condition of the blower working and the cell ventilation system intake being out of action, or switched off, a negative cell pressure can be created that will prevent the opening of cell doors. The cell ventilation system should therefore be interlocked with the blower so as to prevent this occurring.

In the fault condition of the blower being inoperative or the airflow too low due to too small a flow venturi being used, the hot exhaust gas can flow back into the filter and cause a fire. A temperature transducer should be fitted immediately downstream of the dilution tunnel inlet filter and connected to an alarm channel set at about 45°C.

Gas analysers that have to discharge their calibration gases at atmospheric pressure should have an extraction cowl over them to duct the gases into the ventilation extract; such an extract should be interlocked with the analyser, preferably by a flow sensor.

Emissions test procedures

The distinguishing feature of all emissions test procedures is that they all involve transient operation of the engine. They are thus more demanding than any sort of steady state test. They are, however, representative of the conditions of operation of vehicle engines.

A comprehensive survey of current and proposed test procedures would much exceed the scope of this book. It is therefore proposed to describe the two principal current test procedures for passenger cars and to give a brief indication of the scope of other methods and anticipated developments.

Temperature soak areas for legislative testing

A feature of many legislative test procedures is that the vehicle has to soak at a uniform temperature before commencement. The two temperatures stipulated are between 20 and 30°C for 'ambient' tests and −7°C for cold start tests and the soak periods are usually 12 h long. This requirement needs the efficient use of a climatically controlled building space with temperature recording and where vehicles are not subjected to direct solar light. There are several strategies that may be adopted by vehicle test facilities having to deal with multiple vehicles in order to minimize the footprint of the soak area.

For cold start testing, vehicles, additional to the one rigged for the first test, may be parked inside the chassis cell. They may then be moved on to the dynamometer by using special wheeled vehicle skids. ISO refrigerated shipping containers that can be parked close to the cell access doors may be used to soak vehicles providing the transit time, from container to cell, is as short as possible and without engine rotation. In ambient soak areas, it may be possible to use vertical stacking frames to store cars two or three high and test them in order of access.

The European exhaust emissions test procedure

The new NECE urban driving cycle (UDC) was devised to represent city driving conditions while the vehicle is on a chassis dynamometer. It is characterized by low vehicle speed, low engine load, and low exhaust gas temperature and is used in this text as an example of typical legislative drive cycles.

The UDC is used for emission certification of light duty vehicles in Europe. Before the test, the vehicle is allowed to soak for at least 6 h at a test temperature of 20–30°C. The entire cycle includes four ECE segments, Fig. 16.9, repeated without interruption, followed by one EUDC segment, Fig. 16.10.

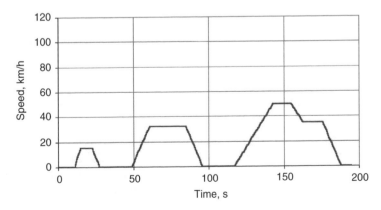

Figure 16.9 *ECE 15 cycle*

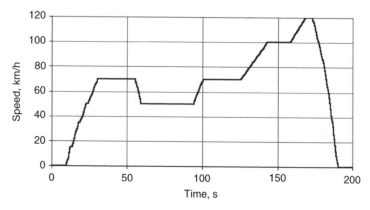

Figure 16.10   *EUDC cycle*

The EUDC (extra urban driving cycle) segment has been added after the fourth ECE cycle to account for more aggressive, high speed driving modes. The maximum speed of the EUDC cycle is 120 km/h. An alternative EUDC cycle for low-powered vehicles has been also defined with a maximum speed limited to 90 km/h.

There have been criticisms of these test procedures because of their lack of severity, in particular of the modest acceleration rates and it has been claimed that they underestimate emissions by 15–25 per cent compared with more realistic driving at the same speed.

The US federal light duty exhaust emission test procedure (FTP-75)

This is a more complex procedure than the European test and is claimed to more realistically represent actual road conditions. The cycle is illustrated in Fig. 16.11 and, in contrast to the European test, embodies a very large number of speed changes. The cycle has three separate phases:

1. a cold-start (505-s) phase known as bag 1;
2. a hot-transient (870-s) phase known as bag 2;
3. a hot-start (505-s) phase known as bag 3.

The three test phases are referred to as bag 1, bag 2 and bag 3 because exhaust samples are collected in separate sample bags during each phase. During a 10-min cooldown between the second and third phase, the engine is switched off. The 505-s driving sequences of the first and third phase are identical. The total test time for the FTP 75 is 2457 seconds (40.95 min), the top speed is 56.7 mph, the average speed is 21.4 mph and the total distance covered is 11 miles.

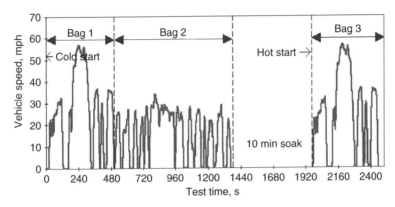

Figure 16.11 *EPA urban dynamometer driving schedule*

Other test procedures

The US procedure (1985) for testing heavy-duty engines prescribes a transient test cycle, while the European (ECE 49) and Japanese (13-mode) procedures use steady state tests. All three procedures measure emissions with the engine removed from the vehicle and installed on a dynamometer. Results are reported in g/bhp/h or g/kWh.

The emissions test procedures for motor cycles and mopeds in force in the USA and in Europe are essentially similar to those described above for passenger cars. Testing is carried out on a single roll dynamometer with inertia simulation, with a clamp to hold the motorcycle upright.

Vehicle evaporative emissions

Most vehicle emissions are produced by the combustion of hydrocarbon-based fuels within internal combustion engines; the exception that is important to the test engineer is the evaporative emissions of a complete vehicle.

These emissions arise because the vehicle both at rest and in motion contains volatile compounds that evaporate differentially under changing solar and atmospheric conditions. In the 1970s, vehicles parked in the sun produced significant amounts of evaporative pollution. This pollution was quantified using a gastight chamber in which the vehicle was placed and subjected to typical temperature variations, while exchanging the polluted air within the chamber with clean air from outside (volume compensation). The resulting test methodology and legislation requires such a chamber, commonly referred to as a SHED (sealed housing for evaporative determination). These are usually purchased as complete integrated systems that are designed to meet legislation.

Typically, the SHED consists of a 'box' fitted with a vehicle access door capable of being closed with a gasproof seal and lined with stainless steel and constructed from materials that are chemically inert at the temperatures used in testing. This box or container together with its external control cabinet is invariably installed within a larger building and connected to the electrical and ventilation services of that building. A typical car SHED unit will measure 6.5 m long by 3 m wide and 2.5 m high and will be equipped with systems to ensure the safe operation while handling potentially explosive gases. Legislation, such as that of the California Air Resources Board (CARB), requires the test vehicle to be subjected to a diurnal temperature range of 20 and 35°C (98/69/EC) or 18.3 and 40.6°C, but some available units extend that range. Special, smaller than standard, SHED units for testing of motor cycles are available. There are also special units that allow refuelling to be carried out within the sealed environment.

The withdrawal of polluted air for analysis and its replacement with treated air in the controlled environment of the SHED requires special control and some explosion-proof air handling equipment.

Major legislation currently requiring SHED testing includes

USA (EPA and CARB)

Hot soak test (40 CFR 86.138-96)
Hot soak evaporative emission test (CARB)
Diurnal evaporative emission test (40 CFR 86.133-96)
Diurnal evaporative emission test (CARB)

Europe (European Union EU2000)

Hot soak test (Directive 98/69/EC)
Evaporative emission test (Directive 98/69 EC)

Implications of choice of site on low level emissions testing

Test engineers not previously concerned with emissions testing should be aware of a number of special requirements that may arise, or have the same importance, elsewhere in other engine or vehicle testing. The first of these is the subject of the test cell site. The problem with measuring exhaust emissions to the levels required to meet the standards set by SULEV is that the level of pollution of the ambient air may be greater than that required to be measured at the vehicle exhaust. Almost any trace of pollution of the air ingested by the engine or used to dilute the concentration of exhaust gas within the measuring process, from exhaust gas of other engines or fumes from industrial plant, e.g. paint and solvents, will compromise the results of emissions tests. Cross-contamination via ventilation ducts must be avoided in the design of the facility and it will be necessary to take into account the direction of prevailing winds in siting the cell. Other than siting the facility in an area having, as far as is possible, pristine air, there are three strategies available to help deal with SULEV level measurement:

1. Reduction of the exhaust dilution ratio by having a heated CVS system and emissions system which prevent water condensation.

2. Refinement or cleaning of the dilution air to reduce the background level of pollutants to <0.1 p.p.m. This supports a CVS system, but is expensive and the plant tends to be large and complex.
3. Use a partial flow system with 'pure' (artificial) air for dilution. These bag mini-dilution systems do not comply with current CVS requirements and require incorporation of an exhaust flow-rate sensor.

Time will tell as to which of these methods become preferred and legislative practice, meanwhile investment in emission laboratories sited in areas with high levels of air pollution would not seem to be advisable.

In addition to the siting of emission laboratories, the following items and trends should be considered by engineers involved in the design and development of test facilities carrying out exhaust emission analysis:

- Since CO_2 reduction requires fuel consumption reduction, the measurement of fuel consumption will become more critical both in terms of mass fuel consumption accuracy and transient flow. Primary and secondary systems will have to be optimized to obtain the best control and measurement accuracy.
- Liquid fuel conditioning (see Chapter 7) will become considerably more demanding. It will be necessary to take great care in the production, storage and handling of fuel and in the avoidance of cross-contamination and deterioration. Control of fuel temperature to better than $\pm1°C$ at the entry to the engine fuel rail is already important and will continue to be so.
- It will become more necessary to condition combustion air, calling for the kind of systems described in Chapter 5. Temperature control of air will be required as a minimum with humidity control becoming more common and requiring a high standard of system integration.
- Since much emissions testing will embrace cold starting and running at winter temperatures there will be an increased need for cells capable of operating at temperatures of $-10°C$ or lower.
- The requirement to test engines fitted with the full vehicle exhaust system will increase the average cell floor area and call for exhaust dilution and extraction ducts (see Chapter 6).
- The running of increasingly complex emission test cycles calls, in many cases, for expensive four quadrant dynamometers and elaborate control systems.
- The requirement for testing onboard diagnostic (OBD) vehicle systems will increase.

The delicate and complex nature of much emissions measurement instrumentation will call for standards of maintenance, cleanliness and calibration at a level more often associated with medicine than with automotive engineering.

Notation

Air to fuel ratio	AFR
Clean Air for Europe (EEC program)	CAFÉ
Compression ignition (engine)	CI
Compressed natural gas	CNG
Critical flow orifice	CFO
Continuously regenerating trap (diesel particle filter)	CTR
Constant volume sampling	CVS
Direct injection	DI
Reference fuel with a total aromatic content of $\times$ wt-%	DIR-x
Diesel particulate filter	DPF
Diesel particulate matter	DPM
Elemental carbon	EC
Environmental Protection Agency (USA)	EPA
Heavy duty	HD
Indirect injection	IDI
Light duty	LD
Organic carbon (bound in hydrocarbon molecules)	OC
Horiba analyser range trade name	MEXA™
Polyaromatic compounds	PAC
Polyaromatic hydrocarbons	PAH
Particles in the size below $10\,\mu$m	PM10
Soluble organic fraction	SOF
Toxic equivalence factor	TEF

Reference

1. *Air Pollution from Motor Vehicles: Standards and Technologies for Controlling Emissions*, World Bank (ISBN: 0821334441).

Further reading

BS 1747 Parts 1 to 13 *Methods for Measurement of Air Pollution.*
BS 4314 Part I *Infra-red Gas Analysers for Industrial Use.*
BS 5849 *Method of Expression of Performance of Air Quality Infra-red Analysers.*
EEC L242 The Motor Vehicles (Type Approval) (Amendment) (No. 2) Regulations 1991 (http://www.legislation.gov.uk/si/si1991/Uksi_19912681_en_1.htm)
EPA CFR 40–86 (http://www.access.gpo.gov/nara/cfr/waisidx_04/40cfr86_04.html)
Horiba (1990) *Fundamentals of Exhaust Emissions Analysis and their Application*, Horiba Instruments Ltd, Northampton.

Useful addresses

National Environment Research Council, Polaris House, North Star Avenue, Swindon, Wiltshire, SN2 1EU, UK

Environmental Protection Agency, 401 M Street SW, Washington DC 20460, USA.

17 Tribology, fuel and lubrication testing

Introduction

Fuels and lubes (F&L) testing is a specialized branch of the engine test industry often represented by publications full of what are to the outsider impenetrable acronyms together with changing roles for national, international and manufacture-specific organizations.

F&L work is supported by laboratories analysing the chemical and physical characteristics of their products and parts of engines before and after strictly prescribed tests carried out either on test rigs or engine test cells. The oil industry has had to react to, and support, engine developments aimed at reducing gaseous emissions; low sulphur diesel and unleaded petrol being obvious examples of product development. There is a continuous development of fuels and lubricants required to prevent the poisoning of more complex catalytic exhaust gas after treatment, while providing longer periods between vehicle oil changes.

Lubricant classification and certification

Up to the early 1980s, the US military played a major role in the development of fuel and lubricant specifications, but since that time and with the proliferation worldwide of interested parties, engine manufacturers, oil companies and chemical industry, a degree of harmonization of test methods has been achieved. Any student of the subject will have to be aware of the acronyms or names of some of the governing or standard setting organizations:

ACEA, the European Automobile Manufacturers Association
ASTM, the American Society for Testing of Materials
ATIEL, Association Technique de l'Industrie Européenne des Lubrifiants
BSI, British Standards Institution
CEC, Coordinating European Council – for the Development of Performance Tests
    for Transportation Fuels, Lubricants and Other Fluids
COFRAC, the French Committee for Accreditation

DIN, Deutsches Institut für Normung (the German Institute for Standardization) is the German national organization for standardization and is that country's ISO member body

EELQMS, European Engine Lubricants Quality Management System

IP, Institute of Petroleum

ISO, International Organization for Standardization is the international standard-setting body composed of representatives from national standards bodies

In Europe, the ACEA details lubrication oil specifications backed by EELQMS standardized tests and performance criteria that have to be met by any product sold to their members. In addition to the ACEA tests, individual member companies have their own test sequences and processes. The nomenclature and ACEA process is rather complex. In descending hierarchy oils are named by sequence, class and category plus a year of implementation designation. Some of the acronyms required to understand the documentation concerning classifications are

DPF   Diesel particulate filter
HTHS   High temperature and high shear rate viscosity
SAPS   Sulphated ash, phosphorus, sulphur (as in 'low SAPS')
SCR   Selective catalytic reduction (NO_x control)
TWC   Three-way catalyst

The current ACEA sequences and classes are: ACEA A/B (combined) for gasoline and light duty diesel engines and ACEA C for gasoline and diesel engines having advanced after-treatment (catalyst compatible).

The category indicates the type of application the lubricant is intended for within its class, there are currently four in the general sequence designated A1/B1, A3/B3, A3/B4 and A5/B5 and three in the catalyst compatible oils, C1 (low SAPS oils), C2 (for vehicles with TWC and DPF) and C3 (high performance gasoline or diesels).

There are four classes in the oils designed for heavy duty diesels designated E2 (general purpose, normal drain intervals), E4 (high rated engines meeting Euro 3 and 4), E6 highly rated engines running on low SAPS fuel) and E7 (highly rated engines with EGR and SCR).

The year of implementation number is placed after the class designation thus: C1-$_{04}$

Using these classes to set the limits of laboratory- and engine-based tests, the ACEA produces tables indicating the minimum allowable quality of oils of the various classes under a system of self-assessment to EELQMS.

Tribology

The science of tribology, launched as a separate discipline with a new name by the Jost Report of 1966, is concerned with lubrication and wear; both are central to i.c. engine operation. It was pointed out some years ago that the difference between a brand new road vehicle and one that was totally worn out was the loss of perhaps 100 g of material at critical points in the mechanism. As far as friction is concerned, the great majority of the power developed by an automobile engine is ultimately dissipated as friction. Even a large slow-speed diesel engine is unlikely to achieve a mechanical efficiency exceeding 85 per cent, most of the losses being due to friction.

As evolution proceeds, different aspects of engine lubrication and wear call for concentrated attention. Some problems, such as excessive bore wear and bore polishing in diesel engines, are largely overcome and sink into the background while others emerge as engine technologies such as common-rail diesel develop, operating conditions become more demanding or emission legislation sets new parameters. The following may be regarded as the main tribology-related problems of particular current concern:

- valve train wear;
- injector life;
- control of lubricant consumption;
- interactions between fuel and lubricant;
- development of synthetic and 'stay in grade' or 'fill for life' lubricants;
- piston design, with particular reference to top land clearance, ring wear, oil control, distortion and reduction of friction;
- the effect on exhaust emissions of lubrication oil constituents.

The pursuit of higher efficiency and reduced friction losses is a continuing activity in all engine development departments.[1,2]

Bench tests

A wide range of standard (non-engine) tests are written into standards such as those of the ASTM, DIN, CEC and IP.

These tests for oils can range from chemical analysis such as ASTM D 1319 (aromatics and olefins content), through physical laboratory tests such as ASTM D 2386 (freezing point), to special rig tests such as DIN 51 819-3 which uses a roller bearing test rig. Similarly, there are series of tests specifically for greases and fuel analysis.

The principle of operation of many of the lubrication oil test machines has not changed for many years but has been updated because, in spite of the fact that the test conditions they impose differ from any to be found in a real engine, their value is their ability to reliably reproduce the classic mechanical wear and/or failure conditions.

The number of analytical and diagnostic tools available to test laboratories (Table 17.1) has increased significantly in the last two decades and range from laser-measuring devices able to differentiate into the realms of nanotechnology, through use of radiotracing techniques of wear particles, to electronic scanning microscopes.

Table 17.1   *Tribology: surface examination techniques*

Visual and microscopic examination
Wear debris quantifying and analysis
Surface roughness measurement
Metallographical sections and etching
Micro-hardness measurement
Scanning electron and scanning tunnelling electron microscopy
Infrared and Raman spectroscopy

After each engine test, whether it is oil- or fuel-related test sequence, the whole engine or that part of it of specific interest will be stripped under laboratory conditions and examined. In addition to the surface examinations described above which look at discolouration, corrosion and wear, fuel tests will examine such effects as the reduction in flow through injector nozzles and ring sticking.

Oil characteristics

Viscosity is a measure of the resistance of a fluid to deform under shear stress. It is commonly perceived as resistance to pouring. It is measured by a viscometer of which there are various designs commonly based on the time taken to flow through an orifice or a capillary tube when at a standard temperature. The SAE 'multigrade' rating known to most motorists has a nomenclature based on flow through an orifice at 0°F and 210°F, with 'W' (standing for winter) after the number for low temperature rating and a plain number for the high temperature rating as in SAE10W40.

Dynamic viscosity

The SI physical unit of dynamic viscosity μ is the pascal-second (Pa·s), which is identical to $1 \, \text{kg·m}^{-1}\text{·s}^{-1}$

The physical unit for dynamic viscosity in the cgs* is the *poise* (P) named after Jean Poiseuille. It is more commonly expressed, particularly in ASTM standards, as *centipoise* (cP). The centipoise of water is almost unity (1.0020 cP) at 20°C.

1 poise = 100 centipoise = 1 g·cm^{-1}·s^{-1} = 0.1 Pa·s.
1 centipoise = 1 mPa·s.

---

* Centimetre gram second system of units.

Kinematic viscosity

Kinematic viscosity is the ratio of the viscous force to the inertial force or fluid density ρ.

$$\nu = \frac{\mu}{\rho}$$

Kinematic viscosity has SI units ($m^2 \cdot s^{-1}$). The cgs physical unit for kinematic viscosity is the *stokes* (St), named after George Stokes. It is sometimes expressed in terms of *centistokes* (cS or cSt).

1 stokes = 100 centistokes = 1 $cm^2 \cdot s^{-1}$ = 0.0001 $m^2 \cdot s^{-1}$

Viscosity index (or VI) is a lubricating oil quality indicator, an arbitrary measure for the change of kinematic viscosity with temperature.

Total base number (TBN)

The TBN of oil is the measure of the alkaline reserve, or the ability of the oil to neutralize acids from combustion tested to ASTM D-2896 by a laboratory method known as potentiometric perchloric acid titration. Depletion of the TBN in service results in acid corrosion and fouling within the engine.

Appendix 17.1, 17.2 and 17.3 show examples of the standards and test methods of the properties of lubricating oils, diesel fuels and gasoline, respectively; for full up to date listings, the websites maintained by the relevant authority is recommended as a starting point.

Reference fuels and lubricants

Any laboratory carrying out fuels and lube testing will have to store and handle correctly reference fuels and oils. Many component inspection results after running engine tests will be based on comparative judgements based on the use of these reference liquids which are expensive and may have limited storage life unless kept in optimum conditions. Connection of reference fuels into test cells is covered in Chapter 7.

Test limits for such criteria as piston cleanliness or sludge accumulation in some engine-based lubrication tests will be based on reference lubricant results that are commonly designated with a CEC reference, such as in the case of CEC RL191 (15W-40).

Designated engines and test regimes in fuel and lube testing

Whereas tribology studies the phenomena associated with wear and lubrication, the classification and specification of all fuels and crankcase lubricants worldwide are based upon standardized engine tests.

There is a complete section of the engine test industry attached to the oil industry using internationally specified engines in exactly determined configurations to run strict test protocols in order to certify engine fuels and lubricants. This use of a prescribed engine builds redundancy and the need for regular updating into the certification process, since the life and availability of any engine is limited as is its suitability to represent the current technology that the fuels and lube industry has to support. At the time of writing, the PSA 2.1 turbo diesel engine XUD11 BTE that is used in several tests such as the 95 h duration lubricant test, CEC L-56-T-98, can be difficult to obtain in some configurations and was due to be replaced in 2006 with the DV4TD engine. Examples of just some of the engines commonly designated in tests are:

- Mercedes M-111 used in the fuel economy test CEC-L-54-T-96 ($C1_{-04}$, $C2_{-04}$, $C3_{-04}$);
- Mack T-8E used in the 'soot in oil' test ASTM 5967 ($E4_{-99}$, $E6_{-04}$, $E7_{-04}$);
- Peugeot TU5JP-L4 used in the 72-h piston and oil viscosity test CEC-L-88-T-02 ($A1/B1_{-04}$, $A3/B3_{-04}$, $A3/B4_{-04}$, $A5/B5_{-04}$).

The approved engines form part of the standard rig which is arranged specifically for dealing with that particular model.

A fuel and lube test stand will typically be configured specifically for a designated engine that will remain in place for prolonged periods, only being removed in whole or part for component inspection after testing. The engine will be mounted on a base-plate, connected to an eddy-current dynamometer and rigged with fuel and/or oil control systems instrumented exactly as required by regulation. The running of the test sequence, the manipulation of the test results and the post-test examination of engine parts, such as valves and injectors, are similarly tightly prescribed.

The major disadvantages of such tests are the cost and time involved and the difficulty, where measurements of such factors as surface wear, deposits, oil consumption and friction are concerned, of obtaining consistent results. The measurements can be much affected by variations in engine build and components, also of fuel in the case of lubrications or of lubrication oil in the case of fuel tests; therefore most fired tests are run with strictly prescribed reference fluids. There are a few internationally recognized specialist test laboratories that provide the service of running such tests and each of the major engine manufacturers have their own special test cells running both public domain test sequences and their own tests on new engine designs.

Motored engine or 'part engine' tests are used for assessing specific components such as valve trains. They are somewhat removed from real conditions, but can

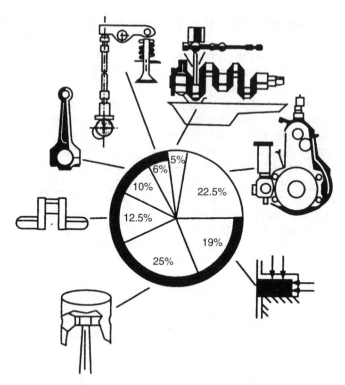

Figure 17.1   *Distribution of friction losses in an i.c. engine*

give a good insight into local problems of wear and lubrication. By the successive removal of components from a motored engine, a good estimate can be made of the contribution made to frictional losses by the different components, see Fig. 17.1.

This book has mentioned some of the very numerous standard tests that are in constant use for assessment and routine confirmation of fuel and lubricant properties. The development, updating and monitoring of these standard tests, not to mention the cost in fuel, components and man-hours, represent a major burden to engine and lubricant manufacturers.

Example of engine tests involving tribology

Measurement of compression ring oil film thickness

A feature of many tribological situations is the extreme thinness of the oil films present, usually to be measured in microns, and various techniques have been developed for measuring these very small clearances.[3] Figure 17.2 shows measured oil film thicknesses between top ring and cylinder liner near top dead centre in a large

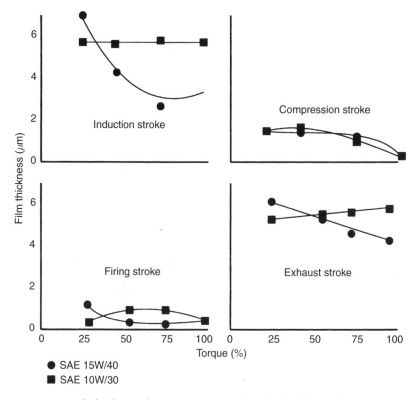

Figure 17.2 *Oil thickness between ring and cylinder liner during four-stoke operation*

turbocharged diesel engine. There is a wide variation in film thickness for the different strokes and an effective breakdown during the firing stroke. It is notable that the thinner oil, SAE 10W/30, gives better protection than a thicker one. This was thought to be the result of higher oil flow to this critical area with the thinner oil and is confirmed by the generally lower rates of wear of aluminium from the piston and chromium from the ring surfaces.

Wear measurements by irradiation

Figure 17.3 shows wear measurement using an irradiated piston ring in an experimental single cylinder diesel engine and indicates a steady wear rate following initial running-in. Wear rate was also found to be sensitive to oil viscosity in this test. This technique is perhaps the only one available for indicating very small increments of wear.

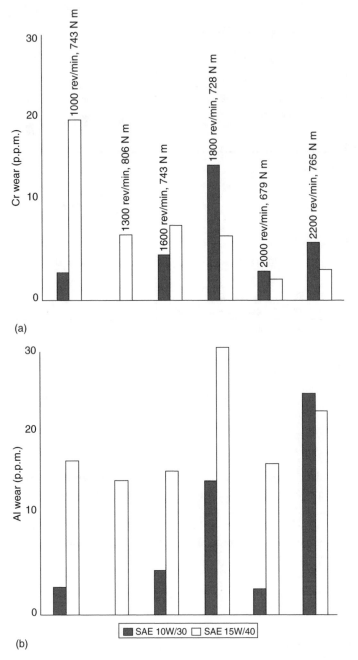

Figure 17.3 *Rates of wear, chromium from ring and aluminium from piston, engine as for Fig. 17.2*

Biofuels

In order to reduce dependency on fossil fuel oil stocks, there are a number of global initiatives intended to increase the use of oil derived from renewable sources (see Chapter 7 concerning storage of biofuels). The European Commission introduced Directive 2003/30/CE which proposes increasing the proportion of energy delivered by biofuels to 5.75 per cent by 2010.

Biodiesel in Europe has to conform to the EN14214 specification and can be derived from a range of animal and vegetable oils, the most common forms are fatty acid methyl ester (FAME) and fatty acid ethyl ester (FAEE). The most common source of biodiesel in Europe is vegetable rape, from which rape-seed methyl ester (RME) is produced. RME is currently more expensive to produce, but has some performance advantages over FAME the most critical of which is low temperature performance. RME biodiesel has a flash point of >150°C which is significantly higher than that of petroleum diesel.

Engine test laboratories will be engaged in testing the emissions and performance of the many possible mixtures and concentrations of these bio-fossil fuels in the coming years. The tendency for some bio-fuels to form wax crystals at low temperatures will require new additives and filter flow testing** in climatic cells; a whole new direction of testing is opening up.

It should be noted that there are worldwide environmental movements encouraging the home production of biodiesel feedstock, the contents of which will be uncontrolled and the performance of which will be highly variable.

Biodiesel is reported to reduce emissions of carbon monoxide (CO) and carbon di-oxide, but it should be remembered that the carbon in biodiesel emissions is within the animal 'carbon cycle' rather than being new carbon released from that tied up in mineral deposits. They are also reported to contain fewer aromatic hydrocarbons but to produce higher emissions of nitrogen oxide NO_x than diesel from petroleum feedstock.

Ethanol (C_2H_6O) can be blended with gasoline in varying quantities; the resulting fuel is known in the United States as gasohol, or gasoline type C in Brazil where there is considerable experience in general use of E20 and E25.

E85 is being used in public service vehicles in the USA; the 15 per cent gasoline being required to overcome cold starting problems experienced with higher ethanol mixtures.

Methanol (CH_3OH) can also be used as a gasoline additive, but is highly toxic and less easy to produce from biological sources than ethanol.

Summary

The fuels and lube test industry, which is a specialized branch of the test industry, has been briefly described. The point to be borne in mind by the test engineer who

---

** Cold filter plugged point (CFPP) is a standard test requirement for winter grade fuels.

may be confronted with a problem of wear or friction is that it is unlikely to be solved on the test bed alone. There exists a vast range of specialized tests and techniques that eliminate a great deal of expensive engine testing and in many cases can clarify aspects of the problem that are simply not susceptible to investigation in the complex environment of a running engine.

For guidance in the tribological aspects of engine performance and testing, the *Tribology Handbook* will be found valuable.[4,5]

The subject of biofuels will become very important to the fuel test and certification industry as political pressure to reduce the dependency on fossil fuels builds.

References

1. *Crankcase Oil Specification and Engine Test Manual* (1987) Paramins, Exxon Chemical, Technology Centre, Abingdon.
2. *Friction and Wear Devices* (1976) American Society of Lubrication Engineers.
3. Moore, S.L. (1987) The effect of viscosity grade on piston ring wear and oil film thickness in two particular diesel engines, *Proc. I. Mech. E.* C184/87.
4. Neale, M.J. (ed.) (1993) *Lubrication: A Tribology Handbook*, Butterworth-Heinemann, Oxford.
5. Neale. M.J. (ed.) (1993) *Bearings: A Tribology Handbook*, Butterworth-Heinemann, Oxford.

Further reading

Automotive Lubricants Reference Book, 2nd edn ISBN: 1-86058-471-3.
BS 2000 Parts 0 to 364, *Methods of Test for Petroleum and its Products*.
Plint, M.A. and Alliston-Greiner, A.F. (1990) Routine engine tests: can we reduce their number, *Petroleum Review*, July.

Appendix 17.1 Listing of major CEC tests and publications

L-02-A-78	Oil oxidation and bearing corrosion
L-33-T-82	Biodegradation of two-stroke outboard engine lubricants
L-38-T-87	Valve train scuffing
L-41-T-93	Evaluation of sludge inhibition qualities of motor oils in a gasoline engine
L-07-A-85	Load carrying capacity of transmission lubricants
L-11-T-72	Coefficient of friction of automatic transmission fluids

L-45-T-93	Viscosity shear stability of transmission lubricants
L-17-A-78	Cam and cylinder wear in diesel engines
L-24-A-78	Engine cleanliness under severe diesel conditions
L-42-A-92	Evaluation of piston deposits and cylinder bore polishing
L-14-A-88	Mechanical shear stability of lubricants containing polymers
L-36-A-90	The measurement of lubricant dynamic viscosity under conditions of high shear (Ravenfield viscometer)
L-37-T-85	Shear stability test for polymer-containing oils
L-39-T-87	Oil/elastomer compatibility test
L-40-T-87	Evaporative loss of lubricating oils
F-03-T-87	Evaluation of gasoline with respect to maintenance of carburettor cleanliness
F-04-A-87	The evaluation of gasoline engine intake system deposits
F-05-T-92	Intake valve cleanliness
M-02-A-78	Internal combustion engine rating procedure
M-07-T-83	The relationship between knock and engine damage
M-08-T-83	Cold weather drivability test procedure
M-09-T-84	Hot weather drivability test procedure
M-10-T-87	Intake system icing test procedures
M-11-T-91	Cold weather performance test procedure for diesel vehicles
M-12-T-91	Representative sampling in service of marine crankcase lubricants
M-13-T-92	Analysis of marine crankcase lubricants
P-108-79	Manual of CEC reference/standardization oils for engine/rig tests
P-221-93	A guide to the definition of terms relating to the viscosity of engine oils
P-222-92	CEC reference fuels manual, 1992, annual report
L-18-U-93	Low temperature apparent viscosity
L-12-U-93	Piston cleanliness

Authority responsible for tests in Europe

The Coordinating European Council for the Development of Performance Tests for
Lubricants and Engine Fuels (CEC)
Madou Plaza, 25th Floor
Place Madou 1
B 1030 Brussels
Belgium

Appendix 17.2 Properties of gasoline: standards and test methods

BSI

BS 2000-12	Methods of test for petroleum and its products

ASTM

D2623-83	Knock Characteristics of Liquefied Petroleum (LP) Gases by the Motor (LP) Method
D2699-84	Knock Characteristics of Motor Fuels by the Research Method
D2700-84	Knock Characteristics of Motor and Aviation-Type Fuels by the Motor Method
D2886-83	Knock Characteristics of Motor Fuels by the Distribution Octane Number (DON) Method
D2885-84	Research and Motor Method Octane Ratings Using On-Line Analyzers
D439-85a	Automotive Gasolene
D3710-83	Boiling Range Distribution of Gasolene and Gasolene Fractions by Gas Chromatography
D240-85	Heat of Combustion of Liquid Hydrocarbon Fuels (General Bomb Method)
D2551-80	Vapor Pressure of Petroleum Products (Micromethod)
D323-82	Vapor Pressure of Petroleum Products (Reid Method)
D2889-81	Vapor Pressures (True) of Petroleum Distillate Fuels
D93-85	Flash Point by Pensky-Martens Closed Tester

Institute of Petroleum

IP34/80	Flash Point by Pensky Martens Closed Tester
IP325/80	Front End Octane Number (RON 100°C) of Motor Gasolene
IP12/73	Heat of Combustion of Liquid Hydrocarbon Fuels
IP236/69	Knock Characteristics of Motor and Aviation Type Fuels by the Motor Method
IP237/69	Knock Characteristics of Motor Fuels by the Research Method
IP15/67	Pour Point of Petroleum Oils
IP171/65	Vapour Pressure – Micro Method
IP69/78	Vapour Pressure – Reid Method

Appendix 17.3 Properties of diesel fuels: standards and test methods

BSI

BS 2000-12	Methods of test for petroleum and its products
BS 2869	Fuel oils for oil engines and burners for non-marine use
BS MA 100	Petroleum fuels for marine oil engines and boilers

ASTM

D613-84	Ignition quality of diesel fuels by the cetane method
D975-81	Diesel fuel oils
D976-80	Cetane index, calculated, of distillate fuels
D189-81	Carbon residue, Conradson, of petroleum products
D524-81	Carbon residue, Ramsbottom, of petroleum products
D2500-81	Cloud point of petroleum oils
D93-85	Flash point by Pensky Martens closed tester
D240-85	Heat of combustion of liquid hydrocarbon fuels (general bomb method)
D97-85	Pour point of petroleum oils
D3245-85	Pumpability test for industrial fuel oils
D129-64	Sulfur in petroleum products by the bomb method
D4294-83	Sulfur in petroleum products by nondispersive X-ray fluorescence spectrometry
D1552-83	Sulfur in petroleum products (high-temperature method)
D1266-80	Sulfur in petroleum products (lamp method)
D2709-82	Water and sediment in distillate fuels by centrifuge

Institute of Petroleum

IP218/67	Calculated cetane index of diesel fuels
IP13/78	Conradson carbon residue of petroleum products
IP34/80	Flash point by Pensky Martens closed tester
IP12/73	Heat of combustion of liquid hydrocarbon fuels
IP41/60	Ignition quality of diesel fuels
IP15/67	Pour point of petroleum oils
IP14/65	Ramsbottom carbon residue of petroleum products
IP107/73	Sulphur in petroleum products
IP276	Total base number

18 Chassis or rolling road dynamometers

Fault tracking under difficulties.

Introduction

This chapter deals with an area of testing activity that is closely allied to the testing of automotive engines and which increased enormously in importance during the last quarter of the twentieth century. It is exclusively concerned with road vehicles and their power-train systems and its expansion has been driven by a number of developments:

- *Technical*: the need to perfect the complex engine, transmission and whole vehicle management systems that are demanded by the modern road user, the economical and repeatable performance of endurance testing, investigations of noise, vibration and harshness (NVH), emissions testing, checking performance under extreme climatic conditions, etc.

- *Legislative*: the need to comply with ever more exacting statutory requirements governing both new vehicles and the after-market in such areas as vehicle safety, brake performance, NVH, electromagnetic compatibility (EMC), fuel consumption and dominating all other tasks, exhaust emissions.

These requirements have called into existence a hierarchy of dynamometers and test chambers of increasing complexity:

- brake testers;
- in-service tuning, ECU remapping;
- end-of-line production testing;
- emissions certification/testing installations;
- mileage accumulation facilities associated with emission testing laboratories;
- anechoic and NVH and EMC test chambers;
- chassis dynamometers within climatic chambers and wind tunnels;
- independent wheel dynamometers.

The major types and uses are discussed later in this chapter.

The road load equation

The behaviour of a vehicle under road conditions is described by road load equations (RLE). The RLE is a fundamental requirement of a true chassis dynamometer which has to resist the torque being produced at the vehicle drive wheels in such a way as to simulate 'real-life' resistance to vehicle motion. It is necessary to have a clear understanding of this control model before studying the various types of machines coming under the heading of chassis dynamometers or rolling roads.

Increasing use of transient testing and road simulation in modern engine test cells means that the following section is relevant to engine test students and engineers.

The RLE is the formula that calculates the change in torque required with change of vehicle speed in order to simulate the real-life performance of a given vehicle; it defines the traction or braking force that is called for under all possible conditions, which include

- steady travel at constant speed on a level road;
- hill climbing and descent;
- acceleration, over-run and braking;
- transitions between any of the above;
- effects of atmospheric resistance, load, towed load, tyre pressure, etc.

The road load equation for a given vehicle defines the *tractive force* or *retarding force, F* (Newton), that must be applied in order to achieve a specified response to these conditions. It is a function of the following parameters:

Vehicle specific:

Mass of vehicle	M	kg
Components of rolling resistance	a_0	N
Speed dependent	$a_1 V$	N
Aerodynamic	$a_2 V^2$	N

External:

Speed	V	m/s
Road slope	θ	rad

The usual form of the road load equation is:

$$F = a_0 + a_1 V + a_2 V^2 + MdV/dt + Mg\sin\theta \tag{1}$$

where $M\, dv/dt$ = force to accelerate/brake vehicle and $Mg\sin\theta$ = hill climbing force.

More elaborate versions of the equation may take into account such factors as wheel spin and cornering.

The practical importance of the road load equation lies in its application to the simulation of vehicle performance. It forms the link between performance on the road and performance in the test department.

To give a feel for the magnitudes involved, the following equation relates to a typical four-door saloon of moderate performance, laden weight 1600 kg:

$$F = 150 + 3V + 0.43V^2 + 1600dV/dt + 1600 \times 9.81\sin\theta$$

The various components that make up the vehicle drag are plotted in Fig. 18.1a which shows the level road performance and makes clear the preponderant influence of wind resistance.

The road load equation predicts a power demand at the road surface of 14.5 kW at 60 mph, rising to 68.8 kW at 50 m/s (112 mph) for a level road.

These demands are dwarfed by the demands made by hill climbing and acceleration. Thus, Fig. 18.1b, to climb a 15 per cent slope at 60 mph, calls for a total power input of $14.5 + 63.2 = 77.7$ kW while, Fig. 18.1c, to accelerate from 0 to 60 mph in 10 s at constant acceleration, calls for a maximum power input of $14.5 + 115 = 129.5$ kW.

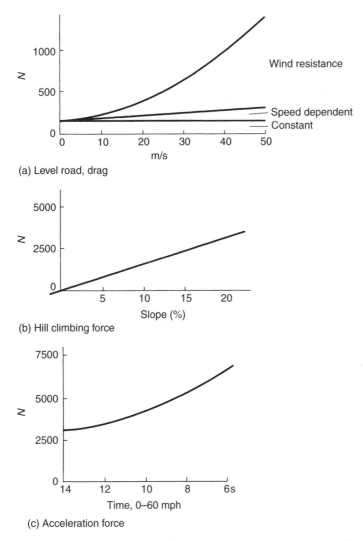

Figure 18.1 *Vehicle drag: saloon of laden weight 1600 kg (a) level road performance; (b) hill climbing; (c) acceleration force*

Genesis of the rolling road dynamometer

The idea of running a complete vehicle under power while it was at rest was first conceived by railway locomotive engineers before being adopted by the road vehicle industry. As a matter of historical record the last steam locomotives built in the UK were tested on multiple roll units, with large eddy-current dynamometers* connected to each driven roll, the tractive force being measured by a mechanical linkage and spring balance. To an observer close to the locomotive on the test rig it must have presented an awe-inspiring sight, running at full power and a wheel speed of 90 mph.

Today the chassis dynamometer is used almost exclusively for road vehicles, although there are special machines designed for fork-lift and articulated off-road vehicles. The advantages to the designer and test engineer of having such facilities available are obvious; essentially, they allow the static observation and measurement of the performance of the complete vehicle while it is, in most respects, in motion.

Before the 1970s, most machines were comparatively primitive rolling roads, characterized by having rollers of rather small diameter, which inadequately simulated the tyre contact conditions and rolling resistance experienced by the vehicle on the road and fitted with various fairly crude arrangements for applying and measuring the torque resistance. A single fixed flywheel was commonly coupled to the roller to give a simulation of a vehicle inertia.

The main impetus for development came with the rapid evolution of emissions testing in the 1970s. The diameter of the rollers was increased, to give more realistic traction conditions, while trunnion mounted d.c. dynamometers with torque measurement by strain gauge load cells and more sophisticated control systems permitted more accurate simulation of road load. The machines were provided with a range of flywheels to give steps in the inertia; precise simulation of the vehicle mass was achieved by 'trimming' the effective inertia electrically.

In recent years the development in digital control techniques and electrical drives has meant that incremental flywheels have been dispensed with in most emission testing designs and been replaced by single (2×2) or double (4×4) roll machines reliant on electrical simulation of all aspects of 'road behaviour'.

Brake testers

These simple machines, commonly installed in service garages and government test stations, are used mainly for the statutory 'in-service' testing of cars and commercial vehicles. They are installed in a shallow pit, and consist essentially of two pairs of rollers, typically of 170-mm diameter and having a grit-coated surface to give a high coefficient of adhesion with the vehicle tyre. The rollers are driven by geared variable speed motors and two types of test are performed. Either the vehicle brakes are fully applied and torque increased until the wheels slip, or the rollers are driven at a low surface speed, typically 2 km/h, and the relation between brake pedal effort and braking force is measured.

The testing of a modern ABS brake system requires a separately controlled roller for each wheel and a much more complicated interaction with the vehicle braking

---

* The National Railway Museum in York holds records and displays the Heenan & Froude specification plate.

system. This calls for computerized control of the tester. In many vehicle management systems, a plug is provided for connection of the brake system to a suitable brake tester computer. Modern 'end of line' brake test machines are used by vehicle manufacturers in a series of final vehicle check stations and therefore have features, such as communication with the test vehicle control systems, that are model specific.

Rolling roads for in-service tuning and assessment

For the typical vehicle maintenance garage, the installation of a rolling road and its installation with a building represents a considerable investment, only justified if the garage has a good market for its expertise in vehicle tuning or other specialized skills.

To cater for this market a number of manufacturers produce complete installations based on small diameter rollers, thus requiring minimum subfloor excavation. This market has tended to be dominated by American suppliers and grew rapidly when the US Environmental Protection Agency (EPA) called for annual emissions testing of vehicles, based on a chassis dynamometer cycle (IM 240).

Road load simulation capability at this level is usually limited to that possible with a choice of one or two flywheels and a comparatively simple control system; most of the testing is concerned with optimizing (tuning) the engine management system and straightforward maximum power certification. The tests have to be of short duration to avoid overheating of engine and tyres.

In-service rolling roads for testing large trucks are confined almost exclusively to the USA. They are based on a single large roller capable of running both single- and double-axle tractor units. The power is usually absorbed by a portable water brake, such as that shown in Fig. 4.1, which may also be used for direct testing of engines, a useful facility in a large agency test facility.

Rolling roads for end-of-line production testing

Test rigs for this purpose range from simple roller sets to multi-axle units with interconnected roll sets and road load simulation. Certain special design features are required and any specification for such an installation should take them into account:

- The design and construction of the machine must minimize the possibility of damage from vehicle parts falling into the mechanism. It is quite common for fixings to shake down into the rolling road, where they can cause damage if there are narrow clearances or converging gaps between, for example, the rolls and lift-out beam, or in the wheel-base adjustment mechanism.
- The vehicle must be able to enter and leave the rig quickly yet be safely restrained during the test. The usual configuration is for all driven wheels to run between two rollers. Between each pair of rolls there is a lift-out beam that allows the vehicle to enter and leave the rig. When the beam descends it lowers the wheels between the

rolls and at the same time small 'anti-climb-out' rollers swing up fore and aft of each wheel. At the end of the test, with the wheels at rest, the beam rises, the restraining rollers swing down, and the vehicle may be driven from the rig.

- To restrain the vehicle from slewing from side to side to a dangerous extent, specially shaped side rollers are positioned between the rolls at the extreme width of the machine where they will make contact with the vehicle tyre to prevent further movement. These rollers can cause problems if a wide variety of tyre profiles are used on test vehicles particularly in the case of low profile tyres.
- Particular care should be taken with the operating procedures if 'rumble strips' are to be used. These are raised strips or keys sometimes fitted to the rolls in line with the tyre tracks, their purpose being to excite vibration to assist in the location of vehicle rattles. However, if the frequency of contact with the tyre resonates with the natural frequency of the vehicle suspension this can give rise to bouncing of the vehicle and even to its ejection from the rig. They can also give rise to severe shock loads in the roll drive mechanism.

Rigs designed to absorb power from more than one axle may have to be capable of adjustment of the centre distance between axles. The complexity of a shaft or belt connection between the front and rear roller sets has been obviated by modern electronic speed synchronization. Special purpose roller rigs for checking wheel alignment sometimes form part of the end-of-line test equipment.

Chassis dynamometers for emissions testing

The majority of chassis dynamometers manufactured worldwide in the last 25 years have been used for some form of emissions testing or homologation.

The standard emissions tests developed in the USA in the 1970s were based on a rolling road dynamometer developed by Clayton and having twin rolls of 8.625-inch (220 mm) diameter. This machine became a *de facto* standard despite its limitations, the most serious of which was the small roll diameter, which resulted in tyre contact conditions much different from those on the road.

Later models used rollers of 500 mm diameter coupled to a four-quadrant d.c. dynamometer with declutchable flywheels to simulate steps of vehicle inertia within which gaps the electrical power operated, Fig. 18.2.

The machine is designed for roller surface speeds of up to 160 km/h and the digital control system is capable of precise road load control, including gradient simulation. Such machines must be manufactured to a very high standard to keep vibration to an acceptable level. It is possible to abuse them, notably by violent brake application which can induce drive wheel bounce and thus very high stresses in the roller and flywheel drive mechanism.

The US Environmental Protection Agency (EPA), which remain the authority defining chassis dynamometers of this class worldwide, have approved machines having a single pair of rollers of 48-inch diameter and 100 per cent electronic inertia

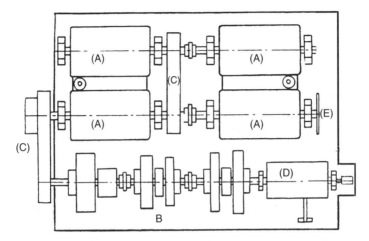

Figure 18.2 *Schematic plan view of four roller (2 × 2 = single driven axle) emission chassis dynamometer of late 1980s design. (A) Four 500-mm rollers; (B) selectable flywheel set; (C) belt drive; (D) d.c. motor fitted with shaft encoder; (E) disk brake for vehicle loading*

simulation. A number of standard designs are now available from the major suppliers and most are of a compact design having a double-ended a.c. drive motor with rollers mounted overhung on each stub shaft. This type of design can produce a standard, pretested package that enables chamber installation time to be minimized.

Mileage accumulation facilities

As a direct result of emissions regulations and the practice of homologation, it has become necessary for manufacturers to check on the change and probable deterioration in emissions performance after the vehicle has been driven for up to 80 000 km. One current legislative trend is to increase the length of this certified 'useful life' of the vehicles emission control system from 50 000 to 100 000 miles or greater.

The cost and the physical strain of using human drivers on test tracks or public roads for driving vehicles the prescribed distances in as little time as possible are high so special chassis dynamometer systems have been developed for running the specified sequences under automatic control, commonly for a period extending to 12 weeks.

Mileage accumulation dynamometers comprise a single large roll, of 48 inches or larger diameter, directly coupled to a d.c. or a.c. motor having sufficient power to run the required repetitive test sequence; for the average saloon car a motor capacity of 150 kW is sufficient.

In order to fully automate the process, the test vehicle is usually fitted with a robot driver. A bewildering range of these devices is available on the world market: for this particular application reliability must be the prime consideration. Some devices can be mounted on the driver's seat and some are capable of also operating the footbrake and engaging reverse gear, but these refinements are not required by current legislative drive cycles. The set-up time of the robot may be appreciable but should be assessed against a test duration of 12 weeks or more. If wear or deterioration takes place in the robot mechanism or in the vehicle controls, for example by change in the clutch 'bite' position, the control system should be capable of automatic rectification, usually by means of periodic 'relearn' and adjustment cycles.

Due to the obvious problems of ventilating such a facility it is invariably located outside, under a simple roof. The control room housing the computerized control and safety systems can occupy a small building at one end of the facility, which in some cases may comprise up to 10 chassis dynamometers.

As the vehicle's own cooling system is stationary, a motorized cooling fan, facing the front grill, is essential. The fan will be fitted with a duct to give a reasonable simulation of air flow under road conditions, and must be firmly anchored. The fan speed is usually controlled to match apparent vehicle speed up to about 130 km/h; above that speed noise and fan power requirements become an increasing problem. Since mileage accumulation facilities usually run 24 h a day, they must be suitably shielded to prevent noise nuisance.

Automatic refuelling systems may be fitted in the facility for each chassis dynamometer unit.

Special purpose chassis dynamometers

Noise, vibration and harshness (NVH), electromagnetic compatibility (EMC) rigs and anechoic chamber dynamometers

Some features of anechoic cell design are discussed in Chapter 3. A critical requirement is that the chassis dynamometer should itself create the minimum possible noise. The usual specification calls for the noise level to be measured by a microphone located 1 m above and 1 m from the centreline of the rolls. Typically, the required sound level when the rolls are rotating at a surface speed of 100 km/h will be in the region of 50 dBA.

To reduce the contribution from the dynamometer motor and its drive system, it is usually located outside the main chamber in its own sound-proofed compartment and connected to the rolls by way of a long shaft (the design of these shafts can present problems; lightweight tubular carbon fibre shafts are sometimes used). The dynamometer motor will inevitably require forced ventilation and the ducting will require suitable location and treatment to avoid any contribution of noise.

Hydrodynamic bearings on quiet roll systems are common, but not without their own problems as their theoretical advantages in terms of reduced shaft noise may not be realized in practice, unless noise from the pressurized supply oil system is sufficiently attenuated.

For a well-designed chassis dynamometer, the major source of noise will be the windage generated by the moving roll surface and this is not easily dealt with. Smooth surfaces and careful shrouding can reduce the noise generated by the roller end faces, but there is inevitably an inherently noisy jet of air generated where the roller surface emerges into the test chamber. If the roller has a roughened surface to simulate road conditions the problem is exacerbated. The noise spectrum generated by the emerging roller is influenced by the width of the gap and in some cases adjustment is provided at this point.

The flooring over the dynamometer pit, usually of steel plate, must be carefully designed and appropriately damped to avoid resonant vibrations. A particular feature of NVH test cells is that the operators may require access to the underside of the vehicle during operation. This is usually by way of a trench at least 1.8 m deep, lying between the vehicle wheels.

Testing for electromagnetic compatibility (EMC)

Worldwide safety standards include regulations governing the acceptable levels of electromagnetic interference produced by road vehicles and the vulnerability of vehicle electronic systems to powerful directional beams of electromagnetic energy, such as may be produced by civil and military aviation systems.

To deal with this problem, a few units have been designed with the chassis dynamometer mounted on a large turntable capable of being rotated through 360°.

Climatic test cells for vehicles

There is a requirement for this kind of facility for development work associated with various aspects of performance under extreme climatic conditions. Subjects include driveability, cold starting, fuel waxing and solar loading, vapour lock, air conditioning and the ubiquitous topic of emissions.

The design of air conditioning plants for combustion air has been discussed in Chapter 5. The design of an air conditioned chamber for the testing of complete vehicles is a very much larger and more specialized problem. Figure 18.3 is a schematic drawing of such a chamber, built in the form of a recirculating wind tunnel so that the vehicle on test, mounted on a chassis dynamometer, can be subjected to oncoming air flows at realistic velocities, and in this particular case over a temperature range from $+40$ to $-30°C$.

The outstanding feature of such an installation is its very large thermal mass and, since it is usually necessary, from the nature of the tests to be performed, to vary the temperature of the vehicle and the air circulating in the tunnel over a wide range

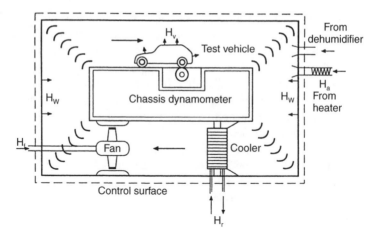

Figure 18.3   *Schematic of climatic wind tunnel facility fitted with a chassis dynamometer*

during a prolonged test period. This thermal mass is of prime significance in sizing the associated heating and cooling systems.

The thermodynamic system of the chamber is indicated in Fig. 18.3 and the energy inflows and outflows during a particular cold test, in accordance with the methods described in Chapter 2, were as follows:

The thermal capacity (thermal mass) of the various elements of the system is defined in terms of the energy input required to raise the temperature of the element by 1°C.

In	.	*Out*	.
H_a make-up air	20 kW	H_r refrigerator circuit	495 kW
H_v vehicle convection/radiation	100 kW	.	
H_r tunnel fan power	154 kW	.	
H_w walls, etc. by difference	221 kW	.	
Totals	495 kW	.	495 kW

C_a air content of chamber, return duct, etc. volume approximately 550 m³, approximate density 1.2 kg/m³, $C_p = 1.01$ kJ/kg.K, thermal mass $= 550 \times 1.2 \times 1.01 = 670$ kJ/deg C

C_v vehicle, assume a commercial vehicle, weight 3 tonnes, specific heat of steel approximately 0.45 kJ/kg.K, thermal mass, say $3000 \times 0.45 = 1350$ kJ/deg C

C_e fan, cooler matrix, internal framing etc. estimated at 2480 kJ/deg C

C_s structure of chamber. This was determined by running a test with no vehicle in the chamber and a low fan speed. The rate of heat extraction by the refrigerant circuit was measured and a cooling curve plotted. Concrete is a poor conductor of heat and the coefficient of heat transfer from surfaces to air is low. Hence during the test a temperature difference between surfaces and wall built up but eventually stabilized; at this point the rate of cooling gave a true indication of effective wall heat capacity. The test showed that the equivalent thermal mass of the chamber was about 26 000 kJ/deg C, much larger than the other elements. Concrete has a specific heat of about 8000 kJ/m^3 deg C, indicating that the 'effective' volume involved was about $26\,000/8000 = 3.2\,\mathrm{m}^3$.

The total surface area of the chamber is about 300 m^2, suggesting that a surface layer of concrete about 1 cm thick effectively followed the air temperature.

Adding these elements together we arrive at a total thermal mass of 30 500 kJ/deg C. Then since 1 kJ = 1 kW-sec, we can derive the rate at which the temperature of the air in the chamber may be expected to fall for the present case.

This shows that the 'surplus' cooling capacity available for cooling the chamber and its contents, H_w, amounts to 221 kW. This could be expected to achieve a rate of cooling of $221/30\,500 = 0.0072$ deg C/s, or 1°C in 2.3 min. However, this is the final rate when the temperature difference between walls and air had built up to the steady state value of about 20°C. The observed initial cooling rate is much faster, about 1.5°C/min. Heating presents less of a problem, since the heat released by the test vehicle and the fan assist the process rather than opposing it.

A further effect is associated with the moisture content of the tunnel air. On a warm summer's day this could amount to about 10 kg of water and during the cooling process this moisture would be deposited, mostly on the cooler fins, where it would eventually freeze, blocking the passages and reducing heat transfer. To deal with this problem, it is necessary to include a dehumidifier to supply dried air to the tunnel circuit.

Special features of chassis dynamometers in climatic cells

Chassis dynamometers intended to operate over the temperature range +40 to −10°C call for no special features, apart from sensible precautions to deal with condensation; for temperatures below this range and particularly those working below −25°C certain special features will be required.

Until the compact 'motor between rollers' ('motor-in-middle' in USA) design of dynamometer became common, the dynamometer motor in climatic cells was often isolated from the cold chamber and operated at normal temperatures. With the new designs two quite different strategies can be adopted to prevent low temperatures causing temperature related variability in the dynamometer system.

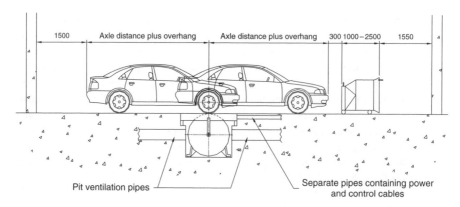

| 1500 | Axle distance plus overhang | Axle distance plus overhang | 300 1000–2500 | 1550 |

Pit ventilation pipes

Separate pipes containing power
and control cables

Figure 18.4 *A single roll set (2 × 2) chassis dynamometer plus road speed following cooling fan, within a cell sized to test rear and front wheel drive cars (AVL Zöllner)*

- The dynamometer pit can be kept at a constant temperature above or below that of the chamber by passing treated (factory) air through the pit via ducts cast in the floor (see Fig. 18.4). This strategy submits portions of the roll's surface to differential temperatures during the static cooldown phase but without reported problems.
- The pit can be allowed to chill down to the cell temperature but the parts of the chassis dynamometer that are crucial to accurate and consistent performance, such as bearings and load cell, are trace heated.

Whatever the layout components exposed to temperatures below 25°C such as rolls and shaft should be constructed of steel having adequate low-temperature strength to avoid the risk of brittle fracture.

Independent wheel dynamometers

A limitation of the conventional rolling road dynamometer is that it is unable to simulate cornering, during which the wheels rotate at differing speeds, wheel spin and skidding. With the increased adoption of electronically controlled traction and braking (ABS), there is a growing requirement for test beds that can simulate these conditions. There are two types of solution:

- *Four roll-set rolling roads*. These machines range from end-of-line rigs consisting of four sets of independent double-roller units, to complex development test rigs, some having steered articulation of each roll unit. The production rigs permit the checking of onboard vehicle control systems, transducers and system wiring by simulation of differential resistance and speeds of rotation. The development of hybrid vehicles may give impetus to the development of independent four-wheel chassis dynamometers since some designs are based on individual wheel drive motors. Electronically controlled transmission systems on 4×4 vehicles also require this type of dynamometer to simulate the road conditions for which they are designed to compensate.

- *Wheel substitution dynamometers.* A major problem faced by engineers carrying out NVH testing on a chassis dynamometer is that tyre noise can dominate sound measurements. One answer is to absorb the power of each drive wheel with an individual dynamometer. If the wheel hub is modified the vehicle can still sit on its tyres, giving approximately the correct damping effect, while the tyre contact and windage noise is eliminated. Alternatively, the wheels may be removed and the individual wheel dynamometers used to support the vehicle. The dynamometers should be four-quadrant machines, to simulate both driving and coasting/braking conditions. Hydrostatic, a.c. and d.c. machines have been used for this application.

The installation of chassis dynamometers

Whatever the type of unit the reader has to specify and install, there are some common planning and logistical problems that need to be taken into account. For single roll machines built within a custom-built chamber, the longitudinal space requirement may have to take account of front- and rear-wheel drive vehicles plus room for tie-down mechanisms and the vehicle cooling fan. A typical layout is shown in Fig. 18.4.

As with engine test cells dealing with the testing of exhaust emissions, chassis dynamometers associated with ultralow emissions work (SULEV) should be built on sites that are out of the drift zone of automotive and industrial pollutants.

Chassis dynamometer pit design and construction details

The minimum requirements, if not the exact specification, of the pit into which a chassis dynamometer is to be installed will invariably be provided by the dynamometer's manufacturer which will include the loading on the pit floor to aid the structural designer. The pit must be built to a standard of accuracy rather higher than is usual in some civil engineering practice since all pit installed dynamometers have a close dimensional relationship with the building in which they are installed; this is particularly true for the pit construction which has to meet three critical dimensional standards:

1. The depth of the pit in relation to the finished floor level of the test cell needs to be held to tight dimensional and level limits. Too deep is recoverable, too shallow can be disastrous.

2. The lip of the pit has to be finished with edging steel that interfaces accurately with the flooring plates that span the gap between the exposed rolls' surface and the building floor.
3. The centre line of the dynamometer has to be positioned and aligned with the building datum and the vehicle hold-down structure that may also be cast into the cell floor. Fine alignment can be achieved by the dynamometer installers, but movement at that stage will be very restricted.

Pit depth

To achieve point 1 above, it is normal practice to make the pit floor lower than the datum dimension and then cast into the floor levelled steel 'sole plates' at the required height minus 5–10 mm to allow some upward adjustment of the machine to the exact floor level. The final levelling can be done by some form of millwright's levelling pads, or more usually levelling screws and shim plates, as shown in Fig. 18.5.

The system described above allows the pit floor to have some slight 'fall' towards a small drainage sump fitted with a level alarm or float operated pump. If the site has a high water table, the pit will have been suitably 'tanked' before final concrete casting to prevent ground water seepage; however, spillage of vehicle liquids and cell washing will drain into the pit so a means of taking these liquids into a foul liquid intercept drain is advisable.

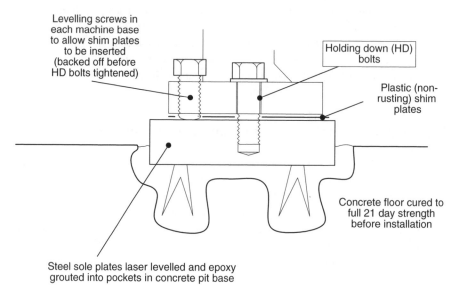

Figure 18.5 *One of the levelled steel plates exactly levelled and cast in the pit floor to support the chassis dynamometer (one per dynamometer HD bolt location)*

The need to run subfloor cable and ventilation ducts, between the control room and the pit and the drive cabinets and the pit requires preplanning and collaboration between the project engineer and builder, it can also complicate the task of creating a watertight seal against ground water. Normally there will be the following subfloor ducts:

- signal and separate small power cable ducts between control room and dynamometer pit;
- high power drive cables and separate signal cable ducts between the drive cabinet room/space and the dynamometer pit;
- pit ventilation ducts which at a minimum will have to include a purge duct for removal of hydrocarbon vapours (see ATEX requirements, Chapter 4).

On the wall of the test cell it is usual to fit a 'break-out' box that allows transducers, microphones or CAN bus plugs to be used for special vehicle-related communications between control room and test unit.

Project managers of any major test facility building are advised to check the layout and take photographic records of all substructure service pipes and steel work before concrete is poured.**

When, due to diminishing access, the dynamometer has had to be located in its pit before the building has been completed and before permanent electrical services are available, it may be vulnerable until building completion to the pit flooding and roll surface damage; provision must be made to guard against both eventualities.

Pit flooring

In most large chassis dynamometer projects, the false floor and its support structure are provided by the dynamometer manufacturer. It may be centrally supported by the dynamometer structure and will have access hatches for maintenance and calibration equipment. The steel or aluminium floor plates will be lifted from time to time and any hatches will need to be interlocked with the control system. It will be clear to readers that unless the pit shape is built to a quite precise shape and size, there will be considerable difficulty in cutting floor plates to suit the interface and that errors in this alignment will be highly visible. It is strongly recommended that the civil contractor should be supplied with the pit edging material by the chassis dynamometer manufacturer and that it is cast into the pit walls under his supervision before final edge grouting. In some cases the pit edging assembly can be fixed with jig beams in temporary place to hold the rectangular shape true while fixing (Fig. 18.6).

** AJM was once told by a steel fixer that his trade was comparable with that of a surgeon in that they both bury their mistakes; henceforth detailed records have been made of prepour structures.

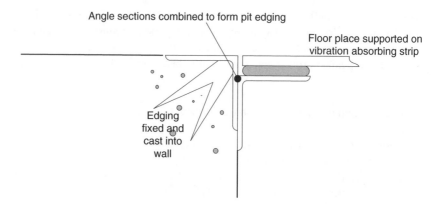

Figure 18.6   *Section through one form of pit edging required to support pit flooring*

Design and installation of variable geometry (4 × 4) dynamometers

Many chassis dynamometers intended for testing four, or more, wheel drive vehicles are designed so that the interaxle distance can be varied so as to accommodate a range of vehicles in a product family. The standard method adopted in 4 × 4 designs is for one, single-axle dynamometer module to be fixed within the building pit and the identical second module to be moved towards or away from the datum module on rails inside the pit. The range of interaxle distance for light duty vehicles (LDV) will be in the range of 2 to 3.5 m, but for commercial vehicle designs the range of movement will be far greater. This range of movement is important in that it tends to determine the design needed to accommodate that section of cell flooring that is required to move with the traversing set of rolls. In the LDV designs where the total module movement is around 1.5 m the moving plates can be based on a telescopic design with moving plates running over or under those fixed. Where the movement considerably exceeds 1.5 m, then the moving floor designs may have to be based on a slatted sectional floor that runs down into the pit at either end, while being tensioned by counterweights or as part of a tensioned cable loop. All these designs can be problematic in operation and require good housekeeping and maintenance standards.

The traversing module is normally moved by two or more electrically powered lead screw mechanisms coupled to a linear position transducer so that a set position can be selected from the control room. For operational reasons, it is recommended that some positional indication, vehicle-specific or actual interaxle distance, is marked on the operating floor.

LDV units are usually traversed on flat machined surface rails having lubricated foot pads fitted with sweeper strips to prevent dirt entrainment; this is very similar to long established machine tool practice and has to be installed with a similar degree of accuracy. Large commercial vehicle rigs often use crane traversing technology rather than that of machine tools with crane rails installed on cast ledges within the pit wall design and the axle modules running within 'crane beams' running on wheels some of which are electrically powered.

Drive cabinet housing

The same precautions relating to the integration of high powered IBGT or thyristor controlled d.c. drives covered in Chapter 10 relate to those fitted to chassis dynamometers.

The drive and control cabinet for high power chassis dynamometers may be over 6 m long, 2.2 m high and, although only 600–800 mm deep, will require space for front and sometimes rear access. These units are heavy with high centres of gravity and vulnerable to damage through damp or dirty atmosphere; if positioned before the building is clean and heated then adequate provision for protection should be made.

Offloading and positioning

Even small machines that are delivered in one unit require lifting equipment to offload, transport and position them in the pit. Forty-eight-inch roll machines or larger units may have to be positioned in already built or partly completed chambers; therefore, the installation will often require specialist equipment and contractors to lift and manoeuvre the machine in a building area of limited headroom and limited floor access as in the case of lined climatic chambers. It is a project phase that is used as an example in the 'Project timing chart' section of Chapter 1, which is recommended reading.

Tyre burst detectors

On all mileage accumulation rigs and others undertaking prolonged automatic test sequences, there should be some form of detector that can safely shut down the whole system in the event of a tyre deflating. The most common form is a limit switch mounted on a floor stand with a long probe running under the car at its midpoint with limit switches fitted one on each side. Any deflation will cause the vehicle body to drop, shutting down the test in a safe manner.

Cable layout

The general treatment of this subject in Chapter 10 applies in the present case, but special attention should be given to the avoidance of electrical interference from power cables with instrumentation lines. In many cases, all the cables running to the chassis dynamometer will at some point run through plastic tubes set within the concrete of the pit walls and plant room floors. These cable tubes should be allocated to either power or signal and should have a drainage slight fall to the pit. After installation is complete, the tubes can be 'stopped' at both ends with acoustic and fire attenuating material.

Loading and emergency brakes

It must be possible to lock the rolls to permit loading and unloading of the vehicle. These consist of either disc brakes fitted to the roll shafts or brake pads applied to the inside surface of one or both rollers. In normal operation these brakes have to be of sufficient power to resist the torques associated with driving the vehicle on and off the rig. Pneumatic or hydraulically powered, they are controlled either by the operational controls (vehicle loading) or by the safety instrumentation (EM stop) and are designed to be normally 'on' and require active switching to be off (machine operational).

It is not considered good practice to rely solely on these brakes to bring the rig to rest in the case of an emergency. They should only be used to supplement the braking effort of the main drive system. In addition there must be a well-thought out 'lock-out' system to ensure that all personnel are outside restricted areas before start-up, particularly important in large facilities where the operator may not be able to see all areas.

Vehicle restraints

The test vehicle must be adequately restrained against fore and aft motion under the tractive forces generated, and against slewing or sideways movements caused by the wheels being set at an angle to the rig axis. It should be pointed out to inexperienced readers that while the vehicle has no momentum driving on a rolling road, the chassis dynamometer does have momentum and it operates in the opposite direction of travel to that which the vehicle would have at equivalent velocity. Sudden braking of the vehicle therefore will cause it to try to jump off the rolls backwards rather than forwards. Vehicle restraint is less of a problem when the powered wheels are located between a pair of rolls rather than resting on a single roller; however, paired rolls are only used for short duration test work because of their limited diameter and hence poor contact pattern. They are generally used for low power end-of-line testing, where they are usually relied on as the only means of restraint so as to allow for rapid loading and unloading of the test vehicles. Where full restraint systems are used, the cell floor must be provided with strong anchorage points. There are many different designs of vehicle restraint equipment, ranging from the use of large pneumatic bags to simple chain or loading strap attachments. Three different types of restraint system may be distinguished:

- A vehicle with rear wheel drive is the easiest type to restrain. The front wheels can be prevented from moving by fore and aft chocks linked by a tie bar. The rear end may be prevented from slewing by two high strength luggage straps with integral tensioning devices; these should be arranged in a cross-over configuration with the floor fixing points outboard and to the rear of the vehicle.
- Front wheel drive vehicles need careful restraint, since any movement of the steering mechanism with the rolls running at speed can lead to violent yawing. Restraints may be similar to those described for rear wheel drives, but with the straps at the front and chocks at the rear. The driver in manned runs should be familiar with the handling characteristics of the vehicle in normal operation, while for unmanned operation the steering wheel should be locked; otherwise, disturbances such as a burst tyre could have serious results. Tyre pressures, usually set above the normal level, should be carefully equalized.
- Four-wheel drive vehicles usual rely on cross-over strapping at both ends, but details depend on fixing points built into the vehicle. Where strong tow points are conveniently supplied they can be used to connect via a short articulated connector to a floor fixed restraint pillar. When using straps, it is good practice to tie the rear end first and then to drive the vehicle slowly ahead with the steering wheel loosely held. In this way, the vehicle should find its natural position on the rolls and can be restrained in this position, giving the minimum of tyre scrubbing and heating.

Guarding and safety

Primary operational safety of chassis dynamometer operators is achieved by restricting human access to machinery while it is rotating. Access to the pit should be hardwire interlocked with the EM stop system and latched EM stop buttons positioned in plant areas remote from the control room, such as the drive room and cell extremities.

The small exposed segment of dynamometer rolls on which the tyres rotate is the most obvious hazard to operators and drivers when the rig is in motion and personnel guards must always be fitted. Such guards are a standard part of the dynamometer system for good safety reasons, but perhaps also because it is the only operational piece of the manufacture's equipment that is visible to the visitor after installation and usually will bear the company logo. Cover plates for the rolls for use when the vehicle is present are also advisable, particularly to prevent surface damage during maintenance periods.

Roll surface treatment

In the past, most twin roll and many single roll machines had no special treatment applied to the roll surfaces, which had a normal finish machined surface. Brake testers and some production rigs have a high-friction surface, which can give rise to severe tyre damage if the machine is used incorrectly. The roll surface of modern single roll machines may be sprayed with a fine-grained tungsten carbide coating, but other common surfaces include emery grit or plain steel.

'Road shells'

For development work, and particularly for NVH development, it is sometimes desirable to have a simulated road surface attached to the rolls. Usually the required surface is a simulation of quite coarse stoned asphalt; but a roughened roll surface greatly increases the windage noise of the rig itself.

Road shell design and manufacture is difficult. The shells should ideally be made up of four or more segments of differing lengths and with junction gaps that are as small as possible and cut helically. It is rarely required to induce low frequency vibrations in the test vehicle; if it is required, it should be generated by randomly distributed variations in roll surface height to avoid possible resonances in the vehicle suspension.

The most usual techniques for producing road surfaces for attachment to chassis dynamometer rolls are as follows:

- Detachable fibreglass road shells having an accurate moulding of a true road surface. These shells are usually thicker than the aluminium type, below, and in both cases the floor plates must be adjustable in order to accommodate them.
- Detachable cast aluminium alloy road shells, made in segments that may be bolted to the outside diameter of the rolls. The surface usually consists of parallel sided pits that give an approximation to a road surface.
- A permanently fixed road surface made up of actual stones bonded into a rubber belt which is bonded to the roll surface.

Most road shells are not capable of running at maximum rig speed because they are difficult to fix and to balance, also they may not be capable of transmitting full acceleration torque. It is therefore necessary to provide safety interlocks so that the computerized speed and torque alarms are set to appropriate lower levels when shells are in use. Road shells also change the effective rolling radius and the base inertia of the dynamometer, requiring appropriate changes in the software constants.

Driver aids

The control room needs to be in touch with the driver in a test vehicle on a chassis dynamometer, either by voice transmission or visually by way of an aid screen which must be easily visible to him.

A major function of such a screen will be to give the driver instructions for carrying out standardized test sequences, for example production test programmes or emissions test sequences. This kind of VDU display is often graphical in form, showing for example the speed demanded and the actual speed achieved. Since in the case of emissions tests, the test profile must follow that prescribed within defined limits, it is usual to include an error checking routine in the software to avoid wasted test time. The screen may also be used to transmit instructions to supplement information conveyed by the two-way voice link.

Fire suppression

For a more general treatment of the subject of fire suppression, see Chapter 4.

The risk of a vehicle fire during chassis dynamometer running is quite high, since the air cooling, even with supplementary fans, is likely to be less than that experienced on the road. Underfloor exhaust systems, in particular, can become very hot and could ignite fuel vapour, whatever its source. A fixed fire suppression system is more difficult to design than in an engine test cell because of both the large size of rolling road cells and the more difficult access to the seat of the fire, which may be within the vehicle body.

All vehicle test facilities should be equipped with substantial handheld or hand-operated fire extinguishers and staff should be trained in their use. One method of fire protection is to fit the test vehicle with a system of the type designed for rally cars which enables the driver or control room to flood the engine compartment with foam extinguishant. Automatic gas-based systems of the type used in some engine test cells are less effective in vehicle cells in view of the greater difficulty in ensuring that all personnel, including the driver, have been evacuated before they are activated. The modern trend is towards water fog suppression systems, which may include discharge nozzles mounted beneath the vehicle and thus near the seat of most potential fires.

There must be a clear and unimpeded escape route for any test driver and the impairment in vision from steam or smoke must be taken into account, particularly in the case of anechoic cells, where the escape door positions should not be camouflaged within the coned surface.

The effect of roll diameter on tyre contact conditions

The subject of tyre rolling resistance has given rise to an extensive literature, which is well summarized in Ref. 1. The bulk of the rolling resistance is the consequence of hysteresis losses in the material of the tyre and this gives rise directly to heating of the tyre. A widely accepted formulation describing the effect of the relative radii of tyre and roll is:

$$F_{xr} = F_x \left(1 + \frac{r}{R}\right)^{\frac{1}{2}}$$

(2)

where:

F_{xr} = rolling resistance against drum
F_x = rolling resistance on flat road
r = radius of tyre
R = radius of drum

Figure 18.7a shows the situation diagrammatically and Fig. 18.7b the corresponding relation between the rolling resistances. This shows that rolling resistance (and hence hysteresis loss) increases linearly with the ratio r/R. For tyre and roller of equal

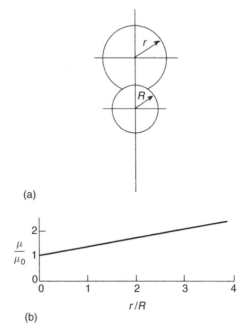

(a)

(b)

Figure 18.7   *(a) Relationship between tyre and chassis dynamometer roller diameters and (b) linear relationship between rolling resistance and the tyre/roll ratio*

diameter the rolling resistance is 1.414 × resistance on a flat road, while for a tyre three times the roller diameter, easily possible on a brake tester, the resistance is doubled. A typical value for the 'coefficient of friction' (rolling resistance/load) would be 1 per cent for a flat road.

To indicate magnitudes involved, a tyre bearing a load of 300 kg running at 17 m/s (40 mph) could be expected to experience a heating load of about 500 W on a flat road, increased to 1 kW when running on a roller of one third its diameter. Clearly the heating effects associated with small diameter rolls are by no means negligible.

Tyres used even for a short time on rolling roads may be damaged by heating and distortion effects, which are aggravated by the fact that the roll will also be heated in the course of the run. Some dynamometer systems are fitted with tyre cooling systems to reduce tyre damage but all tyres used for any but short duration tests should be specially marked and changed before the vehicle is allowed on public roads.

Specification, accuracy and calibration

A typical specification for a chassis dynamometer will include the following information as a minimum:

- base inertia;[†]
- minimum simulated inertia;
- maximum simulated inertia;
- continuous rated power, absorbing mode;
- continuous rated power, motoring mode;
- torque range, absorbing and motoring, sometimes in the form of tractive effort at base speed;
- number of rolls;
- roll diameter with tolerance (typically < ±0.005 inch on a 48-inch roll);
- roll concentricity (typically < ±0.1 mm (0.004 inch) on a 48 inch roll);
- maximum surface speed (km/hour);
- maximum permitted acceleration;
- maximum time to stop from maximum rated speed under regenerative braking;
- type of drive protection system (supply failure during full regenerative load, etc.).

These parameters determine the basic geometry of the rig and the size of motor and its associated variable speed drive system.

[†] It is common practice to use the term 'inertia' to mean the simulated or equivalent vehicle inertia, rather than the moment of inertia of the rotating components. The inertia is thus quoted in kg rather than kgm^2.

Where a purpose of the dynamometer is to carry out homologation tests, these should be clearly specified, together with the range of vehicles to be covered; the acceptance procedure should include performance of these tests.

The essential measurements performed by a chassis dynamometer, equivalent to the measurement of torque and speed in a conventional machine, are roller surface speed and the force (the tractive effort at the roll periphery) transmitted to or from the rolls.

This force is usually not measured directly, but derived from a torque measurement at the motor, performed either in the conventional manner by utilizing a trunnion-mounted motor with torque arm and load cell, or by some form of torque shaft dynamometer. Whatever method is used, there will be an in-built error arising from bearing losses and roll windage: these losses can be measured by running the machine through the speed range with no vehicle present and noting the necessary driving torque, which should be added to the observed torque when absorbing power and subtracted when braking, to give the true tractive force at the roller rim. (This is not the complete answer as there will be a further, unquantified, loss due to roll bearing friction when the rolls are under load.)

Some dynamometers include these corrections in the stated dynamometer constants (see Chapter 8 for a detailed discussion of torque calibration procedures). This is often not a simple matter in the case of chassis dynamometers as the torques involved can be very substantial and space very restricted: the preferred method using graduated weights to apply torque in 10 steps may not be practicable and dynamometers are often supplied with a single mass corresponding to full torque plus a few small weights adding up to about 10 per cent of full scale torque.

The accuracy of calibration demanded and the calibration methods to be used may be included in emissions legislation relevant to chassis dynamometer testing as part of the general tightening of standards.

Equivalent road speed measurement will usually be by means of a high resolution optical encoder on the roll shaft, thus introducing a possible source of error arising from uncertainty regarding true roll diameter, particularly the case with coated rolls. It is common practice to use a so-called 'pi-tape' to obtain the effective diameter by measuring the circumference.

A check on the actual base inertia of the dynamometer is made either by means of a number of 'coast-down' tests or by both accelerating and decelerating the rolls at an approximately constant rate, plotting motor torque and speed in both cases. The latter method is the more accurate since it effectively eliminates the internal drag of the dynamometer motor.

It will be apparent that really accurate measurements of traction/braking force are by no means easy to achieve. Much depends on the quality of manufacture and consistency of performance of the machine, the accuracy of the basic measurements of torque, roller diameter and speed and the capability of the associated computer.

A common operational problem experienced in test laboratories is that after rigging a vehicle on the chassis dynamometer and taking considerable time and effort to connect all of the required emission equipment and transducers, it is then required to perform warm-up and coast-down checks on the dynamometer, during which the vehicle should be removed. The solution is provided on most high end designs by the incorporation of a mechanism that arcs around the drum and lifts the vehicle tyre on two small rollers just clear of the drum surface. On completion of the warm-up routine, the mechanism arcs down, out of contact, and the tyre is returned to its original position.

Limitations of the chassis dynamometer

Like all devices intended to simulate a phenomenon, the chassis dynamometer has its limitations, which are easily overlooked. It cannot be too strongly emphasized that a vehicle, restrained by elastic ties and delivering or receiving power through contact with a rotating drum, is not dynamically identical with the same vehicle in its normal state as a free body traversing a fixed surface. During motion at constant speed, the differences are minimized, and largely arise from the absence of air flow and the limitation of simulated motion to a straight line.

Once acceleration and braking are involved, however, the vehicle motions in the two states are quite different. To give a simple example, a vehicle on the road is subjected to braking forces on all wheels, whether driven or not, and these give rise to a couple about the centre of gravity, which causes a transfer of load from the rear wheels to the front with consequent pitching of the body. On a chassis dynamometer, however, the braking force is applied only to the driven wheels, while the forward deceleration force acting through the centre of gravity is absent, and replaced by tension forces in the front-end vehicle restraints (remember: braking on a rolling road tends to throw the vehicle backwards).

It will be clear that the pattern of forces acting on the vehicle, and its consequent motions, are quite different in the two cases. These differences make it very difficult to investigate vehicle ride and some aspects of NVH testing, on the chassis dynamometer. A particular area in which the simulation differs most fundamentally from the 'real-life' situation concerns all aspects of driveline oscillation, with its associated judder or 'shuffle'.

The authors have had occasion to study this problem in some detail and, while their analysis is too extensive to be repeated here, one or two of their conclusions may be of interest:

- The commonly adopted arrangement, whereby the roll inertia is made equal to that of the vehicle, gives a response to such disturbances as are induced by driveline oscillations, i.e. judder, that are significantly removed from on-road behaviour.
- For reasonably accurate simulation of phenomena such as judder, roll inertia should be at least five times vehicle inertia.
- Electronically simulated inertia is not effective in this instance: actual mass is necessary.

- The test vehicle should be anchored as lightly and flexibly as possible: this is not an insignificant requirement, since it is desirable that the natural frequency of the vehicle on the restraint should be at the lower end of the range of frequencies, typically 5–10 Hz, that are of interest.

It is strongly recommended that investigations of vehicle behaviour under conditions of driveline oscillation should proceed with caution if it is intended to involve running on a chassis dynamometer.

Summary

The subject of rolling roads and chassis dynamometers is a very extensive one, worthy of a textbook in its own right. The application of these machines, at all but an elementary level, is very largely in the field of transient testing and the road load equation. However, the testing techniques employed have a great deal in common with those applicable to engine testing dynamometers, and almost the whole of the present book is relevant.

In this chapter, an attempt has been made to survey the whole range of apparatus that may be described by the general heading of rolling roads/chassis dynamometers, from the simple garage brake tester to complex and very costly climatic chambers and anechoic cells. The design of the associated test cells, vehicle restraint systems, driver communications and safety precautions have been discussed and problems of accuracy and calibration considered.

The subject of the building/dynamometer interface has been covered in some detail because errors discovered at installation are expensive to rectify.

Reference

1. *Tire Rolling Resistance*, Symposium, 122nd Meeting, Rubber Division, American Chemical Society, 1982.

19 Data collection, handling, post-test processing, engine calibration and mapping

Introduction and definition of terms

One of the primary purposes, and in most cases the only purpose, of an engine or vehicle test cell is to produce data. The collection, verification, manipulation, display, storage and transmission of these data should be amongst the prime considerations in the design and operation of any test facility.

It is perhaps the speed of recording, processing, storage of data and the creation of engine 'models' that has been responsible for the most revolutionary changes in the practice of engine testing in the last 10 years.

Test techniques now available are only made possible by the ability of modern computerized systems to create 'virtual engine' models. These models are an essential prerequisite for that part of the engine development process known as calibration.

The process of mapping is part of the process by which engine calibration models are initiated and then developed. In this context, the process refers to the process of creating and testing the three-dimensional control 'maps' held within the engine control unit that determines not only ignition timing and fuel injection at all states of engine speed and torque, but several other engine and vehicle control parameters.

Computerized record systems are no more inherently accurate, nor traceable, than paper systems. All data should have an 'audit trail' back to the calibration standards to prove that the information is accurate to the level required by the user. It is a prime responsibility of the test cell manager to ensure that the installation, calibration, recording and post-processing methods associated with the various transducers and instruments are in accordance with the maker's instructions and the user's requirements.

Bad data are still bad data, however rapidly recorded and skilfully presented.

The traditional approach of test data collection

It should not be assumed that traditional methods of data collection should necessarily be rejected, they may still represent the best and most cost-effective solution in certain cases where budgets and facilities are very limited.

Figure 19.1 shows a traditional 'test sheet'. The sheet records the necessary information for plotting curves of specific fuel consumption, b.m.e.p., power output

UNIT NO.

DATE		CUSTOMER							WORKS ORDER NO.	

DYNAMOMETER TORQUE ARM 265mm

ENGINE VARIABLE COMP	BORE 85mm	STROKE 825mm	CYLINDERS 1	SWEPT VOL. 468 cc	FUEL SHELL 4-STAR	OIL SHELL 20W30
BAROMETER 755mmHg	AIR TEMP. 21°C	AIR BOX SIZE 150 litre	ORIFICE DIA. 18.14 mm	FLOWMETER NO.	FUEL GAUGE 'O'	

POWER kW = $\dfrac{F_N n}{36{,}000}$ b.m.e.p = $7123 F_N$ kN/m²   FUEL Litre/Hour = $\dfrac{180}{1}$ NOTE: COMPRESSION RATIO 7.1

	TACHO rev/min	COUNTER rev	COUNTER sec	rev/min	BRAKE LOAD	POWER	b.m.e.p	1 sec 50/cc	FUEL litre/hour	FUEL litre/kW hr	EXH °C	HEAD cmH₂O	TEMP °C	VOL F/R l/s	EFFY η VOL	MASS F/R g/s	a/l RATIO	REMARKS
			$2x$	n	F_N	kW	kN/m²		litre/hour	litre/kW hr		h_o	T_A	V_a		m_a		
RICH	2000	780		2012	50.5	2.022	360	47.0	3.830	1.357	620	6.15	21	4.02	0.827	6.89	7.38	
		834		2002	67.0	3.726	477	50.0	3.600	0.966		6.15		4.92	0.630	5.89	7.85	
		891		1022	68.0	3.630	484	52.5	3.429	0.945	500	6.13		4.91	0.655	5.88	8.23	
		962		2025	71.0	3.993	506	57.0	3.158	0.791		6.10		4.90	0.620	5.87	8.92	
		1044		1989	74.0	4.089	527	63.0	2.857	0.699		6.05	21.5	4.88	0.629	5.84	9.81	
		1156		2010	74.5	4.160	530	69.0	2.609	0.627		6.00		4.86	0.620	5.82	10.71	
		1304		2019	74.5	4.178	530	77.5	2.323	0.556		5.97		4.85	0.616	5.80	11.99	
		1355		1995	74.5	4.129	530	81.5	2.209	0.535	675	5.93		4.83	0.621	5.79	12.58	
		1560		2035	69.0	3.900	491	92.0	1.957	0.502	685	5.93	19	4.83	0.609	5.79	14.20	
		1615		1988	65.2	3.601	464	97.5	1.846	0.513		5.87		4.81	0.620	5.76	14.98	
		1698		2008	60.0	3.347	427	101.5	1.773	0.530		5.95		4.84	0.618	5.80	15.70	
		1792		2010	52.0	2.903	370	107.0	1.682	0.580	673	6.03		4.88	0.623	5.84	16.66	
WEAK		1908		2000	35.0	1.944	249	114.5	1.572	0.809	682	6.03	19	4.88	0.626	5.84	17.83	

DENSITY OF FUEL 0.75kg/litre

Figure 19.1 *A traditional test sheet*

and air/fuel ratio of a gasoline engine at full throttle and constant speed, with a total of 13 test points. All the necessary information to permit a complete and unambiguous understanding of the test, long after the test engineer has forgotten all about it, is included. The same cannot necessarily be said of computer-acquired test data that has been poorly integrated into a data management system. It is a great temptation to leave the computer to 'get on with the job', without paying sufficient attention to the proper annotation of the information it has generated.

Such simple test procedures as the production of a power–speed curve for a rebuilt diesel engine can perfectly well be carried out with recording of the test data by hand on a pro-forma test sheet. In spite of the low cost of computer hardware there may be little justification for the development and installation of a computerized control and data acquisition system in the engine rebuild shop of, for example, a small city bus company handling perhaps 20 engines per month. If an impressive 'birth certificate' for the rebuilt engine is required, the test sheet may easily be transferred to a computer spreadsheet and accompanied by a computer-generated performance curve.

In the data flow shown in Fig. 19.2, the only post-processing of the test data is the conversion into a delivered engine 'birth certificate' which may be simply a reduced and tidy copy of the original data. The archive is available for use for technical and commercial statistical analysis.

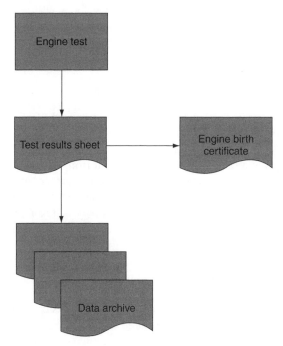

Figure 19.2 *Data flow based on pre-computerized systems*

Chart recorder format of data

Chart recorders represent a primitive stage in the process of automated data acquisition. Multipen recorders, having up to 12 channels giving continuous analogue records in various colours, still have a useful role, particularly in the testing of dispersed systems, such as installed marine installations or of large engines 'in the field' where robustness and the ease with which the record may be annotated are of value.

It is the manner in which the data are displayed as a coloured graph of values against time, now emulated by computerized displays, which is the real value to test engineers who need the clear and immediate indication of trends and the interaction of data channels.

Computerized data recording with manual control

As the number of channels of information increases beyond about 12 and there is a need to calculate and record derived quantities, such as power and fuel consumption, the task of recording data by hand becomes tedious and prone to error.

An operational example, at a slightly higher level of complexity than the previous one, might be a small independent company producing specially tuned versions of mass-produced engines for sports car use. The test cell will be required to produce detailed power curves and will also be used for development work.

Information to be recorded is likely to include

- torque, speed and fuel consumption;
- fluid temperatures and pressures;
- several critical gas and component temperatures.

The total required number of data channels could exceed 40. Typical equipment would be a PC-based data acquisition package with 16 pressure channels, 16 temperature channels, four channels for receiving pulse signals plus torque transducer signal processing and a facility for processing input from a fuel consumption meter.

In this test scenario, no control, either of the test sequence, of the dynamometer or the engine itself, is exercised by the PC; these are under manual control of the operator. A 'snapshot' record of all data channels may be taken and written to disc every time the operator presses a particular key of the test cell keyboard. Each such record will be time and date stamped with a 'time of test' figure derived from the PC's internal clock. As a simple post-processing exercise, the operator is able to call up a power curve or any other preprogrammed graph from any of these data sets using spreadsheet software.

It is entirely possible to adapt a PC data acquisition system at this level to give automatic test sequence control without large additional expenditure, provided the cell is equipped with dynamometer and engine controllers designed to accept external commands.

However, if there is a wide variety of test sequences, not repeated very often, it can take longer to modify or programme the test sequence than to run the tests under manual control.

At all levels of complexity, the appropriate degree of sophistication in the management of the test and the recording of the data calls for careful consideration.

Analogue to digital (ADC) conversion accuracy

Signals from transducers having analogue outputs are digitized, conditioned and linearized in the PC for display and stored in the chosen style and form. Proprietary data collection and signal conditioning cards are commonly available with ADC of 8-, 12- and 16-bit resolution. These should be selected as appropriate to the accuracy and the significance of the signals to be processed.

It is entirely inappropriate and gives a false illusion of accuracy (see Chapter 20) to use 16-bit or higher resolution of ADC for measurement channels having an inherent inaccuracy an order of magnitude greater. For example, a thermocouple measuring exhaust gas temperature may have a range of 0–1000°C and an accuracy of $>\pm3$°C. The 16-bit resolution would be to about 0.016°C, while the entirely adequate 12-bit resolution would be 0.25°C.

The modern standard: full computerized control and data acquisition

Fully computerized test cells are inevitably computer dependent. If the computer and its software are not functioning, then the test bed may be effectively unusable. The degree of dependency is a function of the detailed design of the equipment; some designs will allow manual operation with the computer out of action but many will not. In almost all systems a pseudomanual control mode is available, whereby the operator can start and stop the engine, then change speed and torque via a manually operated panel; however, this mode is invariably fly-by-wire and computer-dependent for all but demand input.

Certain tests, such as those required for engine emission homologation, may be preprogrammed in all details by the control system supplier, but normally the user is provided with a set of software tools to set up the required functions: data collection, labelling, scan rates, logging frequency and display of data. It is vital that training in the efficient use of these tools is given to those in the whole chain of data use.

There should be the ability to set a limited number of channels to higher scan rates than others, but it is possible to place too much emphasis on scan rates when comparing systems. In certain types of transient test logging rates of 1 kHz or more may be required, but for many parameters, fluid temperatures for example, such scan rates are inappropriate and may even, as in the case of engine speed, give misleading 'snapshots' of the value being measured.

Higher than necessary scan rates and greater than necessary levels of analogue to digital conversion can cause unnecessary problems in processing speed and data storage.

'Real-time' display of selected data, often in a graphical format chosen by the operator, is becoming more common as the power of the test bed computer increases; such displays can be very useful when setting up control loops. However, the proliferation of screens showing test data brings with it unique problems of 'information overload' for the operator and some degree of convention and management is required to discipline the use of upward of six screens of information now commonly seen in R&D cells.

Data collection and processing in engine mapping and calibration

In this usage, the term 'calibration' is used to describe the process by which the operational characteristics of one physical engine (more correctly, power-train unit) is changed, through manipulation of all its control mechanisms to suit a target geographical/vehicle/driver combination while remaining within legislative emission limits.

It costs a vast amount of money to develop a new engine and the enterprise is only economically viable if the engine then finds the widest possible field of application. Today this field is essentially worldwide and the developers will hope that their engine will be incorporated in a very wide range of vehicles, manufactured in many different countries, each having their own customer preferences, driving habits and legislative requirements.

The optimized control variables, when found, are stored within the 'map' held in the engine/power train, engine management unit or engine control unit (ECU) and in the vehicle control systems, such as ABS braking and stability control. Thus 'mapping' an engine, outside its intended vehicle, requires that it be run on a test bed capable of simulating not only the vehicle's road load characteristics but also the full range of the vehicle's performance envelope, including the dynamic transitions between states.

In order to reduce the testing time required to produce viable ECU maps, and in order to lower the complexity of the operator's role in the mapping process, the engine test industry is continuing to develop powerful software tools designed to semi-automate the tasks.

During the mapping process, in order to discover boundary conditions, such as 'knock', the engine will be run through test sequences that seek, and then confirm, the map position of these boundaries at a wide range of driver operating conditions. Conditions of excessive temperatures, cylinder pressure and 'knock' require fast detection and correct remedial action. Mapping tools such as CAMEO®, produced by AVL, contain test sequence control algorithms which allow the automatic location of these boundary conditions and can react correctly when they are located.

A typical mapping sequence is that called 'spark sweep' where at a given operating point the spark timing is varied until a knock condition is approached, the timing is then backed off and the knock point approached with finer increments to confirm its map position. The test sequence then sweeps the ignition timing so that a fault condition, such as high exhaust temperature, is located at the other end of the sweep. Given sufficient data, the software can then calculate and extrapolate the boundary conditions for a wide operating envelope without having to run all of it.

Advanced design of experiment (DOE) tools, such as those embedded in products like CAMEO, allow the test engineer to check the plausibility of data, the generation of data models and obtain visualization of system behaviour while the test is in progress.

The ultimate target of the calibration engineer is to be able to set the operating profile of the test engine, then set the sequence running, only to return when the task of producing a viable engine map has been automatically completed; this will be achieved within the lifetime of this edition.

Use of data models in 'hardware-in-loop' testing

Once a viable model of an engine exists it can be used in the testing of engine and vehicle subsystems. Of particular value is the ability to inject fault conditions into the whole-vehicle safety matrix without physically endangering it, or the test engineer.

The various test techniques use the concept of a loop of interactivity in which inputs are injected into either an actual module or a virtual module and the outputs used in a feedback loop. Common examples are known by the following acronyms:

- *HIL*, hardware in the loop, which allows the experimenter to test a real component in a virtual environment in closed-loop mode. Typically the real component is an engine electronic control unit (ECU).
- *SIL*, software in the loop, where software components (e.g. SW of an ECU controller) can run on a test platform.
- *MIL*, model-in-the-loop, where control models can be run in a virtual environment (e.g. hardware such as the ECU is simulated).

Clearly, these test techniques are only as good as the models on which they are based which puts heavy responsibility on the quality control and management of data. However, as the total store of verified data builds, the technique improves and is able to shorten development times in the calibration of vehicle control systems and engine peripherals.

In-vehicle data transfer protocols

The electronically controlled engines of today are installed within electronically controlled vehicles; this has led to the development of various designs of in-vehicle data bus. The CAN bus (controller area network) has come to dominate the automotive industry particularly in Europe. To ensure that CAN bus devices from different suppliers interoperate and that a good standard of EMI and RFI tolerance is maintained, the SAE J1939 family of standards has been developed. Pressure on the data transfer rates beyond the capabilities of the original CAN bus have meant that several other protocols and standards (some optical based) have been developed and it is not clear which will become a future standard. The test engineer dealing with engine ECU and primary vehicle systems such as emission control or transmission control will need to keep up to date with both J1939 systems and any sector specific standards such as SAE J1587/1708 which covers heavy truck systems; these systems will appear in the connection of modules during HIL testing system.

Smart devices and systems: communication and control

The main test bed computer may be linked to smart instrumentation, sending the necessary triggering or control signals and recording the formatted data produced by the device. Linking the test bed control computer with the many types of instrument available is not a trivial task and it is essential, when choosing a main control system, to check on the availability of device drivers and compatibility of third party equipment and data transfer protocols.

Software for production test cells

Computers required to manage production test cells are required to perform a number of specialized tasks, of which examples follow.

Bar code labels or embedded read/write chips are commonly attached to engines arriving for test; these will contain all the information necessary for the test stand software to select the required test procedure with appropriate values, and to create a test 'header' that will form the basis of the engine test certificate. Following the test, the test bed computer may print a bar code label or write test information to the engine chip, which may include fault codes for use in the minor repair stations.

The computer software for production test beds must contain much more information about the end user and the production logic of the plant than would be embodied in the software for an R&D test facility. The identity codes for engines, the procedure for dealing with minor repairs, the pass and fail criteria, integration with production control computers and many other details must be built into the application software at the time of facility design.

Production test cells often require the operator to make visual checks and perhaps make some physical adjustment, such as setting an idling stop, in the course of an otherwise automatic test sequence. An operator may supervise several cells and so will need to be called to a particular unit when his intervention is required, either for normal or for abnormal events. Flashing signals and clear display messages calling for acknowledgement will be required.

There are legal requirements in most countries concerning the archiving of product data

Data processing in cold testing

The vibration signature of rotating equipment has long been used as a diagnostic tool by engineers, particularly in the industrial transmission industry where noise analysis at discrete frequencies allowed tooth contact anomalies to be identified before any physical evidence could be detected.

The availability of ever improved transducers and high-speed data processing has meant that engine developers can now study the vibration patterns produced by engine components forming part of a fully built engine.

A key element of cold testing is the development and maintenance of vast amounts of data relating to the vibration characteristics of a 'good' engine so that the characteristics of a 'bad' engine may be recognized by a computerized system using pattern recognition technology.

While gross build errors, such as valve timing or missing components, are easy to identify, problems with individual components such as fuel injectors will have a more subtle signature and the limits of patterns that are within an acceptable band have to be set empirically.

Typically, cold test is carried out on automotive gasoline and diesel engines where it offers manufacturers considerable advantages over the traditional hot test because it can take a quarter of the time and be carried out within a cheaper infrastructure. When the same vibration signature evidence is sought from a hot test, 'the bang' gets in the way. Cold testing commonly includes

- signature of oil pressure over time to check oil pump and oil circuit integrity;
- 'torque to turn' of the engine over the test sequence gives indication of tight pistons or bearing shells;
- crank and cam timing;
- fuel rail pressure pulsations checking fuel supply and injectors;
- inlet and exhaust airflow and compression check valve operation.

Electrical harness integrity testing may also be carried out at the same test station.

However, there are some shortcomings inherent in the cold test process, two of which are that leaks induced by differential component expansion during the heating of the engine are not evident and the fluid flows are lower than hot test so that manufacturing debris may not be flushed.

It should be noted that in the case of cold testing of diesel engines the fuel injection timing has to be adjusted to prevent the engine running but still allowing fuel to flow, albeit for only a few pulses during the test.

A cold test station does not require much of the test facility infrastructure of a hot test stand. Dedicated ventilation and exhaust systems are absent, hazard containment, fuel system and noise attenuation are simplified, fire risks are lower. The engine is motored by an a.c. dynamometer.

It is the acquired data of the engine's whole life and the application of those data into the recognition of incipient faults in the engine at the time of testing that is the most important asset to the engine builder; such application requires dedicated staff and a policy of continuous improvement.

Anechoic testing and model creation

Modern digital, sound recording, devices and modelling software have reduced the amount of engine or vehicle running in anechoic cells almost to the minimum required to create a sound generation model over the operating range or at the point of interest and then test the results of modifications made based on the analysis of those models.

Human intervention may still be required in order to judge the level of distress or annoyance induced by noises of the various sound patterns recorded. This, as any parent of teenage children knows, is a highly subjective subject. The more difficult part of the development process is to identify the physical modifications to the powertrain system required to change the noise to the acceptable profile that has been produced by manipulation of the original digital recording.

Management of data: some general principles

It cannot be too strongly emphasized that present-day data acquisition systems are capable of storing vast amounts of information that may be totally irrelevant to the purposes of the test or stored in a way that makes it virtually impossible to correlate with other tests.

The first basic rule is to keep a consistent naming scheme for all test channel names, without which it will not be possible to carry out post-processing or comparison of multiple tests. Since most engine and vehicle test work is relative, rather than absolute, this ability of comparison and statistical analysis of results is vital.

Mere acquisition of the data is the 'easy' part of the operation. The real skill is required in the post-processing of the information, the distillation of the significant results and the presentation of these at the right time to the right people in a form they will understand. This calls for an adequate management system for acceptance (rejection), archiving, statistical analysis and presentation of the information.

Data are as valuable as their accessibility to the people that need to use them. The IT manager, in charge of the storage and distribution network of the data, is important in the test facility team and should be local, accessible and involved. One vital part of the management job is to devise and administer a disaster recovery procedure and back-up strategy.

Normally the IT manager will have under his direct control the host computer system into which, and out of which, all the test bed level and post-processing stations are networked. Host computers can play an important role in a test facility, providing they are properly integrated. The relationship between the host computer, the subordinate computers in post-processing offices and the test cells must be carefully planned and disciplined. The division of work between cell computer and host may vary widely. Individual cells should be able to function and store data locally if the host computer is 'off-line', but the amount of post-processing undertaken at test bed level varies according to the organizational structures and staff profiles.

In production testing, the host may play a direct supervisory role, in which case it should be able to repeat on its screen the display from any of the cells connected to it and to display a summary of the status of all cells.

With modern communications it is possible for the host computer to be geograph-ically remote; some suppliers will offer modem linking as part of a 'site support' contract. This ability to communicate with a remote computer raises problems of security and safety since it may be possible, and allowed, for remote staff to cause machinery to operate.

Post-acquisition data processing and reporting

Post-processing and display of data is often carried out using spreadsheet or relational database and graphical software that can read the buffered or stored information directly from the test cell or host computer. It may be important for purchasers of data acquisition systems to check that the data from their system can be converted into a non-specific format, such as ASCII, for importing into their preferred spreadsheets and databases.

A tool widely used by high-end research laboratories is a software suite produced by AVL List called Concerto®, which is specifically designed to deal with the different types of data from test bed, combustion analysis and emission equipment and provides most of the mathematical, statistical and graphical tools required by automotive development engineers. One advantage of such a tool is that security of data is built into the code so that accidental or deliberate changes to data can be seen or prevented.

If the 'customers' of the test facility are not aware of the capabilities of the data acquisition system they will not make optimum use of it. It is essential that they should be given proper training in the capabilities of the facility and its data in order to use it to maximum advantage. This statement will appear to be a truism yet, in the experience of the author, training in the capabilities of modern test equipment and data analysis is often underfunded by purchasers after the initial investment, which leads to skills and use levelling out below optimum.

Physical security of data

The control room of an engine test cell is rarely the ideal environment for storing test data. Rules are required for the management and storage of data, whether locally or at the host computer and access to the various functions must be restricted if the integrity of data and system control is to be preserved.

In most large engine test facilities there are three levels of password enforced security:

1. *Operator.* Typically at this level tests may be started, stopped and run only when the 'header information', which includes engine number, operator name and other key information, has been entered. Operator code may be verified. No alarms can be altered and no test schedules edited, but there will be a text 'notepad' for operator comments, to be stored with the other data.
2. *Supervisor.* On entry of password this level will allow alteration of alarm levels and test schedules, also activation of calibration routines, in each case by way of a form fill routine and menu entry. It will also give access to fault finding routines, such as the display of signal state tables.
3. *Engineer.* On entry of the password, this level is allowed access to debugging routines and language level editing.

Some test facilities, such as those involved with military or motor-sport development can be under constant and real threat to the confidentiality of its data. Physical restriction of access of personnel is built into the design of such test facilities, but modern miniaturized data storage devices and phone cameras seem to conspire to make security of data more and more difficult; it is a management role to balance the threat with sensible working practices.

A note on passwords

Passwords are keys that can be abused and misused. Military practice requires that passwords should be changed at random intervals and this is a sound principle. Such a requirement can be computer generated by the IT management. Increasingly, security card access to secure areas and computers is being used in place of passwords. This all tends to bring with it greater monitoring of the data handling activities of the card holder.

General guidelines for the choice of engine test software

Worldwide, there are probably fewer than 15 proven suites of software that have been developed, and are supported, primarily for the control of R&D engine test beds. Developers requiring the creation and manipulation of 'virtual engine' models have an even more limited choice.

The available proprietary suites of test cell control and data acquisition software, together with the range of engine test hardware they support, can be roughly divided into the same market categories as the users:

1. after-market, engine tuners and engine rebuilders;
2. tier 2 automotive parts suppliers, undertaking endurance testing or quality assurance testing and university departments;
3. OEM, research institutes and international motor sport teams.

The stratification of price, complexity and adaptability matches these categories.

The established suppliers of engine test cell computer systems will be familiar with all the usual safety logic applicable to test cell control and will have included it in the design of their equipment and software. Therefore the prospective user needs to investigate what is successfully used in his peer group and to define his strategy for data management within his operational specification (Chapter 1) before deciding on a supplier shortlist.

The screen displays should be assessed for the ease with which they may be adapted to the user's requirements, their relevance to the tasks to be undertaken, and the clarity of the logic underlying the menu and display hierarchy. It is to be remembered that the almost infinite variety of display that is possible must be managed and used in a safe and auditable manner. Errors and misuse do not merely corrupt data, they can also lead to dangerous failures.

It is important to check that the control equipment can receive and transmit signals to and from associated test equipment of third party suppliers. Special software drivers may have to be written and this work can be complex and expensive. Interfacing with such equipment as exhaust emissions analysers can be very difficult if not planned for when the system is designed.

Finally, the choice of equipment must be reviewed in the light of company policy and the medium-term development strategy for computerization. In particular, consider the extent to which your choice will tie you to a specific software or hardware platform and whether those platforms are sufficiently open or widely available and supported by companies other than the supplier. It is inevitable that modifications to the software in the light of experience will be necessary and the cost of software support and training should be factored into the purchase price (cost of ownership).

Further reading

SAE J1939 standards collection on the web at: *http://www.sae.org/products/ j1939a.htm.*

20 The pursuit and definition of accuracy: statistical analysis of test results

Introduction

This chapter covers what is perhaps the most important subject in the book. Some of the original text carried over from the first edition may appear to some readers as being 'old fashioned'. Furthermore many readers will not be acquainted with the Bourdon pressure gauges mentioned in the text – these are devices in which the effect of the internal forces being measured can be seen to produce physical movement in the instrument linkage.

It is the disconnection from close acquaintance with the forces and temperatures being measured, induced by remote, 'black box' instrumentation, that is an important justification for this chapter. Modern instrumentation, data manipulation, simulation and 'modelling' have tended to obscure questions of accuracy and to give an illusion of precision to experimental results that is often totally unjustified. The intention of this chapter is to point out the dangers and call for a much greater awareness of the problem.

It is a repeated statement of the obvious that the purpose of engine testing is to produce data and that the value of that data, particularly when transformed into 'information', depends critically on its accuracy.

Appropriate accuracy in the test process presents the biggest challenge facing the experimenter, who in addition to a complete understanding of the unit under test (UUT) must master the following skills:

- experience in the correct use of instruments;
- knowledge of methods of calibration and an awareness of the different kinds of error to which instruments are subject, both individually and collectively (as part of a complex system);
- a critical understanding of the relative merits and limitations of different methods of measurement and their applicability to different experimental situations;
- an understanding of the differences between true and observed values of experimental quantities;
- an appreciation that repeatability and accuracy are separate and distinctly different attributes of data;

- an appreciation of the effects on raw data of the signal processing chain;
- an appreciation of data storage and retrieval, and relational database systems.

This is a very wide range of skills, only to be acquired by experience.

The first essential is to acquire, as a habit of mind, a sceptical attitude to all experimental observations: all instruments tend to be liars.

The basics

The digital display of a value gives an illusion of accuracy which can be totally unjustified. The temptation to believe that a reading of, say, 97.12°C is to be relied upon to the second decimal place is very strong, but in the absence of convincing proof must be resisted. An analogue indicator connected to a thermocouple cannot in most cases be read to closer than 1°C and the act of reading such an indicator is likely to bring to mind the many sources of inaccuracy in the whole temperature measuring system. Replace the analogue indicator by a digital instrument reading to 0.1°C and it is easy to forget that all the sources of error are still present: the fact that the readout can now discriminate to 0.05°C does not mean that the overall accuracy of the measuring system is producing a reading within comparable limits.

Analogue display of data has a subtly different message to give to the observer in that it shows how far the measured point is from other values above and below it in addition to the rate and direction of change.

Example: measurement of exhaust temperature of a diesel engine

Temperature measurements are some of the most difficult to make accurately and the following example illustrates many of the pitfalls encountered in the pursuit of accuracy. Figure 20.1 shows a typical situation: the determination of the exhaust temperature of a diesel engine. The instrument chosen is a vapour pressure thermometer, such as is suitable for temperatures of up to 600°C and commonly installed in the individual exhaust ports of medium speed diesel engines. It comprises a steel bulb, immersed in the gas of which the temperature is to be measured, and connected by a long tube to a Bourdon gauge which senses the vapour pressure but is calibrated in temperature.

Let us consider the various errors to which this system is subject.

Sensing errors are associated with the interface between the system on which the measurements are to be made and the instruments responsible for those measurements. In the present case, there are a number of sources of sensing error. In the first place, the bulb of the temperature indicator can 'see' the walls of the exhaust pipe, and these are inevitably at a lower temperature than that of the gas flowing in the pipe. It follows that the temperature of the bulb must be less than the temperature of the gas. This error can be reduced but not eliminated by shielding the bulb or by employing a 'suction pyrometer'. A further source of error arises from heat conduction from the

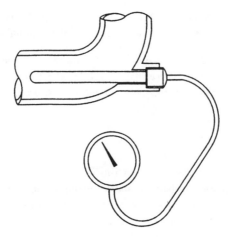

Figure 20.1   *Measurement of exhaust gas temperature by vapour pressure thermometer*

bulb to the support, as a result of which there is a continuous flow of heat from the exhaust gas to the bulb and no equality of temperature between them is possible.

A more intractable sensing error arises from the circumstances that the flow of gas in the exhaust pipe is constant neither in pressure, velocity nor temperature. Pulses of gas, originating at the opening of the exhaust valves in individual cylinders, alternate with periods of slower flow, while the exhaust will also be to some extent diluted by scavenge air carried over from the inlet. The thermometer bulb is thus required to average the temperature of a flow that is highly variable both in velocity and in temperature, and it is unlikely in the extreme that the actual reading will represent a true average.

A more subtle error arises from the nature of exhaust gas. Combustion will have taken place, resulting in the creation of the exhaust gas from a mixture of air and fuel, perhaps only a few hundredths of a second before the attempt is made to measure its temperature. This combustion may still be incomplete, the effects of dissociation arising during the combustion process may not have worked themselves out, and it is even possible that the distribution of energy between the different modes of vibration of the molecules of exhaust gas will not have reached its equilibrium value. As a consequence it may not be possible even in principle to define the exhaust temperature exactly.

We can deal with some of these sensing errors by replacing the steel bulb of Fig. 20.1 by a slender thermocouple surrounded by several concentric screens (Fig. 20.2) and further improve accuracy by the use of a suction pyrometer, Fig. 20.3, in which a sample of the exhaust gas is drawn past the sensor at uniform velocity by external suction. This deals with the 'averaging' problem but still leaves the question regarding the definition of the exhaust temperature open.

Our example includes a Bourdon type pressure gauge and these may be particularly prone to a variety of classic instrument errors.

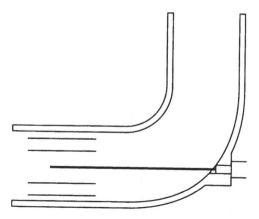

Figure 20.2 *Measurement of exhaust gas temperature by shielded thermo-couple*

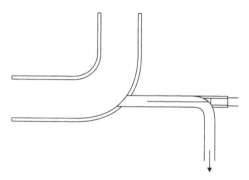

Figure 20.3 *Measurement of exhaust gas temperature by suction pyrometer*

Two of the most common are zero error and calibration error. The zero error is present if the pointer does not return precisely to the zero graduation when the gauge is subjected to zero or atmospheric pressure.

Calibration errors are of two forms: a regular disproportion between the instrument indication and the true value of the measured quantity, and errors that vary in a non-linear manner with the measured quantity. This kind of fault may be eliminated or allowed for by calibrating the instrument, in the case of a pressure gauge by means of a dead weight tester. These are examples of systematic errors.

In addition, the pressure gauge may suffer from random errors arising from friction and backlash in the mechanism. These errors affect the repeatability of the readings.

The sensitivity of an instrument may be defined as the smallest change in applied signal that may be detected. In the case of a pressure gauge it is affected particularly by friction and backlash in the mechanism. The precision of an instrument is defined in terms of the smallest difference in reading that may be observed. Typically, it is possible to estimate readings to within 1/10 of the space between graduations, provided the reading is steady, but if it is necessary to average a fluctuating reading the precision may be much reduced.

Finally, one must consider the effect of installation errors. In the present case these may arise if the bulb is not inserted with the correct depth of immersion in the exhaust gas or, as is quite often the case, if it is installed in a pocket and is not subjected to the full flow of the exhaust gas. If a thermocouple probe that is physically smaller than the bulb is used, some of the vagaries of gas flow temperature measurement may be greater.

A consideration of this catalogue of possible errors will make it clear that it is unlikely that the reading of the indicator will reflect the temperature of the gas in the pipe with any degree of exactness. It is possible to analyse the various sources of error likely to affect any given experimental measurement in this way, and while some measurements, for example of lengths and weight, require a less complex analysis, others, notably readings of inherently unsteady properties such as flow velocity, require to be treated with scepticism. A hallmark of the experienced experimenter is that, as a matter of habit and training, he questions the accuracy and credibility of every experimental observation.

A similar critical analysis should be made of all instrumentation. This example has been dealt with in some detail to illustrate the large number of factors that must be taken into account.

Some general principles

1. Cumulative measurements are generally more accurate than rate measurements. Examples are the measurement of speed by counting revolutions over a period of time and the measurement of fuel consumption by recording the time taken to consume a given volume or mass.
2. Be wary of instantaneous readings. Few processes are perfectly steady. This is particularly true of flow phenomena, as will be clear if an attempt is made to observe pressure or velocity head in a nominally steady gas flow. A water manometer fluctuates continually, and at a range of frequencies depending on the scale of the various irregularities in the flow. No two successive power cycles in a spark ignition engine are identical. If it is necessary to take a reading instantaneously it is better to take a number in quick succession and statistical analyse them.
3. Be wary of signal damping, often entirely sensible; sometimes leading to serious misrepresentation of the true nature of a fluctuating value.
4. A related problem: 'time slope effects'. If one is making a number of different observations over a period of nominally steady state operation, make sure that there is no drift in performance over this period.
5. Events must be recorded in the correct order for the correct causal deductions to be made; the more complex the data streams and number of sources, the more difficult this becomes.

6. The closer one can come to an 'absolute' method of measurement, e.g. pressure by water or mercury manometer or dead weight tester; force and mass by dead weights (plus knowledge of the local value of 'g' (see Chapter 8)), the less the likelihood of error.
7. In instrumentation, as in engineering generally, simplicity is a virtue. Each elaboration is a potential source of error.

Definition of terms relating to accuracy

This is a particularly 'grey' area.[1-3] The statement 'this instrument is accurate to ± 1 per cent' is in fact entirely meaningless in the absence of further definition (Table 20.1).

Table 20.1   *Various terms that are used, often incorrectly, in any discussion of accuracy*

Error	The difference between the value of a measurement as indicated by an instrument and the absolute or true value
Sensing error	An error arising as a consequence of the failure of an instrument to sense the true value of the quantity being measured
Systematic error	An error to which all the readings made by a given instrument are subject; examples of systematic errors are zero errors, calibration errors and non-linearity
Random error	Errors of an unpredictable kind. Random errors are due to such causes as electrical supply 'spikes' or friction and backlash in mechanisms
Observer error	Errors due to the failure of the observer to read the instrument correctly, or to record what he has observed correctly
Repeatability	A measure of the scatter of successive readings of the same quantity
Sensitivity	The smallest change in the quantity being measured that can be detected by an instrument
Precision	The smallest difference in instrument reading that it is possible to observe
Average error	Take a large number of readings of a particular quantity, average these readings to give a mean value and calculate the difference between each reading and the mean. The average of these differences is the average error; roughly half the readings will differ from the mean by more than the average error and roughly half by less than the average error

Statements regarding accuracy: a critical examination

We must distinguish between the accuracy of an observation and the claimed accuracy of the instrument that is used to make the observation. Our starting point must be the true value of the quantity to be measured, Fig. 20.4a.

The following are a few examples of such quantities associated with engine testing:

- fuel flow rate;
- air flow rate;
- output torque;
- output speed;
- pressures in the engine cylinder;
- pressures in the inlet and exhaust tracts;
- gas temperatures;
- analysis of exhaust composition.

By their nature, none of these quantities is perfectly steady and decisions as to the sampling period are critical, see above.

Sensing errors

These are particularly difficult to evaluate and can really only be assessed by a systematic comparison of the results of different methods of measuring the same quantity.

Sensing errors can be very substantial. In the case discussed above regarding measurement of exhaust temperature, the authors have observed errors as great as 70°C, Fig. 20.4b.

Typical sources of sensing error include

1. mismatch between temperature of a gas or liquid stream and the temperature of the sensor;
2. mismatch between true variation of pressure and that sensed by a transducer at the end of a connecting passage;
3. inappropriate damping or filtration applied within the instrument's measuring chain;
4. failure of a transducer to give a true average value of a fluctuating quantity;
5. air and vapour present in fuel systems;
6. time lag of sensors under transient conditions.

Note that the accuracy claimed by its manufacturer does not usually include an allowance for sensing errors. These are the responsibility of the user.

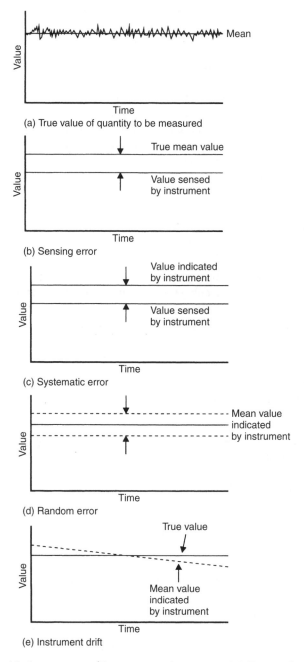

Figure 20.4 *Various types of instrumentation error. (a) True value of quantity to be measured plotted against time – on the same time scale; (b) sensing error; (c) systematic error; (d) random error; (e) instrument drift*

Systematic instrument errors (Fig. 20.4c)

Typical systematic errors include the following:

1. *Zero errors*: the instrument does not read zero when the value of the quantity observed is zero.
2. *Scaling errors*: the instrument reads systematically high or low.
3. *Non-linearity*: the relation between the true value of the quantity and the indicated value is not exactly in proportion; if the proportion of error is plotted against each measurement over full scale, the graph is non-linear.
4. *Dimensional errors*: for example, the length of a dynamometer torque arm may not be precisely correct.

Random instrument errors

Random errors, Fig. 20.4d include

1. effects of stiction in mechanical linkages;
2. effects of friction, for example in dynamometer trunnion bearings;
3. effects of vibration or its absence;
4. effects of electromagnetic interference (see Chapter 10).

Instrument drift

Instrument drift, Fig. 20.4e, is a slow change in the calibration of the instrument as the result of

1. changes in instrument temperature or a difference in temperature across it;
2. effects of vibration and fatigue;
3. fouling of the sensor, blocking of passages, etc.;
4. inherent long-term lack of stability.

It will be clear that it is not easy to give a realistic figure for the accuracy of an observation.

Instrumental accuracy: manufacturers' claims

With these considerations in mind it is possible to look more critically at the statement 'the instrument is accurate to within ± 1 per cent of full scale reading'.

At least two different interpretations are possible:

1. No reading will differ from the true value by more than 1 per cent of full scale. This implies that the sum of the systematic errors of the instrument, plus the largest random error to be expected, will not exceed this limit.

2. The average error is not more than 1 per cent of full scale. This implies that about half of the readings of the instrument will differ from the average by less than 1 per cent full scale and half by more than this amount. However, the definition implies that systematic errors are negligible: we are only looking at random errors.

Neither of these definitions is satisfactory. In fact the question can only be dealt with satisfactorily on the basis of the mathematical theory of errors.

Uncertainty

While it is seldom mentioned in statements of the accuracy of a particular instrument or measurement, the concept of uncertainty is central to any meaningful discussion of accuracy.

Uncertainty is a property of a measurement, not of an instrument:

> The uncertainty of a measurement is defined as the range within which the true value is likely to lie, at a stated level of probability.

It is interesting to note that in most scientific disciplines data, when presented graphically, have each data point represented by a cruciform indicating the degree of uncertainty or error band; this practice is rarely, if ever, seen in engine test data and helps to confirm the unjustified illusion of exactness.

The level of probability, also known as the confidence level, most often used in industry is 95 per cent. If the confidence level is 95 per cent, there is a 19 to 1 chance that a single measurement differs from the true value by less than the uncertainty and one chance in 20 that it lies outside these limits.

If we make a very large number of measurements of the same quantity and plot the number of measurements lying within successive intervals, we shall probably obtain a distribution of the form sketched in Fig. 20.5. The corresponding theoretical

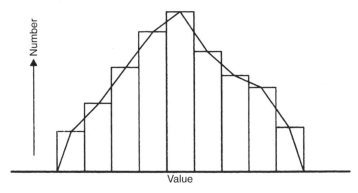

Figure 20.5 *A frequency distribution*

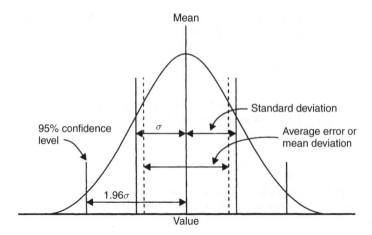

Figure 20.6   *Normal or Gaussian distribution*

curve is known as a normal or Gaussian distribution, Fig. 20.6. This curve is derived from first principles on the assumption that the value of any 'event' or measurement is the result of a large number of independent causes (random sources of error).

The normal distribution has a number of properties shown on Fig. 20.6:

- The *mean value* is simply the average value of all the measurements.
- The *deviation* of any given measurement is the difference between that measurement and the mean value.
- The σ^2 (sigma)^2 is equal to the sum of the squares of all the individual deviations, divided by the number of observations.
- The *standard deviation* σ is the square root of the variance.

The standard deviation characterizes the degree of 'scatter' in the measurements and has a number of important properties. In particular, the 95 per cent confidence level corresponds to a value sigma = 1.96. Ninety-five per cent of the measurements will lie within these limits and the remaining 5 per cent in the 'tails' at each end of the distribution.

In many cases the 'accuracy' of an instrument as quoted merely describes the average value of the deviation, i.e. if a large number of measurements are made about half will differ from the true or mean value by more than this amount and about half by less. Mean deviation = 0.8σ approximately.

However, this treatment only deals with the random errors: the systematic errors still remain. To give a simple example, consider the usual procedure for checking the calibration of a dynamometer torque transducer. A calibration arm, length 1.00 m, carries a knife-edge assembly to which a dead weight of 10 kg is applied. The load is applied and removed 20 times and the amplifier output recorded. This is found to range from 4.935 to 4.982 volts with a mean value 4.9602 volts.

At first sight we could feel a considerable degree of confidence in this mean value and derive a calibration constant:

$$k = \frac{4.9602}{10 \times 9.81 \times 1.0} = 0.050563 \,\text{V/Nm}.$$

The 95 per cent confidence limit for a single torque reading may be derived from the 20 amplifier output readings and, for the limiting values assumed, would probably be about ± 0.024 V, or ± 0.48 per cent, an acceptable value.

There are, however, four possible sources of systematic error:

1. The local value of g may not be exactly 9.81 m/s^2 (see Chapter 8).
2. The mass of the dead weight may not be exactly 10 kg.
3. The length of the calibration arm may not be exactly 1.00 m.
4. The voltmeter used may have its own error.

In fact none of these conditions can ever be fulfilled with absolute exactness. We must widen our 95 per cent confidence band to take into account these probably unknown errors.

This leads straight on to our next topic.

Traceability

In a properly organized test department, careful thought must be given to this matter. In many small departments the attention given to periodic recalibration is minimal. In many more, while some effort may be made to check calibrations from time to time, the fact that the internal standards used for these checks may have their own errors is overlooked. To be confident in the validity of one's own calibrations, these standards must themselves be checked from time to time.

Traceability refers to this process. The 'traceability ladder' for the dead weight referred to in the above example might look like this:

- the international kilogram (Paris);
- the British copy (NPL);
- national secondary standards (NPL);
- local standard weight (portable);
- our own standard weight.

Combination of errors

Most derived quantities of interest to the experimenter are the product of several measurements, each of them subject to error. Consider, for example, the measurement

of specific fuel consumption of an engine. The various factors involved and typical
95 per cent confidence limits are as follows:

- torque ±0.5 per cent;
- number of revolutions to consume measured mass of fuel ±0.25 per cent;
- actual mass of fuel ±0.3 per cent.

The theory of errors indicates that in such a case, in which each factor is involved
to the power 1, the confidence limit of the result is equal to the square root of the
sum of the squares of the various factors, i.e. in the present case, the 95 per cent
confidence limit for the calculated specific fuel consumption is

$$\sqrt{0.5^2 + 0.25^2 + 0.3^2} = \pm 0.58\%$$

The number of significant figures to be quoted

One of the most common errors in reporting the results of experimental work is
to quote a number of significant figures, some of which are totally meaningless –
the *illusion of accuracy* once more; often engendered by the misuse of a common
spreadsheet format. This temptation has been vastly increased by the now universal
employment of digital readouts.

Consider a measurement of specific fuel consumption readings as follows:

engine torque	110.3 Nm
number of revolutions	6753
mass of fuel	0.25 kg

Specific consumption equals

$$\frac{3.6 \times 10^6 \times 0.25}{2 \times \pi \times 6753 \times 110.3} = 0.1923 \, \text{kg/kWh}$$

Let us assume that each of the three observations has the 95 per cent confidence limit specified in the previous paragraph. Then the specific fuel consumption
has a confidence limit of 0.58 per cent, i.e. between 0.1912 and 0.1934 kg/kWh.
Clearly it is meaningless to quote the specific fuel consumption to closer than
0.192 kg/kWh.

Particularly in development work where the aim is often to detect small
improvements this question must be kept in mind.

Absolute and relative accuracy

This is a topic linked with the last one. In a great deal of engine test work – perhaps the majority – we are interested in measuring relative changes and the effect of modifications. The absolute values of a parameter before and after the modification may be of less importance; it may not be necessary to concern ourselves with the absolute accuracy and traceability of the instrumentation. Sensitivity and precision may be of much greater importance. Some test engineers have an unhealthy respect for the attribute of test cell 'repeatability' of results while not considering that this can be almost completely disconnected from accuracy of results.

The cost of accuracy: a final consideration

An engineer with responsibility for the choice of instrumentation should be aware of the danger of mismatch between what he thinks he requires and what is actually necessary for an adequate job to be done. The cost of an instrument to perform a particular task can vary by an order of magnitude; in most cases the main variable is the level of accuracy offered and this should be compared with the accuracy that is really necessary for the job in hand.

Summary

The degree of understanding of the subject of accuracy is perhaps the main criterion by which the professional quality of a test engineer should be judged. A proper understanding requires a wide range of knowledge and the principal elements of this have been described, with a special warning as to the dangers of digital readouts. An example of temperature measurement illustrates the problems and terms relating to accuracy have been defined.

This leads to a discussion of the meaning of the term as used by instrument manufacturers, and different types of error are identified. The mathematical basis of the concept of uncertainty is given and the method of dealing with combined errors is described.

References

1. Hayward, A.T.J. (1977) *Repeatability and Accuracy*, Mechanical Engineering Publications, London. ISBN: 0-7008-0491-9.
2. Dietrich, C.F. (1973) *Uncertainty, Calibration and Probability*, Adam Hilger, London.
3. Campion, P.J. et al. (1973) *A Code of Practice for the Detailed Statement of Accuracy*, HMSO, London.

Further reading

BS 4889 *Method for Specifying the Performance of Electrical and Electronic Measuring Equipment.*
BS 5233 *Glossary of Terms used in Metrology.*
BS 5497 Part 1 *Guide for the Determination of Repeatability and Reproducibility for a Standard Test Method by Inter-laboratory Tests.*
BS 7118 Part 1 and 2 *Linear and Non-linear Calibration Relationships.*
http://www.npl.co.uk/force/faqs/glossary.html
http://www.swan.ac.uk/lis/help_and_training/pdf/standards.pdf

Index